Schleiermacher and Whitehead

Theologische Bibliothek Töpelmann

Herausgegeben von
O. Bayer · W. Härle · H.-P. Müller

Band 125

Walter de Gruyter · Berlin · New York

Schleiermacher and Whitehead

Open Systems in Dialogue

Edited by
Christine Helmer

in Cooperation with
Marjorie Suchocki, John Quiring, and Katie Goetz

Walter de Gruyter · Berlin · New York

♾ Printed on acid-free paper which falls within
the guidelines of the ANSI to ensure permanence and durability.

ISBN 3-11-017992-X

Library of Congress Cataloging-in-Publication Data

Schleiermacher and Whitehead : open systems in dialogue / edited by Christine Helmer in cooperation with Marjorie Suchocki, John Quiring, and Katie Goetz.
p. cm. – (Theologische Bibliothek Töpelmann ; Bd. 125)
Includes indexes.
ISBN 3-11-017992-X (hardcover, 23 × 15,5 cm : alk. paper)
1. Schleiermacher, Friedrich, 1768–1834. 2. Whitehead, Alfred North, 1861–1947. 3. Philosophical theology. 4. System theory. I. Helmer, Christine. II. Theologische Bibliothek Töpelmann ; 125. Bd.
BT55.S35 2004
210'.92'2–dc22
2004013295

Bibliographic information published by Die Deutsche Bibliothek

Die Deutsche Bibliothek lists this publication in the Deutsche Nationalbibliografie; detailed bibliographic data is available on the Internet at <http://dnb.ddb.de>.

Printed in Germany
Cover design: Christopher Schneider, Berlin

Acknowledgements

This collection of essays records a dialogue on thinking anew about system for contemporary theology and philosophy. The dialogue took place at a symposium entitled, "System and Life: Schleiermacher and Whitehead," held on March 6–8, 2003 at the Claremont School of Theology and the Center for Process Studies in Claremont, California. I thank Prof. Dr. Marjorie Hewitt Suchocki for her wholehearted support of the project and for the opportunity to teach a spirited seminar on Schleiermacher and Whitehead with her during the spring term 2003. I also thank Prof. Dr. John Quiring for his patient and detailed organization of the conference and for hospitality at the Center for Process Studies, especially for the chocolates!

My gratitude goes to the authors who each contributed a chapter to this volume. I thank them for having engaged the topic, for having pressed it for its dialogical possibilities, and for having cleared creative paths for constructive systematic thinking.

The financial aspect is not only necessary for the facilitation of academic dialogue, but represents the confidence of the funding institutions in such an endeavor. I am grateful to the Claremont School of Theology, especially to President Philip A. Amerson and to Dean John R. Fitzmier, as well as to the Center for Process Studies, for extending financial grace to both the conference and the book. The final preparation of this volume for publication was done during a sabbatical leave in the spring of 2004. I thank both the Claremont School of Theology for granting me a research leave and the Alexander von Humboldt-Stiftung for generously funding my stay as a Humboldt Research Fellow at the Eberhard Karls Universität Tübingen.

The production of this book is the result of wisdom, expertise, and time, all granted by the following colleagues. My sincere appreciation goes to Dr. K. F. Diethard Römheld, who typeset the articles with an infectious happiness that permeated the electronic communication of this bi-continental book. I thank Dr. Albrecht Döhnert, editor for theology, Jewish studies, and religious studies at Walter de Gruyter, for his enthusiastic guidance of this project from start to finish and for his serious commitment to the international, "bilingual," and open dialogue represented by this collection. And I also thank Prof. Dr. Oswald Bayer, Prof. Dr. Wilfried Härle, and Prof. Dr. Hans-Peter Müller, for kindly consenting to accept this volume in Walter de Gruyter's Theologische Bibliothek Töpelmann series. This book would not have been possible without the editorial efforts of Prof. Dr. Marjorie Hewitt Suchocki, Prof. Dr. John Quiring, and Katie Goetz. I am grateful to them for their generosity of editing spirit, and especially to Katie, for her incredible attention to both detail and the big picture and for sustaining the production by cheerfully mediating all levels of intersubjective dialogue.

The dialogue with Prof. Dr. Robert A. Orsi was an invitation to live. I thank him for the joy of lived novelty that transforms and sustains identity.

Christine Helmer

Abbreviations

Friedrich Schleiermacher

BO	*Brief Outline of Theology as a Field of Study (1811 / 1830).* Trans. Terrence N. Tice. Schleiermacher Studies and Translations, vol. 1. Lewiston: The Edwin Mellen Press, 1990.
CF	*The Christian Faith (1830/31).* Ed. H.R. Mackintosh and J.S. Stewart. Trans. D.M. Baillie et al. Edinburgh: T & T Clark, 1999.
CG	*Der christliche Glaube nach den Grundsätzen der Evangelischen Kirche im Zusammenhange dargestellt (1830/31).* Reproduction of 7th ed. of 1960. Ed. Martin Redeker. De Gruyter Studienbuch. Berlin: Walter de Gruyter, 1999.
Dial(1811)	*Dialektik (1811).* Ed. Andreas Arndt. Philosophische Bibliothek, no. 386. Hamburg: Felix Meiner, 1986.
Dialectic(1811)	*Dialectic or, The Art of Doing Philosophy (1811).* Trans., with Introduction and Notes Terrence N. Tice. American Academy of Religion Texts and Translations Series, no. 11. Atlanta: Scholars Press, 1996.
Dial(1814 / 15)	*Dialektik (1814 / 15)* and *Einleitung zur Dialektik (1833).* Ed. Andreas Arndt. Philosophische Bibliothek, no. 387. Hamburg: Felix Meiner, 1988.
Dial2001	*Dialektik.* Ed. and Introduction Manfred Frank. 2 vols. Suhrkamp Taschenbuch Wissenschaft, no. 1529. Frankfurt a. M.: Suhrkamp, 2001.
DialJ	*Dialektik, aus Schleiermachers handschriftlichem Nachlasse.* Ed. Ludwig Jonas. Berlin: G. Reimer, 1839 [= SW III. 4 / 2].
DialKGA	*Vorlesungen über die Dialektik.* Ed. Andreas Arndt. 2 vols. Berlin: Walter de Gruyter, 2002 [= KGA II. 10 / 1–2].

DialO	*Dialektik: Im Auftrage der Preußischen Akademie der Wissenschaften auf Grund bisher unveröffentlichten Materials*. Ed. Rudolf Odebrecht. Leipzig: J. C. Hinrichs, 1942; reprint, Darmstadt: Wissenschaftliche Buchgesellschaft, 1976.
Hermeneutics	*Hermeneutics and Criticism and Other Writings*. Ed. and trans. Andrew Bowie. Cambridge Texts in the History of Philosophy. Cambridge: Cambridge University Press, 1998.
KD	*Kurze Darstellung des theologischen Studiums zum Behuf einleitender Vorlesungen (1810/1830)*. Ed. Heinrich Scholz. Quellenschriften zur Geschichte des Protestantismus, no. 10. Leipzig, 1910; reprint in Bibliothek Klassischer Texte. Darmstadt: Wissenschaftliche Buchgesellschaft, 1993.
KGA	*Kritische Gesamtausgabe*. Ed. Hermann Fischer, Ulrich Barth, Conrad Kramer, Günter Meckenstock, and Kurt-Victor Selge. 5 parts, until now 21 vols. Berlin: Walter de Gruyter, 1983–present.
LJ	*The Life of Jesus*. Ed. and Introduction Jack C. Verheyden. Trans. S. Maclean Gilmour. Philadelphia: Fortress, 1975; reprint, Mifflintown: Sigler Press, 1997.
On Religion	*On Religion: Speeches to its Cultured Despisers*. Ed. and trans. Richard Crouter. Cambridge Texts in the History of Philosophy. Cambridge: Cambridge University Press, 1996.
Speeches	*On Religion: Speeches to its Cultured Despisers*. Trans. John Oman. Foreword Jack Forstman. Louisville: Westminster/John Knox, 1994.
SW	*Sämmtliche Werke*. Ed. Ludwig Jonas et al. 3 parts, 31 vols. Berlin: G. Reimer, 1834–1864.

Alfred North Whitehead

AI	*Adventures of Ideas (1933)*. New York: The Free Press, 1967.
CN	*The Concept of Nature (1920)*. Cambridge: Cambridge University Press, 1964.
ESP	*Essays in Science and Philosophy*. New York: Philosophical Library, 1947.
FR	*The Function of Reason (1929)*. Boston: Beacon Press, 1958.
Imm.	"Immortality." In *Essays in Science and Philosophy*, 77–96. New York: Philosophical Library, 1947.
MG	"Mathematics and the Good." In *Essays in Science and Philosophy*, 97–113. New York: Philosophical Library, 1947.
MT	*Modes of Thought (1938)*. New York: The Free Press, 1968.
PM	Alfred North Whitehead and Bertrand Russell, *Principia Mathematica*. 3 vols. 2nd ed. Cambridge: Cambridge University Press, 1925–1927.
PR	*Process and Reality: An Essay in Cosmology (1929)*. Corrected ed. David Ray Griffin and Donald W. Sherburne. New York: The Free Press, 1978.
RM	*Religion in the Making (1926)*. New York: Fordham University Press, 1996.
S	*Symbolism: Its Meaning and Effect (1927)*. New York: Capricorn Books, 1959.
SMW	*Science and the Modern World (1925)*. New York: The Free Press, 1967.

Table of Contents

III. System and Contemporary Themes

Indices

Introduction

Christine Helmer, Claremont School of Theology

I. System: A New Look

System is a distinctive feature of Western thought. Whether lovingly sought after, as is the case with the Romantics, or triumphantly found, as is the case with the Absolute Idealists, system since Plato is the goal of Western intellectual striving. Whether in the form of theoretical reason or practical reason, system represents a partial or total comprehension by thought of the reality of particular entities or actions. At the intersection between reason and nature—the "point of indifference," as Schelling would say—the human being is compelled to think a system classifying the reality that confronts it and that it desires to change.

The categorization of reality in thought is the problem of philosophy that takes its point of departure from the simple fixing of universal concepts of experience. How can similar qualities be predicated of different particulars? How can historical movement be conceptualized by interpretative keys of transhistorical value? The philosophical question relating the many to the one and the philosophical-historical question relating particular events to transhistorical significance are the central questions driving philosophy's preoccupation with reason and reality. The narrative of Western philosophical thought can be told from the perspective of these questions. Reason's *raison d'être* is to classify and deduce, as Plato has Socrates admit to being a "lover of division and collection."[1] Reason's content is itself supplied by reality: given in intuition, perceived by the senses, felt by the "organic pole." And the fact that reality can be categorized truly by reason is explained by the necessary relation between the physical and

1 Plato, *Phaedrus*, 266 b. Schleiermacher alludes to reason's capacity to divide and collect in *On Religion*, 20 (speech two).

the mental poles; the inevitable categorization of reality presupposes conditions for that precise possibility. Classification proceeds, presupposing relations between spirit and nature and assuming embeddedness in an organized, interrelated system of real objects of experience. The question regarding reason and reality is the question relating system to life.

The lure to system reveals distinctive features playing into system and its construction. Common exigencies are relatively stable for systems of philosophy and of theology throughout the history of their building. Whether representing a cosmology or a religious worldview, a system of thought aims to grasp the whole. System also locates particulars within a whole that is more than their sum total. The conceptualization of this totality and the particulars related to this totality, furthermore, is accompanied by the question regarding the explanation for the existence of the whole. Most philosophical and theological systems make the move from the world's conditioned existence to its ground; the world's existence is explained by its utter dependence on a "Whence" for whose own existence no explanation can be given. Hence the search for, and in some cases even the rational exhaustion of, the Absolute is a constitutive feature of Western systems. It is also the Absolute that usually serves to unite, as is the case with the German Idealists, and to distinguish between, as is the case with the Jena Romantics, philosophical and theological systems. If the Absolute is the transcendently or transcendentally conceived ground of the world from "a" philosophical perspective, then its determination as the living God reflects "a" theological perspective. God and world function as co-constituents of "the" system.

Together with the God-world relationship, the individual plays a constitutive role in system. The self / world / God relations characterize the coordinates of Western systems that privilege the human individual at the site of metaphysical questioning. The questions concerning the existence of the whole, the existential processes of particulars and their interrelations are questions that human beings are privileged to ask. In the tradition of Continental Idealism, the human individual represents the point of indifference at which spirit and matter coincide. This is the privileged place, though not to the exclusion of other living beings, at which the question of being is raised. Reality begs to be known, and loved, by individuals, and with the rise of historical consciousness, history itself is raised as the question from the perspective of a being that knows itself to be historically constituted. By posing these questions, the human individual reveals onto-anthropological features that drive to system. The human organism unites reason and perception in harmony, mirroring a world that also integrates diversity, spans opposites, and unites poles in order to maximize beauty (as Whitehead would say). While God perpetually injects the world with truth, beauty, and goodness, the individual and the world perpetually change in relation to these lures.

Yet it is precisely the reality of the self / world / God relations that poses the disconcerting question concerning its elusiveness to system's grasp. System has its own cultured despisers. Plato's own early dialogues are characterized by Socratic irony, a moment of realization that reality evades complete determination. The Jena Romantics, particularly Schlegel and the early Schleiermacher, resisted the absolute exhaustion of reality by reason on the grounds that finite reason was not capable of transcending its boundaries in order to comprehend its ground; homesickness, not homecoming, captured the mood in the absence of systematic closure. Furthermore, an ethics acknowledging the ineradicable face of the other refuses subsumption into "the" hegemonic I; the categorical imperative of love inscribed on neighbor protests hate-full totalitarianism. And life itself, with its messy experiences of evil and tragedy, calls systematic prettiness into question. Are claims of redemption and love falsified by empirical instances of their opposite? If reality is a thorn in the flesh of system, then language too deals its deck. Recent despisers deconstruct the myth of Western system by showing that language's infinite movement subverts static opposites. A Heraclitean world resists system, in fact makes impossible the task of system-building, because languages, not metaphysical essences, dance to the tune played by local musicians.

There is merit in the criticism of reason's hubristic elevation over reality. System must be called to humble accountability in the face of life that is jagged, wounded, and raw. Life challenges reason to its breaking-point; can experiences of "horrendous evils,"[2] with their simultaneously shattering and numbing shock effect, be suspended in a meta-narrative that would explain their necessity? Yet reason responds, driven by the desire to answer the "why?" haunting life.

The key theme raised in this volume addresses precisely the accountability of system in the face of life. Whether raised in terms of system's adequacy to empirical data or its applicability to religious pluralism, whether probed in view of the universal's relation to the individual or its embod-

---

2 I borrow the term from Marilyn McCord Adams, *Horrendous Evils and the Goodness of God*, Cornell Studies in the Philosophy of Religion (Ithaca: Cornell University Press, 2000).

iment in a particular perspective, the key question concerns system's account of the ebb and flow of living experience. By posing the question of system's openness to experiential polyvalence, this volume both agrees with the criticism against the rational homeostasis of systematic stagnation and pleads for a new look at system for both the philosophical and theological disciplines. A rehabilitation of system as a viable genre for conceptual thinking is this volume's response to keep system open on the formal grounds of reason's predilection for classification and division, on the cosmological grounds that the universe is ordered by proportion and harmony, and on the material grounds that the ontological truths of beauty and love will prevail over their empirical falsification.

The contemporary context for system's rehabilitation is admittedly shaped by criticism. Yet the stage for a love of system is also set by the contemporary context stemming from a trajectory of Western Enlightenment philosophy. System is currently being mined for its conceptual gold in fields such as sociology, with the work, for example, of Niklas Luhmann,[3] or family systems theory in the study of psychology and religious care and counseling of souls,[4] with work on organic models of system in the biological[5] and philosophical-anthropological sciences,[6] and in philos-

---

3 See the very important book by NIKLAS LUHMANN, *Social Systems*, trans. John Bednarz, Jr., with Dirk Baeker (Stanford: Stanford University Press, 1995).

4 DOROTHY STROH BECVAR and RAPHAEL J. BECVAR, *Family Therapy: A Systemic Integration*, 2nd ed. (Boston: Allyn and Bacon, 1993). I thank Martha Taylor for kindly providing this reference.

5 The most significant and widely acknowledged recent book on open systems in the biological sciences is HIROAKI KITANO, *Foundations of Systems Biology* (Cambridge: Massachusetts Institute of Technology Press, 2001). The designation information theory is commonly used in recent work in this area of study; see esp. SUSAN OYAMA, *The Ontogeny of Information: Developmental Systems and Evolution*, 2nd ed. (Durham: Duke University Press, 2000). I thank Philip Clayton for kindly drawing my attention to these texts.

6 I thank Stephan Grätzel for kindly providing me with the following references to open systems in biology and philosophical anthropology (personal correspondence, April 8, 2004). LUDWIG VON BERTALANFFY first articulated the concept of open system in the biological sciences in his work, *Theoretische Biologie: Stoffwechsel, Wachstum*, vol. 2, 2nd rev. ed. (Bern: A. Francke, 1951), 50: "Ein offenes System ist ein solches, worin Ein- und Ausfuhr von Materialien stattfindet." Bertalanffy's concept of the fundamental openness of the organism was appropriated by Plessner, who initiated the concept of open systems in philosophical anthropology: see HELMUTH PLESSNER, *Die Stufen des Organischen und der Mensch: Einleitung in die philosophische Anthropologie* (Berlin: Walter de Gruyter, 1928). Most recently, the philosopher STEPHAN GRÄTZEL uses this idea as a key determination of the human being in his *Utopie und Ekstase: Vernunft-*

ophy, with Nicholas Rescher being an important voice.[7] Modeling system after the structures and processes of organisms orients thought to be most attentive to the movement of life itself. The organism is the model of system par excellence, for which the whole is more than the sum of its parts, and in which the parts are individually constituted by their relations to each other and to their movement within the whole. With this direction, modern thinking appropriates the Greek notion of system as an organic interconnected whole. The same organic metaphor pervades Schleiermacher's descriptions of system, while Whitehead provides his term for his life's work: the "philosophy of organism."

A system aligned with organic attentiveness resists a criticism launched against a dominant strand of Enlightenment thought. Spinoza's and the Leibnizian-Wolffian philosophy exemplify the rationalist advocacy for system obedient to the geometric method of deductive logic. These systems were conceived according to the Enlightenment identification of system with science; the ground from which the *more geometrico* sprung was fertilized by the rigorous requirements for scientific presentation. With a mathematical rigidity that constructed a mechanistic universe, the rift between the *res extensa* and the *res cognitans* was opened up. The subsequent post-Kantian Idealist search for a foundational philosophy (*Grundsatzphilosophie*) aimed to close the gap, but ended up adopting the category of ontological necessity as the measure of historical development. It was Romantic melancholy that re-opened the unbridgeable chasm between the finite and the infinite, and it was this mood that set the stage of protest against systematic homeostasis. Schleiermacher's own philosophical system bears witness against the rational exhaustion of nature by reason, and his own theological system radically historicizes the contingent standpoint of subject as well as its object. And Whitehead goes even further by proposing a symbiotic model of a God who changes together with the world. Even the classic metaphysical attributes of immutability characterizing the bedrock of system are called into question by his philosophy of organism.

---

offenheit in den Humanwissenschaften, Philosophie im Kontext, no. 1 (St. Augustin: Gardez!-Verlag, 1997), 25–6.

7 Nicholas Rescher, *Cognitive Systematization: A Systems-Theoretic Approach to a Coherentist Theory of Knowledge* (Totowa, N. J.: Rowman & Littlefield, 1979).

II. Open Systems

If current smoke signals on the horizon of some academic disciplines indicate that system is making a come-back, then its rehabilitation for contemporary thought requires a constructive-critical look. This collection of essays represents such a close look at the various formal and material features that play into the concept of system. Conceiving the ingredients of "open systems" is both an experiment in anticipation and an engagement with thinkers from the past who have provided visionary accounts of system. Both Schleiermacher and Whitehead envisaged ways in which systematic thought could better be tailored to suit the results of empirical-scientific and historical research. Both thought through the requirements for systematic thinking, and proposed constructive approaches to system that could be termed "open." The essays in this volume explore the two systems of these thinkers in view of their respective proposed relations between reason and reality. In the process, features are teased out that make system what it is.

One key feature of system held relatively stable since its neo-Platonic conceptualization is its metaphysical commitment. System is the conception of individuals in relation to both a whole, whether understood as *this* world and *the* world (or *this* universe and *the* universe),[8] and to the ground of that whole. Cosmology lends itself to metaphysical reflection concerning the God-world relation. The tension in Western thought is focused precisely on this relation that has its implications for the "openness" of system. On the one hand, the neo-Platonic model explains the common ontological structures dispersed through diverse particulars by a theory of emanation and a metaphysics of participation connecting all reality with erotic glue. On the other hand, a medieval Christian understanding sees all particulars as ideas in the mind of God that are then actualized *ex nihilo* by divine command. The systems represented by Whitehead and Schleiermacher play with this tension in such a way as to magnify intra-worldly openness to motion, freedom, and life. The respective cosmologies are integrally related to metaphysical claims. Whitehead's cosmology is explicitly built up from the metaphysical structure of the actual occasion. Schleiermacher's cosmology reflects a more tentative metaphysics that is presupposed rather than full-blown. Whether explicit or implicit, both thinkers

8 David Ray Griffin specified the Whiteheadian distinction between *this* world / universe and *the* world / universe in a personal correspondence (March 18, 2004).

are similarly committed to working out a God who cannot be conceived without the world.[9]

The second feature is intimately related to the first. Since Kant, no metaphysical claims can innocently avoid epistemological justification. Metaphysics is accountable to the forum of reason. Although Schleiermacher and Whitehead differ on the degree to which speculation is permitted by finite reason, both work out complex theories of perception, thinking, and knowing to give their systems the necessary epistemological rationale. Knowing for both thinkers is given a wide foundation in experience; reality is perceived not only through physical senses but also by intuition and feeling (called "prehension" by Whitehead). Furthermore, the relation between perception in this broad sense and thought is an intimate one. Both thinkers appropriate Leibniz's minimum / maximum continuum between the organic and intellectual pole in order to explain why experience can be categorized by reason at all. The reason for system's potential adequacy to life is given by the monistic epistemological model that itself keeps system open to irritations provided by false schematization and to subsequent testing for the elimination of error. Systemic openness affirms a world that can be experienced in its infinite diversity, so that it can be known.

Metaphysics and epistemology are adjusted to each other by a third feature of system. Criteria must be supplied for any system advocating its truth and truthful relation to reality. Both Schleiermacher and Whitehead keenly recognize the necessity to stipulate criteria. Whitehead stipulates features of system and both rational and empirical criteria for its testing in the opening pages of *Process and Reality*: a system of "speculative philosophy" is "the endeavor to frame a coherent, logical, necessary system of general ideas in terms of which every element of our experience can be interpreted," and tests for its truth consist of rational criteria—coherence and logical consistency—as well as of empirical criteria—adequacy and applicability.[10] These criteria lend to Whitehead's system its openness to be tested against the yields of the natural and social sciences, and the entire range of human experience, including moral, religious, and aesthetic experience. Schleiermacher provides two requirements for knowledge in his philosophical text, the *Dialektik*. These requirements, the agreement of

9 For Whitehead, God can be conceived apart from this particular world, this particular "cosmic epoch," as David Ray Griffin has clarified in a personal correspondence (March 18, 2004).

10 PR 3.

thought with being and the identity of the pathway arriving at this agreement,[11] presuppose a system modeling the totality of reality in a coherent set of interrelations. In his theological work, Schleiermacher expands upon the features stipulated in the *Dialektik*, adding the criteria of coherence and comprehensiveness[12] and appealing to assertoric and divinatory qualities to test the adequacy of a historically located system of theology to the contemporary state of the church (in his case, the Protestant Christian church).[13] The question of testing inevitability opens system up to a stance of respect in the face of other areas of life and its investigation. System is not an imposition onto reality but arises in dialogue with it.

The fitting of a system to reality is a process that is never definitively closed. It is also not the privilege of one exclusive standpoint. Open system invites a plurality of constructions. The commitment of this volume is to multiple perspectives. Yet this commitment cannot be interpreted to mean the avoidance or rejection of questions concerning common features, universal claims, or systematic concerns. Perspectival individuality and common identity are not mutually exclusive options. Rather, the affirmation of plural perspectives is integrally joined to the affirmation of a plurality of co-existing systems. And with a number of perspectives, dialogue can take place. The quest for knowledge is truly intersubjective; even the truth of one's individual system can be contested only in a dialogical context.

III. Schleiermacher and Whitehead in Dialogue

"Systems in dialogue" is the way that this book stages its theme. As a practice in the intersubjective search for knowledge and truth, dialogue presupposes individual starting-points, an agreement concerning a common subject matter under investigation, and the conditions for the possibility of successful communication and understanding. Dialogue invites the exercise of philosophical and hermeneutical strategies in order to locate sameness and difference.

The dialogical stage is set by Schleiermacher and Whitehead, brought into dialogue by scholars of these respective thinkers. Yet both, who lived

11 DialKGA, 423–5 (*Einleitung* 1833, § 5,1)
12 BO §§ 200–1 (101–2). (Page numbers in parentheses.)
13 BO § 202 (102).

almost a century apart, made system their life's métier. Both present, as the authors of the chapters in this volume agree, new options for the construction of system, one in protest against the exhaustive rationalization of reality, the other in acceptance of a worldview shaped by Darwin's and Einstein's discoveries. What it means to have an individual perspective is illustrated by these two different thinkers. What it means to have caught the lure of system, and to have been sufficiently intrigued in order to actualize it, is precisely the focus here. In the dialogue between those who have engaged their thought, the merits of system are teased out in view of the contemporary question regarding the future viability of the systematic genre.

The agreement between Schleiermacher and Whitehead might have to do with respect to their appeals to similar predecessors. Plato and Leibniz tower as giants behind both systems. The love for coherence motivates Schleiermacher and Whitehead to aspire to the creation of a beautiful system, while the love for life drives both to address the relation between the world of becoming, or the truths of fact, and the world of being, or the truths of reason. On the basis of a common philosophical foundation, differences in perspectives can be detected. An adherence to Kantian critical philosophy explains Schleiermacher's resistance to speculation, while a dialogue with Hume and Einstein directs Whitehead to speculation on the processes of actual entities.

Both systems betray common philosophical underpinnings, while the way of moving from philosophy to theology is a significant difference between them. The applicability of Whitehead's system to theology is one to which he gestured in his own major work. A philosophical doctrine of God is a necessary aspect of system, Whitehead argues in working out the primordial and consequent natures of God in relation to world-process. This piece is given explicit theological determination by those process theologians committed to his thought, particularly those included as authors in this volume. Philosophy's fit to theology is not as neat in someone who exemplified them both on their own terms. Schleiermacher strove for distinctive philosophical and theological systems on the basis of what he considered to be distinct fields of study. His philosophical system, on the one hand, carves out reason, its activities, its requirements, goals, and presuppositions for knowledge; the best he can do with the Absolute is to posit a transcendent ground to his system without determining it in any religious way. His theological system, on the other hand, is decisively determined by the historical data teased from the history and state of the Christian church; his ecumenical Lutheran and Reformed theological program arises from contextualized Christian religion and is written for it.

IV. Open Systems in Dialogue

The concept of this collection of essays is designed to facilitate a dialogue between Schleiermacher scholars and Whitehead scholars on the topic of "open system." There are three subdivisions to the book, each concentrating on distinct features at which the systems of Schleiermacher and Whitehead are mined for their contemporary gold. Represented in each of the three subdivisions is both a "monolingual" Schleiermacherian position and a "monolingual" Whiteheadian position as well as two "bilingual" Schleiermacher and Whitehead protagonists who set the dialogue into play.

The first section, "System and Historical Context," identifies the historical-philosophical trajectories feeding into the ways in which Schleiermacher and Whitehead conceived their respective systems. Questions regarding the features of open systems are addressed in this section by considering philosophers with whom both Schleiermacher and Whitehead were deeply familiar (e.g., Plato, Spinoza, Leibniz, Kant) and specific philosophical issues relevant to system-building (e.g., synthetic and analytic judgments). The second section, "System and Categorization," explores key terms in the relation between individual and universal features of system, particularly in view of the relation between these features and their ground. The third section, "System and Contemporary Themes," explores the relevance of open systems in view of issues moving the contemporary imagination.

The keynote essay of this collection is written by Jack C. Verheyden. The essay, "Mapping the Land of Beginning Again," focuses on Schleiermacher's systems of theological and philosophical thought. According to Verheyden, the early Schleiermacher's pitting the "craving of system" against living religion does not preclude the later Schleiermacher from a commitment to the systematic form of academic theology and philosophy. The later theological works aim at a systematic presentation of doctrine that remains open to both the history of a lived religion and to the historical location of any theological articulation. And the philosophical system thematizes the relation of system to being.

The first paper of section one ("System and Historical Context") introduces the central issue at stake in the exploration of the idea of "open system." In "Whitehead at Infinite Speed: Deconstructing System as Event," Roland Faber exposes the dual tendencies in the history of Western philosophy between system's "self-confirmation," based on systematic criteria of reason, and system's "self-relativization," based on reality's subversion of system by its constant elusion of rational comprehension. All four types of system in Western philosophy—cosmological, intellectual, axiomatic, and organic systems—as Faber argues, contain both tendencies. Faber situates Whitehead's unique position in this history between Leibniz and Deleuze, showing that Whitehead represents a balance between the desire for system and the drive to keep system open to process, fluency, and history.

Jacqueline Mariña thematizes the relation between system and judgment formation in her paper, "Schleiermacher between Kant and Leibniz: Predication and Ontology." The two historical lens through which Mariña views Schleiermacher are precedents that Mariña contends do not mix well in Schleiermacher's *Dialektik*. On the one hand, the systematic issue of Leibniz's complete concept is related to his theory of predication, which is based on logical relations between singular concepts. On the other hand, the problematization of the analytic / synthetic distinction in the Kant-Eberhard controversy situates Leibniz's complete concept in relation to Kant's two-source theory of knowledge. Kant's theory explains how synthetic judgments are based on receptivity through sense perception of objects given to experience; hence the concept, for Kant, is informed by real relations between predicates and object. A system that is adequate to experience must take, according to Mariña, Kant's synthetic judgments into account.

In her "bilingual" paper, "Relative Freedoms: The Influence of Spinoza on the Systems of Whitehead and Schleiermacher," Krista Duttenhaver undertakes an exploration of both Whitehead and Schleiermacher in view of Spinoza's legacy: the stress on absolute determinacy as a restriction of an indeterminacy that threatens to undermine system. For both Whitehead and Schleiermacher, as Duttenhaver argues, the element of indeterminacy, whether as the evolution of the actual occasion in Whitehead's case, or as the self-formation of the human individual in Schleiermacher's case, contributes significantly to an understanding of the "openness" in their respective systems. By providing more adequate accounts of reality in terms of dynamic creativity and self-causation, in Whitehead's case, and in terms of the relative freedom and dependence of the finite causal nexus that itself is absolutely dependent on God, in Schleiermacher's case, both Whitehead and Schleiermacher posit relative freedom as constitutive of the indeterminate element that is nevertheless integrally related to a determinate specification of system.

Marjorie Hewitt Suchocki's paper, "System without Certainty," chooses to treat the empirical criteria for the critical testing of a system, for Whitehead, adequacy and applicability, in order to make a case for Whitehead's "open" model regarding the God-universe relation. The model's openness consists precisely in the fact that its adequacy to data can be empirically tested by physics, particularly the W-MAP (Wilkinson Microwave Anisotropy Probe) explorations. By bringing Whitehead's conception of God and the universe into dialogue with the W-MAP, Suchocki argues that a system tested according to its adequacy to data yielded by science, particularly physics, remains principally open to continued revisions that arise as a result of dialogue between the sciences. Such an open system resists systemically certain closure, preferring the modesty of provisionality.

As the first contribution to section two ("System and Categorization"), Anna Case-Winters' paper, "System and Dynamism in Whitehead's Thought: The Category of the Ultimate and the Concept of God," finds the answer to the question concerning the systematic longing for permanence and the acknowledgement of perpetual novelty in Whitehead's theistic-metaphysical system. It is the concept of God that, for Whitehead, is a metaphysically necessary feature of system, guaranteeing, on the one hand, the everlasting truth of the eternal objects and, on the other hand, grounding openness on the creativity that is constitutive of world process. Whitehead's God-world connection is predicated upon the metaphysical relation between God and the world. Yet as nontemporal, God is the world's metaphysical ground, luring a world in process to truth, beauty, and goodness.

In her paper focusing on the relation between the individual and system, Christine Helmer looks at a term more explicitly associated with Whitehead than with Schleiermacher. Nevertheless, in "Novelty and System in Schleiermacher's Thought," Helmer contends that novelty is constitutive of Schleiermacher's privileging of the individual in view of the post-Kantian problematization of knowing the individual. In Schleiermacher's system, knowledge of the individual presupposes structures of identity, such as ethical processes, structures of reason, and regulative-ontological guarantees for the correspondence between the system of thought and the system of the totality of reality. Although these features are significant, Schleiermacher's emphasis is on the dialectical intersubjective procedure that orients the search for knowledge of the individual to the hermeneutical task of isolating individual novelty from shared features of language. The mutual entailment between individual and identity keeps system "open" to novelty.

The constructive question regarding the openness of Whitehead's system for the articulation of a hermeneutical theory is answered by J. R. Hustwit in his "bilingual" paper, "Open Interpretation: Whitehead and Schleiermacher on Hermeneutics." For Hustwit, a Whiteheadian theory of interpretation is "open" to the interpretative event between authorial intention and the reader occasion because of two levels. Firstly, the metaphysic of the actual occasion's concrescence explains how the selection of specific data makes expression more linguistically precise, but does not diminish the influence of the data's less conscious polyvalence in expression. Secondly, Hustwit shows that the relationship between feeling and language in Whitehead's thought grounds interpretative polyvalence in the eternal objects that serve as the ontological parameters for translateability from authorial intention to interpretation.

In his "bilingual" paper, "Systematizing Agency: Toward a Panentheistic-Participatory Theory of Agency," Philip Clayton focuses on the topic of agency in relation to system. Clayton relaxes the tension he sees between agency that introduces novelty into system and system as a schematization of the whole, by marrying Schleiermacher to Whitehead. Although the finite agent is understood differently by Schleiermacher and Whitehead respectively, both complement each other by conceiving a participatory model of the relation between finite and infinite subject and a panentheistic model that allows for the freedom of finite agents. The reciprocal relation between God and finite agent opens system to the reciprocal activity of both.

Julia A. Lamm sets the constructive emphasis of section three ("System and Contemporary Themes") with her paper, "The Force of Dialogue and a Dialogue of Forces: Resources for Open Theological Systems." In this paper, Lamm identifies resources in Schleiermacher's own thought for a robust and contemporary Christian systematic theology. Particularly Schleiermacher's dialogical dialectic and his metaphysic of nature as a system of dynamic individual forces provide new ways to conceive an open system of theology that is motivated by love, is provisional even as it seeks the truth, and is inherently communicative.

The question concerning the relation of contemporary ethics to systematic thought is the theme of David Ray Griffin's paper. In "Feeling and Morality in Whitehead's System," Griffin analyzes elements of Whitehead's theistic naturalism and worldview in order to defend particular moral features that Griffin claims are presupposed in daily life (human freedom, normative values, and ultimate importance) and to argue that the initial aim injected into each unity of the system by God is a universal, metaphysical characteristic of morality. With Whitehead's theistic naturalism, Griffin can explain how an ethical system can be conceived together with a God as an explanation for a universal dimension to ethics, yet this God is one who does not supernaturally intervene in the system in order to actualize an ethical will.

In his "bilingual" paper, "Process and God in Whitehead and Schleiermacher," Eilert Herms provides a comparison between these two thinkers in view of their common rejection of Kant's claim that metaphysics is both impossible and unnecessary. For Herms, both Schleiermacher and Whitehead are committed to working out a theory of being as a theory of the universal conditions of becoming. Furthermore, this theory is pertinent to the central claim of religion: the subject matter of a universal theory of becoming includes the becoming of human persons and the work of God. Thus systematic openness is related to a metaphysical conception of becoming.

The last paper in the volume, by John B. Cobb, Jr., addresses the significant issue of theology and philosophy in relation to religious pluralism. In "Schleiermacher and Whitehead on Religious Pluralism," Cobb sets up two models to account for the plurality of religions. According to Cobb, Schleiermacher (as well as the trajectory influenced by him) determines the essence of religion in general and subsequently uses this determination as a norm by which other religions are evaluated. Whitehead's approach overcomes Schleiermacher's restriction to the Abrahamic religious traditions by providing a number of definitions of religion, such as "the cultivation of the inner life," that succeed, according to Cobb, in both describing many forms of religious life and in encouraging their respective pursuits of goals. Whitehead's model is characterized by its openness to the contemporary reality of religious pluralism that includes both theistic and non-theistic religious, and also political and social expressions.

Mapping the Land of Beginning Again

Jack C. Verheyden, Claremont School of Theology

Arising in the far northwest corner of the state of Texas is the river, originally known in Spanish as *El Rio de los Brazos de Dios*, "The River of the Arms of God." In the early nineteenth century, it was aggressively anglicized by the incoming Texans to simply, "The Brazos." It winds its way through the heartland of Texas to the southeast as the state's largest interior river finally opening onto the Gulf of Mexico.

Many lakes dot the valley of the Brazos, and each year during the winter flood season, millions of acre-feet of water are hurled beyond its banks, only to dry to a rather narrow stream in spots under the torrid summer sun. The river does not move in canyons but often cuts across prairies. And when the next year's floods come and recede, the river is careless about reestablishing itself in last year's path, revising its route perpetually.

The sovereignty of the Brazos area also changed regularly. First dislodging the Native American Indians was the flag of Spain, to be succeeded for a time by the *fleur de lys* on a white shield of France, only to be replaced by the Spanish flag again. But that was in turn succeeded by the flag of the nation of Mexico, which then gave way to the Lone Star flag of the Republic of Texas. Then came the symbol of sovereignty of the United States, but that changed to the flag of the Southern Confederacy for a few years until the Stars and Stripes returned.

The people, continually flocking to the area, said: "We are going to make a new start here," and they drew new boundaries hither and yon. All these things—and others—sometimes brought the name to the area around the Brazos as "the land of beginning again." Can there be a map of such a place? A new one must constantly be re-drawn.

I. The Meanings of System

The theme of "Open Systems" raises the question of terminology with respect to "system." What kind of "map" of the changing reality of experienced life is indicated by this word? Even a casual exposure to philosophical and theological literature uncovers a wide, flexible, and even disparate use of the word system. And one might think that the derivative word "systematic" might be merely the adjectival form of the same meaning. But actually it very often conveys a meaning that indicates a methodical procedure in treating a topic, not that the writer or speaker has "a system." Indeed, "systematic" can mean no more than a classification of items. I do not, of course, have any idea that a few remarks at the outset here will clear up this usage for some proper one, but I do think that we should try to be aware of how we are using the term system.

What are the features that make something a system? Sometimes we hear that being consistent makes a system. I have never found that persuasive. One can write three paragraphs on a topic that are consistent but are hardly a system. Poems can be consistent, but systematic poetry seems a virtual contradiction in terms. We can then introduce the words "rational concepts" into the mix in order to move us along to the type of discourse found in philosophy and theology. This appears more promising but by itself runs into counter currents. For instance, Socrates in the Platonic dialogues proceeds along the line of the exploration of rational concepts, but most interpreters would not hold that he presents a system. Indeed, Plato himself would be, at best, quite an ambiguous representative of formulating a system. Clearly, the dialogues themselves are not in the form of a system. With Aristotle, we reach a much more promising candidate. The Aristotelian system would not be a phrase that many would contest. But for a substantial number, Aristotle's method of procedure would still prevent the constitution of a system.

If we say that a system is a body of philosophical or theological thought, then Plato and Aristotle would certainly qualify. And surely in the subjects of philosophy and theology, one frequently comes across the series of bodies of thought referred to throughout history as a variety of systems. But, then again, opposing currents are raised. Søren Kierkegaard has a body of thought that has been much studied and analyzed during the twentieth century. But is it a system? Certainly he did not think so; indeed, his body of thought contains probably the most famous attack upon and rejection of "system." Nietzsche from a very different perspective is hostile to system in his body of thought. Then to jump across the Atlantic, William James does not show much enthusiasm at all for system.

In addition to "rational concepts," I would say that, in general, the idea of a philosophical system has at least two main features. One such feature is the notion of a single structure of conceptual implication or relevance that applies across a variety of areas or fields.[1] In this respect, the idea of system in philosophy stands opposed to the kind of thought which would explain topics piecemeal with concepts that are without pertinence to others. One could divide philosophical inquiry into specialized fields—for example, logic, metaphysics, and ethics—each which can be pursued independently of each other and, therefore, would not constitute a system. The second main feature of system is "completeness." A system is an articulated totality that displays the form of unity.[2] It is not a fragment of the topic treated.

The opening pages of Whitehead's *Process and Reality* articulate these two features.

> Speculative Philosophy is the endeavour to frame a coherent, logical, necessary system of general ideas in terms of which every element of our experience can be interpreted. By this notion of "interpretation" I mean that everything of which we are conscious, as enjoyed, perceived, willed, or thought, shall have the character of a particular instance of the general scheme. Thus the philosophical scheme should be coherent, logical, and, in respect to its interpretation, applicable and adequate. Here "applicable" means that some items of experience are thus interpretable and "adequate" means that there are no items incapable of such interpretations.[3]

Whitehead's words about a "logical, necessary system" go beyond the two features I mentioned above. Whitehead points to an area of system that has been especially controversial in the history of philosophy, even though he does not explicitly reflect on the specific issue. It appears to have arisen in modern philosophy in the seventeenth century with Descartes. Much philosophy in this century was modeled after the pattern of mathematics and mathematical logic, and spoke of deducing truth from certain ultimate axioms. In this way, system is seen as something that results from a single, fundamental principle. Descartes' "I think, therefore I am" establishes a

---

1 LOUIS MINK, *Mind, History, and Dialectic: The Philosophy of R. G. Collingwood* (Bloomington: Indiana University Press, 1969), 245.

2 TOM ROCKMORE, *Before and After Hegel: A Historical Introduction to Hegel's Thought* (Berkeley: University of California Press, 1993), 123.

3 PR 3.

fixed point from which to build a system of knowledge. This characteristic of system carries through the seventeenth into the eighteenth century.

German classical philosophy, the period generally characterized as lasting from Kant to the later Schelling, brings the discussion of system to the highest pitch, I believe, in the history of philosophy. Kant's revamping of philosophy means that he takes up the idea of system in the *Critique of Pure Reason*. He says in the Preface to the second edition of the *Critique* (1787) that "The critique is a treatise on the method (of the science of metaphysics), not a system of the science itself."[4] It is a task of speculative reason to set down in advance the entire outline of a system of metaphysics. Kant states that he is not writing such an outline, and acknowledges his work as merely a propadeutic for such an undertaking. At the end of the *Critique*, in the section in which he discusses the architectonic of pure reason, "Transcendental Doctrine of Method," Kant has some important words to say on system. Kant says, "By an architectonic I mean the art of systems. Since systematic unity is what first turns common cognition into science, that is, turns a mere aggregate of cognition into a system, architectonic is the doctrine of what is scientific in our cognition as such."[5] Along with the major point relating systematic unity to the aggregate of common knowledge, this passage sharply brings out the close alliance in the German language of this era between "system" and "science." In Kant's context, the two terms are defined in relation to each other. Kant continues the passage, speaking of reason's concept of the form of a whole and the purpose of this whole structuring sense congruent with that purpose. Such conceptual wholes can be both empirically and practically grounded, in which case they yield technical unity, the kind of unity in artistic craft. These wholes can also be grounded a priori in conformity with an idea, and this is what is termed science. The purposes, then, do not await empirical legitimation.[6] It is this type of conceptual whole that is an architectonic unity.

Kant's view of system sets the stage for the first great post-Kantian system, that produced by Johann Gottlieb Fichte. Fichte wanted to both show the unity of the first two critiques and determine the transcendental foundation for the system not provided by Kant's prolegomena in the 1794 *Foundation of the Entire Science of Knowledge*. This supposedly tran-

4 IMMANUEL KANT, *Critique of Pure Reason*, trans. Norman Kemp Smith (London: Macmillan, 1929), B xxiii (25). (Page numbers in parentheses.)
5 Ibid., B 860 (653).
6 Ibid., B 861 (653).

scendental system became the most discussed system for the next several years. It appears to take very seriously the separation of the fundamental idea grounding a system from the empirical mentioned in Kant's discussion of architectonic unity. Fichte's absolute principle is the self positing itself as infinite activity. The system follows from this foundational principle (*Grundsatz*) as a series of deductions.

II. The Trajectory of System in Schleiermacher's Publications

This review of Kant's comments on system and Fichte's production of a deductive transcendental system are important background for Schleiermacher's statements about system. In his early twenties, Schleiermacher's letters express an antipathy to the idea of system. He writes, "I continue to remain aloof from craving a system, a system which would provide me with definitive answers to all the questions which can be posed, answers which at the same time would also be congruent with the remainder of my knowledge .... I have always believed that critical analysis and examination, as well as listening to all witnesses and parties, is the only way which enable us to achieve certainty."[7] Schleiermacher does not intend to lose himself in the "deserts of metaphysics." Remarks such as this one antedate Fichte's 1794 production of a system.

The appearance of the *Speeches on Religion* in 1799, however, provides clear evidence for the atmosphere of the orientation toward transcendental system extant in the 1790's. The *Speeches* are not an academic type of writing but an example of a rhetorical and literary work, originally intended for the romantic salon of Henriette Herz, known as the Wednesday Club. It is Schleiermacher's first book and was spurred into composition by the event of being hounded by his fellow Wednesday clubbers. Schleiermacher had, in fact, just turned thirty and had not yet published a book!

This work is most famous for its sustained argument concerning the distinctive nature and source of the religious in human life. The religious element cannot be subsumed under some other component of human nature, whether that be theoretical knowledge or moral action. Religion neither originates in the pure impulse to know, nor does it know anything of the deducing and connecting that characterize metaphysics or systems. There

7 Friedrich Schleiermacher, *Aus Schleiermacher's Leben in Briefen*, vol. 1 (Berlin: G. Reimer, 1860–1863), 78. I have followed the translation of Gerhard Spiegler, *The Eternal Covenant: Schleiermacher's Experiment in Cultural Theology* (New York: Harper, 1967), 39–40.

is, therefore, a pervasive polemic against system in the *Speeches* as something that is antithetical to the life of religion.

I need to dwell on this important point. The rejection of system is not merely a passing aspect of the book. The theme appears in the first speech and can still be found in the fifth and last speech. The initial critical remarks come in respect to the systems of theology. You romantics, Schleiermacher suggests, despise such things because you have not found religion there and that is because it is not there.[8] The speaker then broadens his criticism to philosophy. These systems come from academic institutions, which are the habitations and nurseries of systems and their dead letters.[9] "[R]eligion ... by its whole nature is just as far removed from all that is systematic as philosophy is by its very nature inclined toward it."[10]

The Fichte-like system echoes between the lines when the speaker asks in the second speech, what does your metaphysics or transcendental philosophy do? System classifies and divides, "and deduces the necessity of what is real while spinning the reality of the world and its laws out of itself."[11] This is not a realm into which religion should venture. "A system of intuitions? Can you imagine anything stranger? Do views, and especially views of the infinite, allow themselves to be brought into a system?"[12] Schleiermacher turns to the astronomical starry sky. He claims that this infinite chaos is the most suitable and highest symbol of the religious because every point there represents a world. Those who know their religion and its essence will utterly subordinate every apparent connection to the particular, and will not sacrifice the smallest part of this particular to such connections.[13] Philosophy tries to subsume all knowing under one common knowledge. This mania for system rejects what is different, even when it is quite conceivable and true, because it would spoil the systematic ranks and disturb the beautiful connections by claiming its place.[14] "Whoever only thinks systematically... inevitably circumscribes himself."[15] The religious life, essentially understood, provides the mind with unlimited freedom.

The withering fire of the *Speeches* continues on, but there is one text, which strikes a different note and should not be omitted here. Late in the

8 *On Religion*, 13 (speech one).
9 *On Religion*, 13–14 (speech one).
10 *On Religion*, 14 (speech one).
11 *On Religion*, 20 (speech two).
12 *On Religion*, 26 (speech two).
13 *On Religion*, 27 (speech two).
14 *On Religion*, 27–8 (speech two).
15 *On Religion*, 28 (speech two).

second speech, we find a paragraph articulating a chaos, now not representing points that are worlds in themselves, but only a buzzing confusion. This confusion is infused with the infinite and gives rise to a fetish and idols. At another level, this matter moves into the universe, presenting itself as a multiplicity without unity, as a manifold, whose constant conflicts mark its manifestations. This level is higher than the previous one, but higher still is the third level at which the universe manifests itself as totality. At this apex, conflict is again united; this is a step that arrives at "unity in multiplicity, as system and thus for the first time [the universe] deserves its name."[16] The movement of this paragraph adumbrates propositions 7 and 8 of *The Christian Faith*.[17] In these paragraphs, Schleiermacher discusses philosophy of religion as the movement of the religious consciousness in a history from fetishism through polytheism to monotheism. Unity in multiplicity is parallel with monotheism, which Schleiermacher approves.[18] Since he speaks of this unity "as a system," this passage stands as a positive sign amidst the other discourse of the *Speeches* on system.

To list the reasons Schleiermacher advances in the *Speeches* for rejecting systems, the following should be included. (1) Religious life is too immediate as it is lived out to be grasped by a system because the infinite is too close. Systems require "distance" to be concocted. This distance is a result of abstractions and convolutions; inferences and drawn out conclusions of system mean that the fountain of life is squeezed dry. (2) A system drives toward that which is universal, its tentacles reaching out prematurely for connections. Such a feature turns away from individuality and so smothers it. (3) The term system indicates for Schleiermacher a rigid and inflexible body of knowledge that forces everything into a certain mold. System therefore strangles that which is new and different. (4) Schleiermacher rejects the transcendental system that deduces reality from an intellectual principle.

Schleiermacher's follow-up companion piece to the *Speeches*, the *Soliloquies*, was published the following year in 1800. The language is more robustly romantic and perhaps, also the content. These two books, along with the *Confidential Letters Concerning Schlegel's Lucinde*, also published in 1800, as well as his associations in the circles of Romanticism, earned Schleiermacher a reassignment from his chaplaincy post in Berlin to the

---

16 *On Religion*, 52 (speech two).

17 CF §§ 7–8 (31–9). (Page numbers in parentheses.)

18 CF § 8,4 (37–8).

boondocks at Stolpe. In this setting, he again took up the study of Christian theology, but two other matters called his attention and efforts to publication. One of these works was the translation of Plato into German, a project stretching over the decades, and a long work, *The Basic Features of a Critique of Previous Ethical Theory*, published in 1803 when Schleiermacher was thirty-four years of age. In his massive volume of 1870, *The Romantic School*, Rudolf Haym sees these two endeavors as a landmark event in the movement of Romanticism.[19] Here one of its leaders has appeared in writings of disciplined precision and scientific rigor, and such a thing is a startling change against the backdrop from which these writings come. "The transition out of the flowing form, which his ideas had had during the time of the common life with his poetic friends, into the rigid form of strict scientificness... appear as an undertaking accomplished at a pace of almost breakneck speed."[20]

Haym thinks that as a production in style, the *Critique* is open to many criticisms. In spite of the problems, Haym thinks that the work marks the beginning of a new epoch because he sees the revolutionary spirit of Romanticism entering into a new arena. Schleiermacher's book represents the attempt to direct the feelings and needs of the imagination into the rational structure of system. It emphasizes the role of individuality as a significant measure for the success of ethical systems. Haym sees Hegel's *Phenomenology of Spirit*, published three years later, as a kind of marching companion to Schleiermacher's *Critique* because the *Phenomenology* is a combination of poetry and science. It is, however, one thing for Haym to point out this historic role in the history of the Romantic school, and it is another matter to understand what this new style means for the kind of objections to system that Schleiermacher advances in the *Speeches*. This is particularly the case since he publishes a second edition of the *Speeches* in 1806 in which almost all of the texts referred to in 1799 are still present. Of course, one must not overlook the fact that the subject of the *Speeches* is "religion," while that of the *Critique* is ethical theory.

Schleiermacher does not present his own work in the *Critique* as a system, but he does lift up system as a criterion for evaluating the reviewed systems of ethics. The attack of the *Speeches* is not the whole story about Schleiermacher on this subject. The remainder of Schleiermacher's books, with the exception of the continuing Plato translations and the introduc-

19 Rudolf Haym, *Die Romantische Schule*, 5th ed. (Berlin: Weidmann, 1870).
20 Ibid., 927.

tions to the dialogues, are all theological: *Christian Eve*, the *Brief Outline of the Study of Theology*, the *Critical Essay on the Gospel of Luke*, the two editions of *The Christian Faith*, and the *Sendschreiben*, or *Open Letters to Dr. Lücke*. The *Brief Outline* and *The Christian Faith* display Schleiermacher's theological system. In terms of his publications, it is these that one has to utilize if one is to understand Schleiermacher as an advocate of system for his own work.

III. The Christian Faith according to the Principles of the Protestant Church Presented in Connection

It should be said from the onset that the language of system in theology has a different function than that of system in philosophy. The reason for this difference is that Christianity in its intellectual form is frequently referred to as a system. In theology, one is not inherently dealing with different fields of thought in their interrelationships, but rather with one field. The field, however, addresses many different topics: God and the triune nature, creation, the human as the image of God, sin, Jesus Christ and redemption, the church, and eschatology, are said to be facets of the Christian system. To say, therefore, that *The Christian Faith* presents a system of theology does not in itself say much at all. This book is preeminently a systematic presentation. Indeed, it is the most systematic statement of Christian doctrine that I have ever read, and I have read a representative number of books on the subject. I do not make this observation strictly as a compliment, by the way. Traditional theology did not set out to accomplish the rational drive for unification that appeared in the seventeenth and in the eighteenth centuries. Augustine, as a one time teacher of rhetoric, would intentionally digress because that was the rhetorical ideal at the time. His method does not make him appear very systematic now, even though I doubt that few would question his power of intellect. Thomas Aquinas has a *Summa* or two, which would probably qualify him as having a system. But does Luther have a system? Hardly. Calvin organized the facets of Christian doctrine most influentially in a single Protestant work, *Institutes of the Christian Religion*, but many modern readers do not find him to be systematic. It is possible for Calvin to elaborate doctrinally more than one side of an issue. But a reader, and particularly a new reader, might respond by pointing out that Calvin contradicts what he claimed some pages before. Calvin would reply that he is presenting Christian doctrine according to the Bible. If both positions are found in scripture, then that is the way, Calvin says, that he is going to present them. He is not going to take the Holy Spirit to school and try to tell God how it should be done! The modern drive for rational unification has a difficult problem with the lack of this interest in system in much traditional thought. In the 1920's, Karl Barth was critical of system in theology because it lifted up the drive of the human subject in such a manner as to get in the way of letting the proper object of theology appear. And he thinks that Schleiermacher led the way in the early nineteenth century for this approach.[21]

If one looks at *The Christian Faith* as a particularly intense theological system, what are the features that bring this out? One feature that does not bring it out is the word system itself. Schleiermacher may have used it plentifully in 1803 when discussing other systems of ethics, but he does not continue this practice for this book eighteen and twenty-seven years later. The English reader will not notice this fact because the word is freely splashed through prominent places in the Introduction to the book.

Take the famous proposition 19 that says: "Dogmatic Theology is the science which systematizes the doctrine prevalent in a Christian Church at a given time."[22] But the German according to my literal eye says, "Dogmatic theology is the science of connections prevalent in a Christian society at a given time."[23] The most prominent use of system calls for special comment. This is the word *Glaubenslehre*, which literally means "doctrine of faith," but is rendered again and again as "system of doctrine." In the four pages of table of contents, the English translation lifts up in full capitals "the first part of the system of doctrine, and the second part of the system of doctrine" where the German says *Glaubenslehre*.

I think that Schleiermacher has a certain resistance to the word system. He had protested this word so vehemently in the *Speeches* and he continued this polemic in 1806 and in 1821. In *The Christian Faith*, Schleiermacher does not absolutely reject the word. Proposition 20 utilizes the word system,[24] and proposition 28 utilizes the term systematic. Proposition 28 and its explanation are important for my theme. The proposition reads,

---

21 Karl Barth, *The Theology of Schleiermacher: Lectures at Göttingen, Winter Semester of 1923–24*, ed. Dietrich Ritschl, trans. Geoffrey W. Bromiley (Grand Rapids: Eerdmans, 1982), 190.

22 CF § 19, proposition (88).

23 "*Dogmatische Theologie ist die Wissenschaft von dem Zusammenhange der in einer christlichen Kirchengesellschaft zu einer gegebenen Zeit geltenden Lehre.*" CG § 19, proposition (119).

24 CF § 20, proposition (94).

"The dialectical character of the language and the systematic arrangement give dogmatics the scientific form which is essential to it."[25] "Dialectical" means language that is technically correct and can be used in the arena of public knowledge where poetic and rhetorical expressions do not suffice. Dialectical language thus gives rise to the scientific formation appropriate to dogmatics. In doing this, it draws on philosophy, a realm of much diversity. No philosophies are excluded from dogmatics except those which do not distinguish God from the world in any way, that do not admit any opposition between good and evil, and do not differentiate the spiritual and the sensible.[26] The more frequently that philosophical systems change within these boundaries, the more frequently do revolutions in dogmatic language occur. These revolutions are unavoidable when a philosophical system has become out of date, i. e., when thinking no longer conforms to its type. But these revolutions often take place at an earlier stage when theologians have been caught up with intense zeal by a rising philosophical system, and who hope that the new system will be better fitted than any previous one to make an end to all divisions in the theological interpretation of faith. "Others again, by the spectacle of that very zeal, are made apprehensive lest one particular philosophical system should set itself up as lord and judge in theological matters."[27] But the apprehension is just as unfounded as the hope.

At this point in the Introduction, Schleiermacher gives his definition of the systematic aspect in his theology. Language alone cannot produce the scientific formation because even the clearest proposition reflecting a most definite idea is not stable enough in meaning until it has been placed in a certain framework. He writes, "But this is the very essence of systematic arrangement, that by comprehensive co-ordination and exhaustive subordination each proposition should be brought into a perfectly definite relation with all others."[28] A dogmatic construction is capable of the above if it reflects a self-contained whole. Schleiermacher gives us a specification of system and systematic in propositions 20 and 28, and the meaning he advances shows an important overlap with another word that he uses much more freely and prominently in the Introduction: *wissenschaftlich*. The word is usually translated into English as "scientific," with the well known incongruities in the attached meanings between German and English con-

25 CF § 28, proposition (118).
26 CF § 28,1 (118).
27 CF § 28,1 (119). (I have revised the translation slightly.)
28 CF § 28,2 (120).

cerning the word "science." Scientific language can be explained in a way quite parallel to the "dialectical" language mentioned above and it executes the connected arrangement just seen in that which is systematic.

Doctrinal propositions possess a two-fold value—a churchly one and a scientific one.[29] The churchly value comes from the capacity of a statement to express the self-consciousness of the human being's relation to God, "the religious affections of the inner life" or soul. Does a specific utterance of faith accurately represent the content of that which is Christian? Unless it does so, one cannot get further in theology. The primary norm is whether or not something has a churchly character, which is to say, whether it belongs to the religious community of Christians.[30] And the reference to Jesus Christ must be as prominent in dogmatic propositions as it is in the Christian religious consciousness itself.

The scientific value of propositions comes from what has already been mentioned: the definiteness of the concepts employed and their connection with each other in passing beyond the poetical and the rhetorical. Scientific value is found in the fruitfulness of propositions, that is, in their capacity to harmonize with others. Those statements which most open up and can combine with the largest range of other religious statements are to be preferred because they contribute to an interrelated whole. The more the whole is fittingly interrelated, the greater the confidence one may have that the doctrines are correct. The interrelated whole, that is, the system, provides insight into the truth. And the understanding that is thereby attained has consequences for those responsible for guiding the Christian religious community. Leadership of the Christian church is augmented and enhanced by this understanding.

The scientific approach is a word that Schleiermacher is more comfortable with in his *The Christian Faith*, although it has the same function as the systematic approach. In proposition 27, he discusses three types of doctrinal constructs.[31] One construct emphasizes the appeal to scripture throughout. In this case, confessional documents retire into the background as does the homogeneity or coordination of the religious statements. In contrast with this scriptural dogmatic, there is the symbolical dogmatic which is both based on the confessional documents and interprets scripture through these documents and their authorized spokesmen.

29 CF § 17, proposition (83).
30 CF § 17,1 (83–4).
31 CF § 27 (112–8).

Finally, there is the third type that Schleiermacher calls the scientific dogmatic. This kind starts "from some recognized principal points" and intends to "make everything clear by its orderly sequence, the parallelism of its component parts, and the coherence which the individual propositions have to one another."[32] Here scriptural citation and confessional documents play a role, but they recede into the background. One issue must be clear: the above "principle points must be none other than the fundamental facts of the religious self-consciousness conceived in a Protestant spirit."[33] That is, the norm of churchly criticism must be respected.

It is not difficult to see where Schleiermacher's own production fits. When Schleiermacher describes the scientific dogmatic, he is describing his own system; his organization and interrelation of doctrines reaches an extraordinary pitch.

This is nowhere more emphatically shown than in Schleiermacher's discussion of the attributes of God. Let me try to illustrate the systematic dedication to arrangement around certain principle points. In proposition 30 of the Introduction, Schleiermacher writes that he will divide Christian doctrines into three aspects.[34] There is the primary form, which describes the Christian religious self-consciousness as experienced human life. There are also two forms of doctrine which derive from this religious self-consciousness. One of these two forms addresses the attributes and actions of God, the other focuses on certain qualities or characteristics of the world. These two forms are correlates in these respective subjects to the human religious statements. It is possible to make statements about God, as it is also possible to make claims about certain characteristics of the world that are not based on human religious self-consciousness. These statements can be based on science, whether metaphysical science or natural science.[35] And they may or may not be true in any particular case. But they do not belong in theological doctrine because they do not derive from the religious states of inner life as the correlates of those states.

The principle here should be noticed because it is indicative of the strictness of Schleiermacher's systematic procedure. It is not that metaphysical accounts of God are wrong. It could be that they depict many accurate aspects about God. This depiction, however, does not imply their viability

32 CF § 27,4 (117). (I have revised the translation slightly.)
33 Ibid.
34 CF § 30, proposition (125).
35 CF § 30,2 (126).

for inclusion in Christian doctrine. This principle also holds for Schleiermacher's own theological reflection. In his philosophizing about the relation of thought and being, he finds it necessary to postulate a transcendent ground that is their unity if knowledge is to transpire.[36] His arguments in the *Dialektik* progress to the conclusion that this transcendent ground is living, and he eventually names this ground, God. God as the transcendent ground of thought and being, however, does not enter the system of doctrine. Indeed, Schleiermacher's discussion of the transcendent ground has been taken as a kind of rational argument for the existence of God. In response to this interpretation, proposition 33 must be consulted. This proposition states that the recognition of the feeling of utter dependence as a universal element of life completely replaces such arguments.[37] There is no mention that such arguments are not rationally successful any more than that they can be. They simply are not appropriate for the system of the doctrine of faith.

Since statements are so closely correlated in this manner with their basis in human self-consciousness, the arrangement of the order of treatment of the relevant human states also defines where the systematic location of the doctrine of God is elaborated. For instance, in his statement of doctrine, Schleiermacher begins with the religious consciousness in its nature as defining religion generally, a dimension that is contained in every kind of this consciousness. To attempt to illustrate this point, let us take a statement of didactic language that is not scientific. In the 1806 edition of the *Speeches* in the speech on "The Essence of Religion," Schleiermacher makes the following statement. "The sum total of religion, then, is to feel all that moves us in our feeling, in the supreme unity of it all, as one and the same, and to feel all that is individual and particular as mediated only through that unity—that is, to feel our being and life as a being in and through God."[38] Such a statement does not reflect any particular concrete religion, but could be included in many.

The first part of the doctrine of faith presents a discussion of the religious consciousness as founded on the feeling of utter dependence. This feeling of utter dependence is a feeling that is logically presupposed, so to speak, by various religions. The utter dependence of the human in its world is ex-

36 DialO 302.

37 CF § 33, proposition (133–4).

38 Friedrich Schleiermacher, *On Religion: Speeches to its Cultured Despisers*, trans. Terrence N. Tice (Richmond: John Knox, 1969), 94 (speech two).

pressed in Christian doctrine by statements concerning the creation and the preservation of the world. These statements have a correlate in God who does create and preserve. Such attributes of God express divine causality, eternity, omnipresence, and absolute spirituality. These are not all that can be said about God, by any means, but the discussion is confined to the human state that has been indicated. Other affirmations about God must wait upon further discussion of human states concerning sin which correlate with the attributes of divine holiness and justice. The account of this aspect of God comes ninety-three pages after the initial group. After a three hundred and sixty page long discussion of the consciousness of grace, we encounter an elaboration of the attributes of love and wisdom.

It would be extremely difficult to find anything like this in the history of Christian thought prior to Schleiermacher. Schleiermacher's system probably embodies system in theology most rigorously, if system is defined as Schleiermacher defines it in terms of the order and arrangement of the parts in an interrelated whole that is executed on the basis of a few principle points. I think, however, that his arrangement is too systematic. In my opinion, his delay of the discussion of the attributes of love and wisdom to the last few pages of the long work is a tactical mistake that he could and should have avoided. As it stands, the "metaphysical attributes" overwhelm the "ethical ones," and the unique status of love as defining the essence of God is seriously blunted in the reader's understanding.

In light of the scientific and systematic drive of his doctrinal theology, how does Schleiermacher's position relate to his protest in 1799 in the *Speeches*? Is the building of systems completely opposed to the spirit of religion?[39] Schleiermacher had first objected to the distancing of the intellectualized systems from the immediacy of religious life. *The Christian Faith* opens with a discussion in propositions 3, 4, and 5 with an account of feeling in the economy of human selfhood with its distinctions between feeling, knowing, and doing.[40] Schleiermacher links feeling with immediate self-consciousness and by this word, he intends to contrast religious feeling with other levels and aspects of consciousness. On the one hand, since he is determining self-consciousness, he does not mean the unconscious or what we might term "stream of consciousness." These elements of consciousness are too unfocused. On the other hand, Schleiermacher also distinguishes immediate self-consciousness from ideas of the reason-

39 *On Religion*, 113 (speech five).
40 CF §§ 3–5 (5–26).

ing mind, even if these ideas are of one's own person. Such thinking is more at the control of the intellect than the "immediate existential relationship" which religious feeling attests.[41]

I cannot here fully depict Schleiermacher's analysis of religious self-consciousness. The feeling of utter dependence which defines the unity of selfhood and is the same thing as "being conscious of being in relation to God"[42] is not found in reflective or objective self-consciousness. Schleiermacher's contention in paragraph 4 is important for understanding its relation to the first rejection of system above. On the one hand, Schleiermacher neither constructs rational concepts nor carries out deductions and inferences. He is using scientific language to describe a primal experience. On the other hand, Schleiermacher's drive to interconnect various expressions of religious convictions is not descriptive in that fashion. The harmonious whole that results provides a reference to true understanding. In that respect, there is a collision between *The Christian Faith* and the *Speeches*. Insight from an interrelated whole is not congruent with the symbol of infinite chaos.

The second protest against system is its capacity to smother individuality. *The Christian Faith*, as virtually all of Schleiermacher's writing, advocates distinctive individuality whether it addresses immediate self-consciousness or the religions of history. In that sense, his theological system is an advocacy of individuality rather than of the strictly universal. A pervasive principle of Schleiermacher's thought is distinctiveness within the framework of relatedness.

The third protest is that a closed system forces the new into a certain mold. I am not sure that any system can completely avoid this feature. William James was bothered a good deal by the quality of closure characterizing system. Schleiermacher thinks in terms of evolution, not revolution. He does not embrace sharp breaks and reversals in the interpretation of life. Such a developmental outlook is perhaps easier to associate with a system. Schleiermacher's own theological work demonstrates the potential for including the new and different within system. Schleiermacher's famous definition of dogmatic theology, "Dogmatic theology is the science which connects the doctrine prevalent in a Christian church society at a given

---

41 Friedrich Schleiermacher, *On the Glaubenslehre: Two Letters to Dr. Lücke*, trans. James Duke and Francis Fiorenza, AAR Texts and Translations, no. 3 (Chico, Calif.: Scholars Press, 1981), 40.

42 CF § 4, proposition (12).

time,"[43] affirms that doctrine has to be continually rewritten as the church moves along through time and as it is faced with new challenges and issues. This impetus became something of a truism in twentieth-century theological thought, but it was certainly not such when Schleiermacher said it. Through the centuries, the Christian churches have not looked upon doctrine as changing. Schleiermacher indicates a flexibility here that partially deflects the protest of the *Speeches*.

The fourth protest concerns the transcendental system that deduces the principles of the whole of reality from a fundamental principle. It is a philosophical matter that goes beyond the system of *The Christian Faith*. I hold that Schleiermacher never changed his mind on this aspect of polemic against system.

IV. The Relation of Theological System to Schleiermacher's Philosophy

In this section I address a subject that has occasioned much interest and discussion: the relation of Schleiermacher's theological system to his philosophy. For the purpose of this paper, I will consider some aspects of this subject. I have written this essay to reflect Schleiermacher's own published works, which are overwhelmingly theological. His most widely read philosophical book, the *Soliloquies*, bypasses system and is consonant with the *Speeches*. Yet Schleiermacher has gained a considerable reputation as a philosopher. In fact, he is the only person since the Enlightenment who, as a preacher and a theologian, has had separate chapters devoted to his thought in distinguished histories of philosophy.[44]

Scheiermacher's reputation as a philosopher is largely due to his posthumously published lecture materials—manuscripts, outlines, and student notes. I do not know of another figure who has earned such philosophical renown on the basis of such unstable material. In one area of philosophy, hermeneutics, Schleiermacher is regarded as one of the classically defining figures of the fundamental trajectory of this area of thought even though he never published anything at all on the subject! His lectures in the area of

43 CG § 19, proposition (119). (My translation.)

44 Frederick Copleston, *A History of Philosophy*, vol. 7 (London: Burns, Oates, and Washbourne, 1965); Friedrich Ueberweg, *Grundriss der Geschichte der Philosophie*, vol. 4 (Berlin: E. S. Mittler and Sohn, 1923).

philosophy cover an astoundingly varied number of topics, but three have been especially prominent. In addition to the hermeneutics, the philosophical works include philosophical ethics, which in English might better be described as the philosophy of human culture, and the dialectic. The latter is concerned with the nature of thought and its attempt to know. In the process of moving from thinking to knowing, the thinker comes into conflict with others concerning the object of thought. The dialectic addresses the resolution of intersubjective conflict by prescribing the move through dialogue to the reconciliation of differing judgments. There is a distinct Platonic background to Schleiermacher's dialogical-dialectical procedure. The philosophical influences of Spinoza, Leibniz, Kant, and Schelling on Schleiermacher can be rehearsed in this connection, but the most important philosopher for Schleiermacher is the master of the Academy.

The question that arises is the relation between the theological books that Schleiermacher published to the philosophical lecture materials that he did not publish. I will cite one work here whose position I consider so extreme that my comments will indicate the direction of my own orientation. In 1941, Richard Brandt published *The Philosophy of Schleiermacher*.[45] The work represents a full inspection of the philosophical division of Schleiermacher's *Collected Works* and much of the theological corpus. The final hundred-page chapter of the book is entitled "The System of Philosophy and Theology."[46] Brandt takes up various topics under this rubric. First, he assumes that Schleiermacher has *one system* embracing various areas; whether a topic is either philosophical or theological does not preclude a mixing of one with the other. In section III above, I discussed Schleiermacher's theological intention to describe immediate self-consciousness, feeling, or the realm of the soul or heart as the object of his reflection. The *Dialectic*, as an example of Schleiermacher's philosophical work, is concerned with objective thought and how it can advance toward knowledge. The utter dependence of the religious consciousness is never mentioned in the *Dialectic*. This is no small item when one sets out to weave all theological and philosophical points into one fabric as Brandt does. Furthermore according to Brandt's interpretation, in Schleiermacher's single system we are admitted into what Schleiermacher truly thinks in the philosophical lectures not in his theological publications, which tend to express Schlei-

45 Richard Brandt, *The Philosophy of Schleiermacher: The Development of his Theory of Scientific and Religious Knowledge* (New York: Harper, 1941).

46 Ibid., 200.

ermacher's views of the Christian community in a way not necessarily his own standpoint. It is rather in his philosophical works that Schleiermacher speaks more freely about his own position.[47] Philosophy makes claims about reality, rather than about feeling.[48] If discussions of a religious object are to be valid, they must be made on metaphysical ground.[49] In this way, Brandt attempts to correct Schleiermacher's theology by using the philosophical lectures.

There are a number of reasons that speak against Brandt's interpretation. The first reason is the sheer fact that Schleiermacher worked for years to publish in multiple editions his theological system in the *Brief Outline* and *The Christian Faith*. Brandt's approach disregards this fact, and readily sacrifices what is said there in favor of continuously reformulated philosophical lecture materials. I do not think this makes any sense at all.

The second reason is given in a letter that Schleiermacher wrote to the philosopher Friedrich Heinrich Jacobi in 1818.[50] Written during Schleiermacher's fiftieth year, the letter formulates the relation of theology to philosophy in a striking and illuminating manner. Jacobi had written a piece indicating that he philosophized with his rational understanding and in so doing, considered himself a pagan.[51] With his religious feeling, however, he considered himself a Christian even though his reason negated this admission.

Schleiermacher responds that he is also a philosopher with his reason that is an independent and original activity. He is furthermore a believer with his religious feelings and specifically a Christian believer. He does not want to be entangled in contradiction and aims to maintain a balance in both spheres. Balance comes about by means of an alternating ascending on one side and a descending on the other. Such oscillation exhibits the general form of all finite existence. Understanding and feeling are like the two foci of the ellipse characterizing his being. "And I have the entire fullness of my earthly life in this complementary swinging between them. Consequently, my philosophy and my dogmatics are firmly resolved not to contradict each other, but precisely for this reason both of them, also, must always remain incomplete; and for as long as I can recall they have always been

47 Ibid., 250.
48 Ibid., 289–90.
49 Ibid., 297.
50 Friedrich Schleiermacher, *Aus Schleiermacher's Leben in Briefen*, vol. 2 (Berlin: Walter de Gruyter, 1974), 349–53.
51 Ibid.

mutually in accord with each other, and have always more approached each other."[52] In light of this passage, I do not think that Brandt's idea of one system can appropriately describe Schleiermacher's thought. An alternative interpretation is that Schleiermacher has systems. He does not conflate the different systems and he does not see any of them as complete.[53] For Schleiermacher, philosophy and theology can neither be united with each other, nor can the various areas of the philosophical lectures be brought together into an interrelated system.

Schleiermacher's claim that his philosophy and his dogmatics do not contradict each other requires some careful differentiation to be sustained. Interpreters of the relation of the lectures on dialectic to the system of theology often tend to overlook the precise differentiation between the two disciplines. An example of such an interpretative error is a book written by John Thiel on the relation between God and the world.[54] The author is not heavy handed in the manner of Brandt, but falls prey in one instance to a similar revision of Schleiermacher's theology by philosophy on the issue of the creation *ex nihilo*. The *Dialectic* treats the relation of thought to being. Schleiermacher's sustained argument follows the movement of thought that eventually reaches the principle of "the world not without God, God not without the world."[55] Thought that is proper can only take place vis-à-vis being. From this result, Schleiermacher concludes that the creation *ex nihilo* should either be rejected or that "there is little distinction whether it is said that God has made the world out of eternal matter or that God has made the world out of nothing."[56] According to Schleiermacher, the thinking mind cannot think the creation *ex nihilo,* and therefore changes creation into a formation of matter. Thiel concludes his discussion: "This most basic principle of Schleiermacher's epistemology annuls even the possibility of thinking God without or apart from the world. The radical difficulty Schleiermacher finds with the doctrine of creation *ex nihilo* is that it is unthinkable."[57] For Thiel, Schleiermacher could omit creation from his

52 Ibid., 350.

53 Although Schleiermacher produced two editions of his system of theology, the production of the second edition, extensively rewritten seven years after he had finished the first edition, shows that Schleiermacher considered the first edition to be incomplete.

54 John Thiel, *God and World in Schleiermacher's Dialektik and Glaubenslehre: Criticism and the Methodology of Dogmatics* (Las Vegas: Peter Lang, 1981).

55 DialJ 167 (A 225). (My translation.)

56 DialJ 417 (A 183). (My translation.)

57 Thiel, *God and World in Schleiermacher's Dialektik and Glaubenslehre*, 182.

system, and thereby achieve his systematic-theological purpose solely by the doctrine of preservation.

There is one important problem raised by the quote just cited. If the passage is taken strictly on Schleiermacher's terms, it is not only the creation *ex nihilo* that cannot be thought, but God also cannot be thought by proper thinking. Creation *ex nihilo* in traditional thought has not been held to be the usual kind of conception of the thinking mind. From early times, theology recognized that concerning this subject thought it was attempting to indicate something beyond the conditions of normal thinking. Augustine understands the doctrine on the one hand, as a negation of any divine status attributed to the world, and on the other hand, as a negation of the idea that God formed the world out of some already extant stuff.[58] For Augustine, time only arises with the act of creation. Augustine's joking reply to the question of what God was doing prior to creation was that God was preparing hell for the overly curious. This joke is a famous illustration of Augustine's awareness of the limits of the finite mind. Over a thousand years later, Calvin also used Augustine's reply but this time, it is not advanced as a joke!

Schleiermacher does not see the issue in Christian doctrine as one of proper thinking in relation to being. Rather, Christian doctrine is a reflection on the immediate self-consciousness of religious feeling, which is the feeling of utter dependence. The doctrines of creation and preservation are conceived in relation to the systematic role that the feeling of utter dependence plays in Schleiermacher's system; they are not posited by objective consciousness. It is from New Testament texts such as: "God created the world and everything in it" (Acts 17:24[59]) and: "It is faith that enables us to see that the universe was created at the command of God, so that the world we see did not arise out of matter" (Heb. 11:3) that Christians have confessed that: "I believe in God the Pantocrator" (Apostles' Creed). In propositions 39, 40, and 41 of *The Christian Faith*, Schleiermacher argues that the doctrine of creation is to be explained especially with a view to the exclusion of every alien element.[60] For Schleiermacher, the religious self-consciousness of utter dependence on infinite being contradicts every

58 AUGUSTINE, *Confessions and Enchiridion*, ed. Albert C. Outler, Library of Christian Classics, vol. VII (Philadelphia: Westminster Press, 1955), book 12, ch. 7; cf. book 11, ch. 12.

59 The Christian Bible is cited according the *The Holy Bible: Revised Standard Version* (New York: T. Nelson, 1952).

60 CF §§ 39–41 (148–56).

representation of the origin of the world; the world's origin excludes every element from an understanding of its sole origination from God. Theology's task is to reflect on the significance of this claim, and not on objective thought in relation to being. Philosophy and theology do not constitute a single system. The two disciplines cannot be mixed together in a manner that ignores their different goals. And they do seem able to contradict each other, at least on this particular point.

In a 1933 article entitled "Schleiermachers politische Ethik als Spiegel seines Denkens," Horst Stephan writes about Schleiermacher: "What characterizes him, consequently, is not a finished, static system, but the never completed wrestling with a connected, uniform reasoning through of reality. Unity stands as an ever new task at the end of his thinking, never in the beginning as a given magnitude when he starts to reflect."[61] I had not read widely in Schleiermacher's lectures when I first read that article forty years ago, but I have found through the years that Stephan was right.

As a lecturer, Schleiermacher approached his material without a settled map of a system. One of his students, David Friedrich Strauss, once said Schleiermacher's style in the lecture hall was to set everything for his listeners as problems to be thought through again before them.[62] To take notes was like the attempt to photograph—with old daguerreotype process—a dancer in full motion. Very difficult to do! And that is not a promising style for the presentation of a system. Unity was always a new task at which to aim; Schleiermacher's situation, therefore, is somehow parallel to the challenge of the one always attempting to chart the land around the Brazos: to map anew the land of beginning again!

---

61 Horst Stephan, "Schleiermachers politische Ethik als Spiegel seines Denkens," *Zeitschrift für Theologie und Kirche* NF 14 [41] (1933): 321. (My translation.)

62 David Friedrich Strauss, *Gesammelte Schriften*, ed. E. Zeller, vol. 5 (Bonn, 1877), 9. Cited in Richard R. Niebuhr, *Schleiermacher on Christ and Religion: A New Introduction* (New York: Scriber, 1964), 137.

I

System and Historical Context

Whitehead at Infinite Speed

Deconstructing System as Event

Roland Faber, University of Vienna

What is a system? It was apparently Johann Micraelius who first defined the term "system" in 1653.[1] Short and precise, the definition introduces the problem of systems: "*Systema est compendium, in quod multa congregatur.*"[2] "System," Micraelius writes, is a "compound" of "many" in respect of their "congregation." *Unity* is the aim, but the basic fact on which this definition focuses is *multiplicity*. "System" by definition is not an idea but a *problem*. It presents the problem of a multiplicity that evidently exists beyond unity and stubbornly resists any act of unification. Through history, therefore, interest in "systems" has arisen from a lack of unity or the loss of its evidence—when dualism prevails as in Descartes or chaos reigns as in Nietzsche, when epistemology appeals not just to unity but interprets unity as certainty, when an epoch is afraid of uncertainty and seeks a foundation.[3] Whitehead, however, was neither intimidated by uncertainty[4] nor shocked by multiplicity—even of systems.[5] In Whitehead's

---

1 Cf. Geo Siegwart, "System," in *Enzyklopädie Philosophie und Wissenschaftstheorie*, ed. Jürgen Mittelstraß, vol. 4 (Stuttgart: Metzler, 1996), 184. This definition still reflects the Greek *systema*: from *syn histemi*—"to stand / put together"; cf. Otto Ritschel, *System und systematische Methode in der Geschichte des wissenschaftlichen Sprachgebrauchs und der philosophischen Methodologie* (Bonn: Karl Georgi, 1906).

2 Johann Micraelius, *Lexicon philosophicum terminorum philosophis usitatorum* (Jena, 1653), 1053.

3 Cf. Karl Steinbacher, "System / Systemtheorie," in *Europäische Enzyklopädie zu Philosophie und Wissenschaften*, ed. Hans-Jörg Sandkühler, vol. 4 (Hamburg: Felix Meiner, 1990), 501.

4 PR 4.

5 AI 159.

thought, systems begin as an *assemblage*[6] and—I may add—systems essentially end their respective processes in new assemblages rather than in clear-cut crystal structures. Whitehead's approach to system is *process*, and its appearance is—*event*. That is the thesis of this chapter.

I. Whitehead and System in Historical Perspective

Whitehead is widely considered to be a systematic thinker.[7] But very little attention has been paid to the historical development of the concept of system in order to understand more precisely what is meant by "system" and in what sense Whitehead's thought might be "systematic."[8] Indeed, a glance at the history of the concept of system reveals why: it is a complicated, un-systematic story of fragments, leaps, fissures, and folds in a confusing, sheer unmanageable landscape of persons, streams of thought, and self-reinterpreting traditions.[9] Nevertheless, to determine Whitehead's unique place in history and, conversely, his contribution to the determination of what a system might be, it is necessary to gain a historical perspective *on* Whitehead.

How can we grasp the historical perspective? Two orientation marks shall guide us. First, any historical view on philosophy is speculative. It seeks connections by interpreting history as a whole composed of perspectives and differences. Hence, it will produce a *heuristic* model for the sake of clarity and direction while losing some of the richness that retreats into a vague background. Second, the model for the historical introduction to Whitehead chosen here is that of gaining a perspective *on* Whitehead rather than seeking the origins and causal relations *of* Whitehead's system. The latter, I think, is not really useful unless we are prepared to pay attention to the impact Whitehead has had on subsequent history.

---

6 MT 3.

7 Cf. Nicholas Rescher, *Process Metaphysics: An Introduction to Process Philosophy* (Albany: SUNY Press, 1996); cf. Willi Oelmüller, Ruth Dölle-Oelmüller, and Carl-Friedrich Geyer, *Diskurs Metaphysik*, 2nd ed. (Paderborn: Schöningh, 1995), 58.

8 Cf. Michael Hampe, "Whiteheads Metaphysik und das philosophische Selbstverständnis der Gegenwart," in *Prozeß, Gefühl und Raum-Zeit: Materialien zu Whiteheads "Prozeß und Realität,"* ed. Michael Hampe and Helmut Maassen, vol. 1, Suhrkamp Taschenbuch Wissenschaft, no. 920 (Frankfurt a. M.: Suhrkamp, 1991), 18–19.

9 Cf. Fritz-Peter Hager and Christian Strub, "System," in *Historisches Lexikon der Philosophie*, ed. Joachim Ritter and Karlfried Gründer (Basel: Schwabe, 1998), 824–56.

For these reasons, I will contextualize Whitehead within the main streams of the history of the concept of system and contrast his thought with philosophies that exhibit a cross-section through this history. In this way, these philosophies can function as navigational lights on the infinite landscape of system-building. The particular models I use to contrast Whitehead's uniqueness are those of Leibniz and Deleuze. Leibniz emphatically understood his monadology as a system, and Whitehead acknowledged it as the basis for his own thought.[10] Deleuze, on the other hand, saw Leibniz and Whitehead to be in line with his own project of nomadology.[11]

A consideration of their thinking indicates that the three philosophies converge according to an interest in system.[12] While they diverge from each other by being very different manifestations of their respective conceptions of system, they relate to each other by intense mutual reconsideration of each other's thought. In his reconstruction of the history of philosophical systems, Whitehead understood Leibniz's philosophy as *the* one system next to his own organic philosophy.[13] Because of the universality of Leibniz's philosophy, whose system ranged "from mathematics to divinity, and from divinity to political philosophy," Whitehead suggested that someone should write a book on "The Mind of Leibniz."[14] We might be tempted to understand Whitehead's own philosophy as fitting that description.[15] Deleuze, in fact, wrote a book on the mind of Leibniz, *The Fold*, that transforms Leibniz's monadology into a Deleuzian nomadology.[16] Surprisingly, at the center of his book, Whitehead appears—as

---

10 PR 80; cf. NICHOLAS RESCHER, "Leibniz and the Concept of System," *Studia Leibnitiana* 13 (1981): 117–8. For Whitehead's remarks on the relation of Leibniz's monadology to his own philosophy, cf. STASCHA ROHMER, *Whiteheads Synthese von Kreativität und Rationalität: Reflexionen und Transformationen in Alfred North Whiteheads Philosophie der Natur*, Alber Thesen, no. 13 (Freiburg: Alber, 2000), 85–90.

11 GILLES DELEUZE, *The Fold: Leibniz and the Baroque* (Minneapolis: University of Minnesota Press, 1993).

12 Cf. GEORGE R. LUCAS, *The Rehabilitation of Whitehead: An Analytic and Historical Assessment of Process Philosophy* (Albany: SUNY Press, 1989), 9; for Deleuze: JOHN MARKS, *Gilles Deleuze: Vitalism and Multiplicity* (London: Pluto, 1998), 78–90.

13 SMW 155; cf. CHRISTOF KANN, *Fußnoten zu Platon: Philosophiegeschichte bei A. N. Whitehead*, Paradeigmata, no. 23 (Hamburg: Felix Meiner, 2001), 204.

14 MT 3.

15 Cf. ROHMER, *Whiteheads Synthese von Kreativität und Rationalität*, 85.

16 Cf. GILLES DELEUZE and FÉLIX GUATTARI, *A Thousand Plateaus* (Minneapolis: University of Minnesota Press, 1987), 351–423. For an introduction to Deleuze's thought,

the event of the transformation itself.[17] When Deleuze asks: "What is an event?," he is really asking *the* question concerning the essence of systems, and he considers Leibniz and Whitehead on their way to his own answer: that a system is not concerned with eternity—the self-same—but with a world that allows for the production of novelty, a world that is creation.[18] This is Deleuze's frame of reference for systems and, hence, for Whitehead's system: if creativity is ground, system appears as event.[19]

Nevertheless, all three systems differ in their ability to integrate novelty and wholeness. In Leibniz, novelty appears as the private subjectivity of monads that form a universal system. Since these monads have no horizontal relations, but are organized by the power of God to harmonize compossibilities for the best of all possible worlds, their pre-established harmony paralyzes any real movement for the sake of a holographic system.[20] Whitehead, however, understands the relation of actual entities—and this was his great systematic advance in Deleuze's eyes—as a process of *prehensions of prehensions*, that is, of (mutual) relations in real time: an asymmetric, moving immanence. Hence, these relations "become" internally, and their "divergent series" create "endlessly bifurcating paths" of "bifurcations, divergences, incompossibilities, and discord" in a chaotic world.[21] In this "chaosmos," even "God desists of being a Being who compares worlds and chooses the richest compossible," but "becomes Process, a process that affirms incompossibilities and passes through them."[22] For Deleuze, finally, singularities create a *chaosmic* system in which they move on a plane of immanence, integrating new constellations of togetherness over and over again. In his last book *What is Philosophy?*, Deleuze calls this process "a pure Event, a hecceitas, an entity."[23] His nomadic system "is defined by *the inseparability of a finite number of heterogeneous components traversed by a point of absolute survey at infinite speed.*"[24] Now, systems

---

cf. CHRISTIAN JÄGER, *Gilles Deleuze: Eine Einführung*, Uni-Taschenbücher, no. 1985 (Munich: Wilhelm Fink, 1997).

17 DELEUZE, *The Fold*, 76–82.

18 Cf. Ibid., 81.

19 Cf. FRIEDRICH BALKE, *Gilles Deleuze*, Einführungen, no. 1090 (Frankfurt a. M.: Campus, 1998), 23–50.

20 Cf. ROHMER, *Whiteheads Synthese von Kreativität und Rationalität*, 85–90.

21 DELEUZE, *The Fold*, 81.

22 Ibid.

23 GILLES DELEUZE and FÉLIX GUATTARI, *What is Philosophy?* (New York: Columbia University Press, 1994), 21.

24 Ibid. (Italics in the original.)

are actualized and transformed into events. And this is the perspective from which Deleuze, as Whitehead's historical precedent, perceived Whitehead.

II. The Complex History of the Concept of System

The concept of system has a rich and puzzling history which itself cannot be grasped by any system. It is, to quote Deleuze, a history of "bifurcations, divergences, incompossibilities, and discord."[25] In studying its complexities, however, we can (more heuristically than systematically) discern some of its threads and distinguish between important lines of argumentation (see Table One).

Table One: Typology of Systems

		monadic	*organic*	*nomadic*	
cosmological system	universe ‹······	··········	··········	········›	multiverse
intellectual system	construction ‹······	··········	··········	········›	deconstruction
axiomatic system	axiom ‹······	··········	··········	········›	incompleteness
organic system	organism ‹······	··········	··········	········›	chaos
		Leibniz	**Whitehead**	**Deleuze**	

The Greek *systema* means a whole that consists of parts that "stand together" as a result of "putting them together."[26] A system is a unity of differences. In antiquity, system appeared without philosophical specification as medical, musical, and political collections, harmonies, and functional parts of wholes.[27] Plato, on the one hand, appealed to system in order to understand the dialectic of unity and plurality, Aristotle, on the other hand, for animal organisms. First effectively used in Stoic philosophy, "system" was either a synonym for "cosmos," the well-ordered and organic whole of the world, or a part of logic, naming the "conclusion" (*logos*) the system of its conditions.[28] The career of the term "system"

---

25 Deleuze, *The Fold*, 81.

26 Cf. Fritz-Peter Hager, "System: I Antike," in *Historisches Lexikon der Philosophie*, ed. Joachim Ritter and Karlfried Gründer (Basel: Schwabe, 1998), 824; cf. Johannes Michael Dittmer, *Schleiermachers Wissenschaftslehre als Entwurf einer prozessualen Metaphysik in semiotischer Perspektive: Triadizität im Werden*, Theologische Bibliothek Töpelmann, no. 113 (Berlin: Walter de Gruyter, 2001), 87–94.

27 Cf. Wolfgang Schrader, "System, Systemtheorie," in *Lexikon für Theologie und Kirche*, 3rd ed., vol. 9 (Freiburg: Herder, 2000), 1217.

28 Cf. Hager, "System: I Antike," 824.

did not begin until the fourteenth century with astronomical cosmology. There it was used for the *systema mundi*, the world-structure, thereby—as in Copernicus—distinguishing the cosmos *itself* as a system of reality from hypotheses *on* the structure of the cosmos.[29] Modern usage surprisingly leads us back to the Protestant theologian Philip Melanchthon who used the term to mean the intellectually coherent presentation of Christian doctrines.[30] From the 1600's on, the idea of "system" spread to every discipline that understood its own subject in terms of a methodologically controlled project of research, seeking knowledge and unity.[31]

The most important leap in the history of system was made in reaction to a loss: the loss of the unity of *scientia* and *sophia*, of science and metaphysics, of intellectual construction and empirical experiment.[32] At this point, "system" was elevated as *the* epistemological notion generating *truth*. With Descartes' *Discours de la Méthode*, system becomes an a priori intellectual construction of the world's unity that expresses truth—and truth as certainty.[33] Descartes' *fundamentum inconcussum* is the *ego cogito*. From this point on, the idea of intellectual system was based on three equations: unity is fundament; fundament is substance; and substance is subject. The Cartesian equations were threaded through all the important philosophical conceptions of the following two centuries: from Spinoza's *natura naturans* to Leibniz's *monas monadum*, from Kant's "transcendental subject" to Fichte's "absolute ego" and Hegel's "absolute spirit," from Schelling's "ground of existence" to Heidegger's "history of Being."[34]

Nevertheless, another train of thought resisted the three equations. The English empiricists, Bacon, Locke, and Hume, rejected any rationalistic foundation—ontologically and epistemologically. These three philosophers were seen as "enemies of system" who demanded that empirical plurality prevail over constructed unities. The French Enlightenment philoso-

---

29 Cf. SIEGWART, "System," 183–4.

30 Cf. STEINBACHER, "System / Systemtheorie," 500.

31 Cf. CLEMENS TIMPLER, *Metaphysicae systema methodicum* (Hanau, 1606).

32 Cf. ROWANNE SAYER, *Wert und Wirklichkeit: Zum Verständnis des metaphysischen Wertbegriffs im Spätdenken Alfred North Whiteheads und dessen Bedeutung für den Menschen in seiner kulturellen Kreativität* (Würzburg: Ergon, 1999), 45.

33 Cf. CHRISTIAN STRUB, "System: II System und Systemkritik in der Neuzeit," in *Historisches Lexikon der Philosophie*, ed. Joachim Ritter and Karlfried Gründer (Basel: Schwabe, 1998), 829–30. REINHARD LAUTH, *Descartes' Konzeption des Systems der Philosophie*, Quaestiones: Themen und Gestalten der Philosophie, no. 12 (Stuttgart: Frommann-Holzboog, 1998).

34 Cf. SCHRADER, "System, Systemtheorie," 1218–9.

phers, d'Alembert, Malebranche, and Voltaire, also criticized rationalist and subjectivist systems with an empiricism that claimed that a system could only be true a posteriori and not deduced from a priori categories. Nietzsche's post-idealist critique of the "will to system" provokingly uncovers systems as amoral projections of subjective interests tyrannizing phenomena and societies.[35] In the twentieth century, Heidegger's *Seinsvergessenheit*, Meinong's empirical a priori of knowledge, and Hartmann's clash between problems and systems set the stage for a new, but again primitive meaning of system: to be "a collection having a regular relation between its members" (Peirce)[36] or a "provisorial collection of knowledge" (Neurath).[37] In late-modern philosophy, for example in Adorno's negative dialectic, in Rorty's ironic self-relativity, and in Derrida's deconstructionism, we again see the bifurcation of truth and certainty. The notion of a system is thereby transformed into a fragmented constellation of history or difference that liberates the plurality of oppressive unifications.[38]

In addition to cosmological and intellectual systems, two other concepts of system must be explored: axiomatic and organic systems. Beginning with Euclid's geometry, axiomatic systems, on the one hand, built mathematical and logical structures on basic axioms, definitions, and rules. In history, these systems intersected with cosmological and intellectual systems and, at times, merged together, as is the case with Spinoza's universal system constructed *more geometrico*.[39] Organic systems, on the other hand, were developed in reaction to the subjectivist principle of the Cartesian tradition in German Idealist philosophy. They were nevertheless already based on Plato's organic cosmology.[40] Two interesting characteristics distinguish organic systems. First, organic systems have no external reasons; all grounds of system interact with one another and with the whole so as to prepare an immanent mutuality of reasons (Fichte). Second, organic systems exist only in interaction with their environment. This line of thought extending beyond Idealism was explored by Schleiermacher.[41] It also stands behind

---

35 Cf. STRUB, "System: II System und Systemkritik in der Neuzeit," 847–8.

36 CHARLES S. PEIRCE, *Collected Papers*, vol. 4, *The Simplest Mathematics* (Cambridge: Harvard University Press, 1933), 517 [=§ 621].

37 Cf. STRUB, "System: II System und Systemkritik in der Neuzeit," 849.

38 Cf. Ibid., 851.

39 Cf. SIEGWART, "System," 184–8.

40 Cf. STEINBACHER, "System / Systemtheorie," 500–5.

41 Cf. GUNTER SCHOLZ, *Die Philosophie Schleiermachers*, Erträge der Forschung, no. 217 (Darmstadt: Wissenschaftliche Buchgesellschaft, 1984), 62.

the modern autopoetic system theories from von Bertalanffy and Luhmann to Maturana and Prigogine.[42]

In scanning this rich history, we can discern four streams of conceptualization—cosmological, intellectual, axiomatic, and organic systems—and two directions of interest—self-confirmation and self-relativity (see Table One). Cosmological systems began with the Stoic inquiry into the cosmos but led to cosmological theories of a plurality of universes (A. Linde). Intellectual systems began with the pre-Socratic quest for the logos of the unity of reality but led to the deconstruction of logocentrism (Derrida).[43] Axiomatic systems began with Euclidean geometry but led to Gödel's proof of the non-axiomatic richness of mathematical truth.[44] Organic systems began with the self-organization of biological organisms but revealed the chaotic core of every living system.

III. Whitehead's Concept of System in Historical Context

Whitehead's concept of system is complex. He uses the term "system"[45] and similar terms like "theory,"[46] "structure,"[47] "scheme,"[48] or "organism,"[49] in different contexts scattered throughout his works. Whitehead, like Leibniz and Deleuze, appealed to all four strands of system's conceptualization mentioned above and to both of their directions. He was interested in a "cosmological construction"[50] and understood philosophy as the "critique of cosmologies,"[51] yet he was convinced of the mutability of natural laws and of the multiplicity of "cosmic epochs."[52] He under-

---

42 Cf. Ibid., 502–4. On the concept of autopoiesis, cf. Humberto Maturana, *Erkennen: Die Organisation und Verkörperung von Wirklichkeit: Wissenschafttheorie*, Wissenschaft und Philosophie, no. 19 (Braunschweig: Friedrich Vieweg, 1982), 157–69.

43 Cf. Wolfgang Welsch, *Vernunft: Die zeitgenössische Vernunftkritik und das Konzept der transversalen Vernunft*, Suhrkamp Taschenbuch Wissenschaft, no. 1238 (Frankfurt a. M.: Suhrkamp, 1996), 247–53.

44 Cf. Roger Penrose, *The Emperor's New Mind: Concerning Computers, Minds, and the Law of Physics* (New York: Penguin Books, 1991), 99–148.

45 FR 69.

46 FR 75.

47 AI 192.

48 PR 3.

49 PR 7.

50 PR 341.

51 SMW xxi.

52 PR 91–3.

stood philosophy as a "speculative" endeavor and he created a "coherent, logical, necessary system of general ideas"[53] governing all possible worlds, yet he never believed that he had actually created a system that revealed these "metaphysical first principles."[54] With his colleague Bertrand Russell, he developed an axiomatic system for mathematics in logical terms,[55] yet he knew—probably in view of Gödel's proof—that there were always mathematical and logical true notions beyond systems.[56] He constructed a "philosophy of organism,"[57] yet had deep respect for the chaotic power of life.[58]

This is a paradox: Whitehead was a thinker who fervently strove for system, but radically knew of the impossibility of systematization,[59] or at least the limited value of any systematic endeavor.[60] When compared with Leibniz's crystalloid universe of monadic harmony and Deleuze's chaos of nomadic virtuality, Whitehead occupies a unique place in the history of the concept of system because he affirms systematization as organic process[61] that, at the same time, confirms and deconstructs system.[62]

The strongest affirmation of systematization appears in texts in which Whitehead explores a "system of general ideas"[63] that, in its character, still seems to be based on axiomatic thought.[64] A system requires—as Whitehead claims in *The Function of Reason*—clarity, logical consistency, and a logical scheme,[65] or—as in *Process and Reality*—coherence, logical consistency, and necessity.[66] Whitehead's system thus far might be seen as a clone of Spinoza's *ars geometricum*, Leibniz's *calculus*, or Kant's "transcendental categories" as necessary self-explanations of pure reason.[67] Whitehead's

---

53 PR 3–4.
54 PR 4; cf. KANN, *Fußnoten zu Platon*, 85–94.
55 PM.
56 MT 2.
57 PR 7.
58 PR 104–5.
59 PR 4, 7–8.
60 AI 159; cf. HAMPE, "Whiteheads Metaphysik," 10–31.
61 SMW 155.
62 MT 2–3; cf. ROLAND FABER, *Prozeßtheologie: Zu ihrer Würdigung und kritischen Erneuerung* (Mainz: Grünewald, 2000), 201–12, 357–72.
63 PR 3.
64 Cf. FELIX D'SOUZA, *Whitehead and Religion: The Significance of his Conceptual Scheme* (Leuven: unpublished Ph. D. dissertation, 1989); cf. RESCHER, "Leibniz," 122.
65 FR 5.
66 PR 3.
67 Cf. BOWMAN L. CLARKE, "Logic and Whitehead's Criteria for Speculative Philosophy," *Monist* 65 (1982): 517–31; cf. DAVID RAY GRIFFIN et al., *Founders of Constructive Post-*

concept of system, however, differs dramatically from both axiomatic and transcendental necessity because, for Whitehead, "system"[68] as a "scheme of thought"[69] is nothing but a "reservoir of potential developments"[70] that, in its abstraction from experience, does not reflect reality or pure reason, but creates imaginations by which speculative reason furthers *life*.[71] For this reason, the general scheme of ideas covering all possible experiences[72] has an *empirical* basis and aim.[73] It cannot be exhaustively discussed by theoretical terms of logic, coherence, and necessity, but permanently requires empirical feed-back.[74] Hence, Whitehead's concept of system is that of "interpretation"[75] that essentially includes *infinite* revision.[76]

In consequence, other texts by Whitehead reflect on the limited value of systems. "Systems, scientifically and philosophically, come and go."[77] The "discordance of competing philosophical systems is a factor essential for progress"[78] and hinders "dogmatic finality."[79] One can never "produce that final adjustment of well-defined generalities which constitute a complete metaphysics."[80] Now it is clear: philosophy—as most of this history of its systems shows[81]—is "misled by the example of mathematics."[82] Philosophy articulates no clear-cut structures, but offers only "tentative

---

modern Philosophy: Peirce, James, Bergson, Whitehead, and Hartshorne (Albany: SUNY Press, 1993). For critical discussion, cf. CATHERINE KELLER and ANNE DANIELL, eds., *Process and Difference: Between Cosmological and Poststructuralist Postmodernism* (Albany: SUNY Press, 2002), VII–XII, 2–8.

68 FR 69.

69 FR 76.

70 FR 72.

71 Cf. KANN, *Fußnoten zu Platon*, 95–100; cf. AI 105.

72 PR 3.

73 PR 4.

74 PR 3; cf. SAYER, *Wert und Wirklichkeit*, 73–148.

75 PR 3.

76 Cf. HANS POSER, "Whiteheads Kosmologie als revidierbare Metaphysik," in *Whiteheads Metaphysik der Kreativität: Internationales Whitehead-Symposium Bad Homburg 1983*, ed. Friedrich Rapp and Reiner Wiehl (Freiburg: Alber, 1986), 105–26.

77 AI 159.

78 AI 144.

79 AI 145; cf. PR 8; cf. CLARKE, "Logic and Whitehead's Criteria for Speculative Philosophy," 519.

80 AI 145; cf. FABER, *Prozeßtheologie*, 201–24.

81 SMW 29.

82 PR 8; cf. MICHAEL WELKER, "Whiteheads Vergottung der Welt," in *Whitehead und der Prozessbegriff: Beiträge zur Philosophie Alfred North Whiteheads auf dem Ersten Internationalen Whitehead-Symposion 1981*, ed. Harald Holz and Ernest Wolf-Gazo (Freiburg: Alber, 1984), 249–72.

formulations of the ultimate generalities."[83] Philosophy appeals to *intuition*[84] that limits systematization in relation to the universe that cannot be fathomed.[85] Therefore, Whitehead deconstructs philosophical systems[86] as a "series of footnotes to Plato."[87] For Whitehead, Plato is the "greatest metaphysician," not because of his "fragmented system"—he was "the poorest systematic thinker"—but because of his "depth of metaphysical intuition."[88] Indeed, when Whitehead envisions a philosophical system, he invokes seven Platonic notions (Ideas, Physical Elements, Psyche, Eros, Harmony, Mathematical Relations, and Receptacle[89]) that by no means have axiomatic value, but rather, they seem to be dramatic characters on the stage of a play that is the universe's process.

A third group of texts explains *why* the true "topic of philosophical research" is, for Whitehead, not the construction of system but a deconstruction of "undiscovered limitations" of intuition.[90] At this point, we must deconstruct two limitations: that a system is a universal hypothesis about reality rather than about an *actuality*, and that it comprises reality with finality rather than being related to *infinity*. Firstly, Whitehead's concept of system departs from a notion of system as an abstract structure and connotes concrete actuality. Every actuality is a "system of all things,"[91] "the universe in this synthesis."[92] Beyond the actual movement of the universe to which they are related as potentials of their respective actualizations, structures contribute nothing of significance. Hence, Whitehead appeals to a "principle of empiricism." By this he means that the concrete movement of the actual world "is not discoverable by abstract reason," but "rests on an empirical basis."[93] His "ontological principle" demands "that actual entities are the only *reasons*."[94] Structures may appear as axiomatic "ab-

---

83 PR 8.
84 PR 22.
85 AI 144; cf. ROLAND FABER, "'The Infinite Movement of Evanescence'—The Pythagorean Puzzle in Plato, Deleuze, and Whitehead," *American Journal of Theology and Philosophy* 21 (2000): 171–99.
86 PR 19.
87 PR 39.
88 AI 166; cf. KANN, *Fußnoten zu Platon*, 36–61.
89 AI 275.
90 AI 145.
91 PR 36.
92 PR 28; cf. FABER, *Prozeßtheologie*, 101–200.
93 SMW 178.
94 PR 24.

stract hierarchies,"[95] but they are irrelevant apart from their inclusion in the becoming of actualities.[96] Secondly, systematization is a process of *concentration*; it bifurcates the universe into both centers of interest and a wide background of de-centered relations to form a finite structure against the infinite background of the universe.[97] In *Modes of Thought*, Whitehead's criticism of the concept of system rests on this difference: all systems are narrow in relation to the importance of the infinite background that is the universe.[98] Hence, "system" is an ever-moving *assemblage*. Philosophy is an endless process of assembling[99]—deconstructing, reconstructing, delimiting.[100]

In situating Whitehead's new concept of a *fluent system* between Leibniz and Deleuze, we are drawn to an important passage in *Process and Reality*, in which Whitehead distinguishes between static "systems of thought" that he identifies with the Cartesian tradition and his own appeal to fluency that he relates to Bergson.[101] While the static system, in appealing to "permanence," shapes a "clear-cut philosophy" that analyzes "the world in terms of static categories" by the "subordination of fluency" to substance, Bergson criticizes this approach as a "spatialization" of the infinite movement of the universe in its "inescapable flux."[102] Moreover, while Leibniz's system of monads, although built on organizing activity[103] and universal relationalism,[104] is essentially static and fundamentally lacks process, fluency, and history, Bergson thinks that the universe is a "moving whole" that, in its *wholeness*, is a never finished, open-ended process.[105] In agreement with Whitehead and Bergson, Deleuze's "transcendental empiricism" takes *actuality* to be the transcendental condition of reality as an ever-moving, an ever-assembling, an ever-diverging, and an ever-differentiating process.[106]

---

95 SMW 167.
96 Cf. Thomas Beelitz, *Die dynamische Beziehung zwischen Erfahrung und Metaphysik: Eine Untersuchung der spekulativen Philosophie von Alfred North Whitehead im Interesse der Theologie* (Frankfurt a. M.: Peter Lang, 1991), 191.
97 AI 159.
98 Cf. CN 12–13.
99 MT 2–3.
100 Cf. Dittmer, *Schleiermachers Wissenschaftslehre*, 423–4, 483–4, 498–9.
101 PR 209.
102 Ibid.
103 SMW 155.
104 PR 80.
105 Cf. Gilles Deleuze, *Cinema 1: The Movement-Image*, 4th ed. (Minneapolis: University of Minnesota Press, 1996), 10–11.
106 Cf. Balke, *Gilles Deleuze*, 29–33; cf. Deleuze, *The Fold*, 76–82.

For Deleuze, "system" is an *event* that holds its parts together by generating a fluent structure. It is comprised by a *singular wholeness* that moves over the actual togetherness of its elements with *infinite speed*.[107] When seen in the context of this discussion, Whitehead's concept of a fluent system *is* a process of togetherness[108] appealing to "actuality," "activity," "creativity," "process," and "organism."

IV. The History of Intellectual Systems

In the history of intellectual systems, two decisions, made in the thirteenth and fourteenth centuries, were of decisive significance. Firstly, Aristotelian philosophy of science was introduced to "first philosophy"—something Aristotle did not do. Aristotle understood science to be (syllogistic) knowledge based on reasons. But his metaphysics, although treating ultimate reasons and first principles, never named reasons axiomatically.[109] In this regard, Aristotle still followed Plato who, in his own first philosophy, was aware of the hypothetical character of all knowledge of the ultimate; hence with the *Timaeus*, Plato explored a *mythology*, a *possible history* to be told.[110] In the high Middle Ages, however, axiomatic onto-theology began to prevail.[111] Secondly, the Aristotelian distinction between knowledge from experience and a higher knowledge of metaphysics led to the bifurcation of speculative reason and empirical reality. Descartes' dualism of the *res cogitans* and *res extensa* already reflected the loss of a coherent connection between necessary principles and empirical experience.

Descartes significantly changed the character of intellectual systems because he understood metaphysics as the knowledge of reasons of *certainty*, grounding the conscious reasoning of the self-reflecting mind in the *ego cogito*.[112] Beginning with Descartes, *subjectivity* became the paradigm for

---

107 Cf. Deleuze and Guattari, *What is Philosophy?*, 10–19; cf. Rescher, "Leibniz," 121: "Systematicity indicates the presence of unity amidst variety."

108 PR 22.

109 Cf. Ludger Honnefelder, "Der zweite Anfang der Metaphysik: Voraussetzungen, Ansätze und Folgen der Wiederbegründung der Metaphysik im 13./14. Jahrhundert," in *Philosophie des Mittelalters: Entwicklungslinien und Paradigmen*, ed. Jan P. Beckmann et al. (Hamburg: Felix Meiner, 1996), 167–71.

110 Cf. Plato, *Timaeus*, 29 d; cf. Kann, *Fußnoten zu Platon*, 42–3; cf. Otfried Höffe, *Aristoteles* (Munich: Beck, 1996), 139–59, esp. 109.

111 Cf. Honnefelder, "Der zweite Anfang," 171.

112 Cf. Strub, "System: II System und Systemkritik in der Neuzeit," 829.

the foundation of truth.[113] With subjective reason as the foundation, *unity* is sought prior to all differences and with its character as certainty, a priori truth is sought prior to all empirical reality.[114] The most concise manifestation embodying these postulates was Spinoza's system of the one, ultimate substance—the *natura naturans*—with its infinite number of attributes and an infinite variety of finite modes that are actualized with axiomatic necessity.[115] Leibniz, the first philosopher who used the term "system," detected the inconsistency in Spinoza's monism to be a consequence of Descartes' dualism. So, against Descartes and Spinoza, Leibniz conceived subjectivity in relational and individual terms. His system of internally related monads organized reality as a system of *organic* subjectivity.[116] With his axiomatic system, however, Leibniz proposed a pre-established harmony between all monads prior to all empirical reality.[117] Later, Kant drew on Leibniz's idea of system as the ultimate intellectual architecture of truth in his own investigation into the transcendental, a priori structure of the human mind. Nevertheless, the ultimate basis of Kant's transcendentalism is a dualist equation: the condition for the possibility of the experience of reality *is* the transcendental structure of pure reason.[118] As a result, Kant could neither sufficiently argue for the freedom and the individuality of subjectivity, which he dissolved both into the general postulates of practi-

---

113 Cf. MARIA-SIBYLLA LOTTER, *Die metaphysische Kritik des Subjekts: Eine Untersuchung von Whiteheads universalisierter Sozialontologie*, Studien und Materialien zur Geschichte der Philosophie, no. 43 (Hildesheim: Olms, 1996), 144–62; cf. PR 80.

114 Cf. SCHRADER, "System, Systemtheorie," 1218; cf. RESCHER, "Leibniz," 121–2 who distinguishes Spinoza's axiomatics based on geometry from Leibniz's axiomatics based on *calculus*.

115 Cf. HEINRICH ROMBACH, *Substanz, System, Struktur*, vol. 2 (Freiburg: Alber, 1966), 13–15; cf. KANN, *Fußnoten zu Platon*, 155; cf. HUBERTUS GEZINUS HUBBELING, "Whitehead and Spinoza," in *Whitehead und der Prozessbegriff: Beiträge zur Philosophie Alfred North Whiteheads auf dem Ersten Internationalen Whitehead-Symposion 1981*, ed. Harald Holz and Ernest Wolf-Gazo (Freiburg: Alber, 1984), 375–85.

116 Cf. ENNO RUDOLPH, "Zu Leibniz's Anfangsgründen der Naturphilosophie und Phänomenologie," in *Naturphilosophische Studien*, vol. 2, *Positionen zur Descarteskritik*, Texte und Materialien der Forschungsstätte der Evangelischen Studiengemeinschaft, no. 22 (Heidelberg: Forschungsstätte, 1986), 9–32; cf. IDEM, "Substance as Function: Ernst Cassirer's Interpretation of Leibniz as Criticism of Kant," in *Philosophy, Mathematics and Modern Physics: A Dialogue*, ed. Enno Rudolph and Ion-Olimpiu Stamatescu (Berlin: Springer, 1994), 235–42; cf. ROMBACH, *Substanz, System, Struktur*, vol. 2, 299–304.

117 Cf. STRUB, "System: II System und Systemkritik in der Neuzeit," 829–31.

118 Cf. Ibid., 835–8; cf. RITSCHEL, *System und systematische Methode*, 69–74; cf. ROHMER, *Whiteheads Synthese von Kreativität und Rationalität*, 77–83.

cal reason and the generality of the transcendental subject. Nor could Kant solve the problem of empirical reality, which he subsumed into a construction of pure reason.[119] His system is *idea*—a priori rational unity.

While Kant assumed the *identity* of unity and system, Fichte and Hegel asked the question as to *why* reason requires systematization. While the answer integrated unity with system, it began to deconstruct the Cartesian foundation on which it was based. Now, the *internal* unity of the system is the reason for systematization, and therefore, unity becomes the *self-explanation* of the system. Its structure is a circle of the reciprocal foundation of its reasons, whereby all reasons explain each other and are explained by the system they ground. Here, axiomatic foundation is abandoned for organic togetherness that is life and history.[120] In life, the abyss of negation cannot be subsumed under any foundational unity. While Fichte still understood system's self-explanation as the process of the negation of the "ego" by the "non-ego," Hegel transformed this subjective process into the objectivity of the "absolute concept" of which "process" is the dialectic of *the negation of the negation* of any simple unity.[121]

With Nietzsche, Schelling, and Heidegger, a new paradigm surfaced. While on the one hand, they criticized the German Idealists' totality of systematization, on the other hand, they insisted on individuality, freedom, and difference. Nietzsche criticized Kant's identification of subjectivity with systematic reason and exposed it as the will to power. Systems are constructions of the irrational will of subjects in line with their individual interests.[122] Nietzsche's criticism of system resulted in the loss of an integral concept of *the* world, but it gained a new symbol for ineffable individuality—"eternal return." Schelling deconstructed rationalism by claiming the groundlessness of existence. For Schelling, existence has no essentiality beyond process; so even in God there is a "ground"—the "*Ungrund*"—as a groundless chaos of potentiality that does not form any

---

119 Cf. Reiner Wiehl, "Whiteheads Kant-Kritik und Kants Kritik am Panpsychismus," in *Natur, Subjektivität, Gott: Zur Prozeßphilosophie Alfred North Whiteheads*, ed. Helmut Holzhey, Alois Rust, and Reiner Wiehl, Suhrkamp Taschenbuch Wissenschaft, no. 769 (Frankfurt a. M.: Suhrkamp, 1990), 198–239; cf. Scholz, *Die Philosophie Schleiermachers*, 53.

120 Cf. Strub, "System: II System und Systemkritik in der Neuzeit," 841.

121 Cf. George Kline, "Begriff und Konkreszenz: Über einige Gemeinsamkeiten in den Ontologien Hegels und Whiteheads," in *Whitehead und der Deutsche Idealismus*, ed. George R. Lucas, Jr. and Antoon Braeckman (Bern: Peter Lang, 1990), 145–61; cf. Ritschel, *System und systematische Methode*, 74–90.

122 Cf. Strub, "System: II System und Systemkritik in der Neuzeit," 845–6.

system prior to the history of existence.[123] The earlier Heidegger, in his existential analysis of *Dasein*, still followed Kant's transcendental project. For the later Heidegger after his turn (*Kehre*), however, it is not *Dasein* that projects *Sein*, but *Sein* that "gives" being. According to Heidegger's ontological difference, the ground of the world is the "gift" beyond any system. It is formless nothingness, that is, nothing for itself or beyond the process of the world. It is ultimately neither substance nor subject, but groundless history / becoming / process.[124]

The last stage that deconstructed the identification of system with subjectivity, certainty, and rationality is reached with Benjamin's *disruptive constellation* and de Saussure's *de-centered system of differences*, finally leading to Foucault's structuralism, Deleuze's post-structuralism, and Derrida's deconstructionism.[125] Benjamin, on the one hand, understands systematization to be an act of political totalitarianism that uses continuity as the power by which the rich terrorize history. To liberate the victims, system's continuity must be shocked. The method of "shock" is discontinuity; it creates surprising constellations of disjunctive pluralities like stellar constellations.[126] De Saussure, on the other hand, understands system as *text*, as a de-centered structure of differences in an infinite process of signification.[127] Finally, on this basis, Deleuze, Foucault, and Derrida developed the concept of system as an endless process of differentiation, without any foundation or center or unity. Derrida calls it *différance*—the "name" for differentiation without ground.[128]

A survey of the history of intellectual systems draws a full circle from Plato to post-modern philosophy. "Becoming reality" cannot be analyzed into a coherent system; it must be understood as a *living whole* beyond

---

123 Cf. HORST FUHRMANS, "Einleitung," in FRIEDRICH WILHELM JOSEPH VON SCHELLING, *Über das Wesen der menschlichen Freiheit*, ed. Horst Fuhrmans (Stuttgart: Reclam, 1995), 38; cf. JOSEPH BRACKEN, *The Divine Matrix: Creativity as a Link between East and West* (New York: Orbis, 1995), 45–8; cf. DITTMER, *Schleiermachers Wissenschaftslehre*, 159–69, esp. 123.

124 Cf. WELSCH, *Vernunft*, 141–64.

125 Ibid., 851.

126 Cf. ROLAND FABER, "Messianische Zeit: Walter Benjamins 'mystische Geschichtsauffassung' in zeittheologischer Perspektive," *Münchner Theologische Zeitschrift* 54 (2003): 68–78.

127 Cf. STRUB, "System: II System und Systemkritik in der Neuzeit," 851; cf. DITTMER, *Schleiermachers Wissenschaftslehre*, 48–50.

128 Cf. WELSCH, *Vernunft*, 259–74.

structure, form, and logic; it is an *organism*.[129] Plato names this reality the *khora*, Derrida identifies this formless wholeness with *différance*, and Deleuze envisions it to be an "Omnitudo" in which innumerable planes cut through chaos as *creative events*. Both *khora* and *différance* elevate "system" / "formless wholeness" beyond any totalizing unity, foundational certainty, and necessary reason.[130]

From the rise of systems in the seventeenth century to their deconstruction in the twentieth century, we discern *four levels of shift* (see Table Two / vertical axis). Regarding the level of *ultimate interest* in systems, shift A: the reduction of disjunctive plurality to all-including *unity* yields to the liberation of *plurality* that is always richer than any oppressive unity. Regarding the level of the *foundation* of systems, shift B: the uncovering of the ultimate *ground* as a moment of interpretation itself; on a "*plane* of immanence," every reason is a moment in the interpretative structure. Deleuze calls it rhizomatic structure.[131] Regarding the level of *togetherness* within the system, shift C: *necessity* is dissolved into *contingency*; this is best seen in the shift from the monadic harmony in Leibniz to the nomadic distribution in Deleuze.[132] Regarding the level of *mind / world-relation*, shift D: early philosophies thought of intellectual systems as *descriptions* of the systematic character of reality according to criteria of harmony, coherence, and continuity; later systems, however, lost the illusion of an access to reality without *interpretation*.

V. Whitehead's Fluent System in Historical Context

For heuristic purposes, I distinguish between *four paradigms* of intellectual systems: *substance*, *subject*, *process*, and *difference* (see Table Two / second horizontal axis). The paradigm of substance was perfected by Spinoza's *natura naturans* and was deconstructed by Nietzsche's eternal return. The paradigm of subject began with Descartes' *ego cogito* and was perfected

---

129 Cf. Kann, *Fußnoten*, 117–30; cf. AI 119–20 in relation to Plato's *Sophist*, 247 d–e; cf. Dittmer, *Schleiermachers Wissenschaftslehre*, 152.

130 Cf. Jacques Derrida, *Passions, Sauf le nom, Khôra* (Paris: Éditions Galilée, 1993); cf. Deleuze and Guattari, *What is Philosophy?*, 35.

131 Cf. Deleuze and Guattari, *A Thousand Plateaus*, 3–25.

132 Cf. Ibid., 351–423.

by Kant's transcendental subject.[133] The paradigm of process developed under the reign of subjectivity in Hegel's absolute dialectic. It was organically de-subjectified by Schleiermacher's hermeneutics[134] and finally lost in Heidegger's "ontological difference."[135] The paradigm of difference began with de Saussure's infinite signification and led to Derrida's *différance*.[136] At the *three folds* of the paradigms, however, three philosophers highlight the changes they respectively influenced (see Table Two / third horizontal axis). Leibniz integrates substance and subject; Deleuze transforms process into difference; and Whitehead fuses subject and process.[137] Precisely this shift from subjectivity to process, and moreover, the fluency of all paradigms in Whitehead's shift may reveal his historical importance. Whitehead deconstructs substantiality for subjectivity, but transforms subjectivity into process that, as essentially beyond subjectivity, anticipates difference.[138] I will explicate Whitehead's place in history by locating his "*process*" as a mirror to Leibniz's "*harmony*" and Deleuze's "*chaosmos*" in view of all four levels of intellectual systems (see Table Two / first horizontal axis).[139]

---

133 Cf. WIEHL, "Whiteheads Kant-Kritik," 198–239; cf. ROHMER, *Whiteheads Synthese von Kreativität und Rationalität*, 79.

134 Cf. DITTMER, *Schleiermachers Wissenschaftslehre*, 421–2; cf. INGOLF U. DALFERTH, "Subjektivität und Glaube: Zur Problematik der theologischen Verwendung einer philosophischen Kategiorie," *Neue Zeitschrift für systematische Theologie* 36 (1994): 18–58.

135 Heidegger's "ontological difference" as process beyond subjectivity was explored by the *il y a* of Levinas; cf. COLIN DAVIS, *Levinas: An Introduction* (Cambridge: Polity Press, 1996).

136 Cf. WELSCH, *Vernunft*, 259–74; cf. MANFRED FRANK, *Das individuelle Allgemeine: Textstrukturierung und -interpretation nach Schleiermacher*, Suhrkamp Taschenbuch Wissenschaft, no. 544 (Frankfurt a. M.: Suhrkamp, 1985), 54.

137 Cf. WELSCH, *Vernunft*, 355–71; for Whitehead cf. MARIA-SIBYLLA LOTTER, "Subject—Superject: Zum Verhältnis von Privatheit und Öffentlichkeit," in *Natur, Subjektivität, Gott: Zur Prozeßphilosophie Alfred North Whiteheads*, ed. Helmut Holzhey, Alois Rust, and Reiner Wiehl, Suhrkamp Taschenbuch Wissenschaft, no. 769 (Frankfurt a. M.: Suhrkamp, 1990), 169–97, and HANS-CHRISTIAN LUCAS, "Substanz—Subjekt—Superjekt: Überlegungen zu Hegels und Whiteheads Abwendung von der Substanzmetaphysik, besonders im Hinblick auf Spinoza und Aristoteles," in *Whitehead und der Deutsche Idealismus*, ed. George R. Lucas, Jr. and Antoon Braeckman (Bern: Peter Lang, 1990), 67–82.

138 Cf. ROLAND FABER, *Gott als Poet der Welt: Anliegen und Perspektiven der Prozesstheologie* (Darmstadt: Wissenschaftliche Buchgesellschaft, 2003); cf. IDEM, *Prozeßtheologie*, 117–43, 168–200, 357–88.

139 Cf. CATHERINE KELLER, "Process and Chaosmos: The Whiteheadian Fold in the Discourse of Difference," in *Process and Difference: Between Cosmological and Poststruc-*

Table Two: Paradigms of Intellectual Systems

(three folds)			*harmony*	*process*	*chaosmos*
(four levels of shift)	(four shifts)				
ultimate interest	A. unity	‹······	······	······	······› plurality
foundation	B. ground	‹······	······	······	······› plane
togetherness	C. necessity	‹······	······	······	······› contingency
mind/world-relation	D. description	‹······	······	······	······› construction
(FOUR PARADIGMS)	SUBSTANCE → SUBJECT → ← PROCESS ← DIFFERENCE				
(three folds)			**Leibniz**	**Whitehead**	**Deleuze**

Leibniz's monadology, on the one side of the "mirror," integrates the paradigms of "substantiality" and "subjectivity" into the pre-established "harmony" of interrelated monads.[140] This "system" has implications at all four levels.

A. ***Unity / plurality***: While Spinoza's substance-monism devalues the plurality of individual beings to mere modes of ultimate substance,[141] Leibniz understands reality to be an indispensable plurality of internally related monads.[142] Their individual unity is internally indivisible and accounts for their spiritual subjectivity, which, in its substantiality, is essentially relational, ecological, and organic.[143] Apart from individuality, there is no unity—not even God's—that can comprise plurality. God only pre-stabilizes a harmony of interrelations by creating a compossible world.

B. ***Ground / plane***: In the reign of substance, ground is unity that devalues plural and relational reality. Leibniz, in contrast to Spinoza, however, discriminates between ground and unity. Although God is still the metaphysical principle of the system, unity is a harmony of plurality.[144]

---

turalist Postmodernism, ed. Catherine Keller and Anne Daniell (Albany: SUNY Press, 2002), 55–72.

140 Cf. RESCHER, "Leibniz," 117–22.

141 Cf. ROMBACH, *Substanz, System, Struktur*, vol. 2, 42.

142 Cf. Ibid., 9–98, 299–394.

143 Cf. ERNST R. SANDVOSS, *Geschichte der Philosophie*, vol. 2, Suhrkamp Taschenbuch Wissenschaft, no. 4441 (Frankfurt a. M.: Suhrkamp, 1989), 213–7; cf. ROMBACH, *Substanz, System, Struktur*, vol. 2, 358–74; cf. IDEM, *Whiteheads Synthese von Kreativität und Rationalität*, 86–7.

144 Whitehead notes a relation between "the Receptacle, according to Plato, the Void, according to Lucretius, and God, according to Leibniz" (AI 135).

C. *Necessity / contingency*: Every monad *is* the universe because monadic relations are internal and universal: "*un miroir vivant de l'univers.*"[145] But this harmony of monads generates a crystalloid structure excluding real process. Monads are mere temporal images of eternal structures. Only God's decision to create the best possible world is contingent.

D. *Description / construction*: Although Leibniz characterizes his *Système nouveau de la nature et de la communication des substances* constructively as "*mon système*," he is convinced that the *systema harmoniae praestabilitae* describes reality itself.[146]

Deleuze, on the other side of the "mirror," is *the* philosopher of immanence who, working through Hume, Kant, Bergson, Nietzsche, Spinoza, Leibniz, Whitehead, and Foucault, transfers "process" into his paradigm of "difference."[147] For this reason, Foucault characterized the twenty-first century to be a Deleuzian century.[148] Deleuze's position on system differs dramatically from Leibniz's.

A. *Unity / plurality*: Deleuze defines reality as a process of differentiation. No unity can ever comprise difference.[149] While metaphysics is concerned with unity, difference frees from oppressive metaphysics.[150] Difference emphasizes alterity over against substantial or subjective unification. Identity beyond oppressive unity, then, appears as the heterogeneity of pre-individual singularities. Plurality is basic, and unity is the movement of nomads on a plane.[151]

145 Johannes Hirschberger, *Geschichte der Philosophie*, vol. 2 (Freiburg: Herder, 1980), 155.

146 Cf. Rescher, "Leibniz," 114–7.

147 For the first time that "difference" appears in Deleuze's work, see Gilles Deleuze, "La conception de la différence chez Bergson," *Les études bergsoniennes* 4 (1956): 77–112. Cf. Friedrich Balke and Joseph Vogl, eds., *Gilles Deleuze: Fluchtlinien der Philosophie* (Munich: Fink, 1996).

148 Cf. Keller, "Process and Chaosmos," 58.

149 Cf. Todd May, "Difference and Unity in Gilles Deleuze," in *Gilles Deleuze and the Theater of Philosophy*, ed. Constantin V. Boundas and Dorothea Olkowski (New York: Routledge, 1994), 33–50, esp. 149.

150 Cf. Gilles Deleuze, *Difference and Repetition* (New York: Columbia University Press, 1994); cf. Keith A. Paerson, *Germinal Life: The Difference and Repetition of Deleuze* (London: Routledge, 1999).

151 Cf. Balke, *Gilles Deleuze*, 43–6, 82–3.

B. *Ground/plane*: Where all structure is based on immanent differentiation without allowing for reference to "reality" beyond the process of infinite signification, any appellation to a transcendent ground vanishes. Ground is distributed between "singularities" on a "plane of immanence."[152] Ground is surface.[153] Vertical, transcendent, or hierarchical structures are transformed into flat cuts through chaos, allowing for ever-changing constellations.[154] Instead of imaging structure as vertical-like trees with roots, structures constantly change their internal connections like ever-moving and ever-renewing "rhizomes."[155]

C. *Necessity/contingency*: No necessities restrict rhizomatic structures; the contingently moving constellations of nomads gather to momentary identities that render their "components inseparable."[156] A nomad is an event that comprises its elements with sudden presence: "*infinite through its survey or speed but finite through its movement that traces the contour of its components.*"[157] No pre-formation of possibilities determines these events; only virtuality contingently actualizes their singularity.[158]

D. *Description/construction*: When system is event, reality is constructed. The event of system, however, is not an illusion because there cannot be any knowledge of a reality "under construction" prior to its creation.[159]

Between Leibniz and Deleuze, Whitehead, the "mirror" of the four paradigms, invents the *organic* system of "process" (see Table One) that deconstructs substantiality as the monistic oppression of plurality,[160] as the du-

---

152 Alain Badiou, "Gilles Deleuze, The Fold: Leibniz and the Baroque," in *Gilles Deleuze and the Theater of Philosophy*, ed. Constantin V. Boundas and Dorothea Olkowski (New York: Routledge, 1994), 51–72.

153 Cf. Faber, "The Infinite Movement," 197.

154 Cf. Deleuze and Guattari, *What is Philosophy?*, 42–3; cf. Jäger, *Gilles Deleuze*, 260–71.

155 Cf. Deleuze and Guattari, *A Thousand Plateaus*, 16.

156 Cf. Deleuze and Guattari, *What is Philosophy?*, 19.

157 Ibid., 21.

158 Cf. Michael Hardt, *Gilles Deleuze: An Apprenticeship in Philosophy* (Minneapolis: University of Minnesota Press, 1995), 16–17; cf. Constantin V. Boundas, "Deleuze—Bergson: An Ontology of the Virtual," in *Deleuze: A Critical Reader*, ed. Paul Patton (Oxford: Blackwell, 1996), 81–106.

159 Cf. Hardt, *Gilles Deleuze*, 18.

160 PR 7.

alistic bifurcation of mind and world,[161] and as the pluralistic isolation of unrelated atoms.[162] Conversely, the organic process integrates individuality with solidarity, atomicity with connectivity, and novelty with continuity.[163] Whitehead's position is unique in that it reconstructs Leibniz's static structure that essentially lacks process, but critically corrects Deleuze's orgiastic chaos that profoundly lacks rationality (see Table Two / four shifts A to D).[164]

A. *Unity / plurality*: Whitehead's actual occasions uniquely unite their prehensive relations in a "space" of possibilities.[165] His "category of the ultimate" takes this process to be *a rhythm of the revolution of unity and plurality powered by creativity.*[166] "The many become one, and are increased by one."[167] Unity is *atomistic*: inseparable in unification but quantified in space-time[168]; and it is *relational*, creating "prehensive" unification.[169] Plurality is *ultimate*, the condition for becoming "one"; and it is *final*, a process of multiplication by the perishing of entities.[170] Whitehead follows Leibniz's relational net of individual entities,[171] but other than Leibniz, Whitehead's system does not freeze relations into a pre-established harmony.[172] Whitehead transforms the crystalloid system of windowless monads into a temporal process of momentary occasions that become and perish, that—in their causal nexus—exhibit change, and that—as "one" universe—are flux.[173] With Deleuze, Whitehead protects the ultimacy of plurality. Other than Deleuze, however, whose universe

---

161 RM 105–7.
162 SMW 155; cf. PR 46; cf. MG 106.
163 PR 7, 40, 56, 67; cf. William Christian, *An Interpretation of Whitehead's Metaphysics*, 2nd ed. (New Haven: Yale University Press, 1967), 155–74.
164 Cf. Deleuze, *The Fold*, 79; cf. Faber, "Evanescence," 194–8.
165 Cf. Faber, *Prozeßtheologie*, 117–54.
166 Cf. Reto Luzius Fetz, *Whitehead: Prozeßdenken und Substanzmetaphysik*, Symposion, no. 65 (Freiburg: Alber 1981), 130–3.
167 PR 21.
168 PR 285.
169 Cf. Charles Hartshorne, "Whitehead's Revolutionary Concept of Prehension," in *Creativity in American Philosophy* (Albany: SUNY Press, 1984), 103–13.
170 Cf. PR 21; cf. Deleuze, *The Fold*, 76.
171 RM 108.
172 PR 80; AI 133.
173 PR 48, 80; cf. Deleuze, *The Fold*, 76–82; cf. Christof Kann, "Zur Leibniz-Kritik bei A. N. Whitehead," in *Leibniz und Europa: VI. Internationaler Leibniz-Kongreß* (Hannover: Leibniz-Gesellschaft, 1994), 392–9.

tends to vanish in chaos,[174] Whitehead revives Plato's *khora* as formless unity that does not oppress plurality, but is the world's "medium of intercommunication."[175] With the ultimacy of plurality and the aim of momentary unity, Whitehead's ultimate metaphysical principle, "creativity," anticipates both Derrida's *différance* and Deleuze's "differenciation," because the process of becoming and perishing differentiates the *khora* and thereby generates multiplicity.[176]

B. ***Ground/plane:*** Whitehead exercises extreme care in discerning metaphysical principles as "internal constituents" for a "self-sufficient explanation of the flux of the world"[177] from "eminently real" hierarchical grounds[178] that are "essentially transcendent and only accidentally immanent."[179] No external ground may be "conceived as the foundation of the metaphysical situation with its ultimate activity."[180] In consequence, metaphysical principles are not "axiomatic certainties from which to start."[181] Hence, Whitehead consistently distinguishes his ultimate metaphysical principle, creativity, from God as its primordial exemplification.[182] He thereby radically disentangles ground from unity—a move that ends directly in Deleuze's radical immanence.[183] If the metaphysical ground of Whitehead's

---

174 Cf. DELEUZE and GUATTARI, *What is Philosophy?*, 50–1; cf. DOROTHEA OLKOWSKI, "Nietzsche's Dice Throw: Tragedy, Nihilism, and the Body without Organs," in *Gilles Deleuze and the Theater of Philosophy*, ed. Constantin V. Boundas and Dorothea Olkowski (New York: Routledge, 1994), 119–40.

175 AI 134; Cf. KANN, *Fußnoten*, 126–30.

176 Cf. FABER, *Poet*, 161–70; cf. CATHERINE KELLER, "Introduction: The Process of Difference, the Difference of Process," in *Process and Difference: Between Cosmological and Poststructuralist Postmodernism*, ed. Catherine Keller and Anne Daniell (Albany: SUNY Press, 2002), 16–18; cf. LUIS G. PEDRAJA, "Whitehead, Deconstruction, and Postmodernism," in *Process and Difference: Between Cosmological and Poststructuralist Postmodernism*, ed. Catherine Keller and Anne Daniell (Albany: SUNY Press, 2002), 73–90; cf. BOUNDAS, "Deleuze—Bergson," 82–3.

177 AI 121.

178 PR 342.

179 AI 121; FABER, *Prozeßtheologie*, 275–8.

180 SMW 179.

181 PR 13.

182 SMW 179; PR 7; cf. ALLISON H. JOHNSON, "Some Conversations with Whitehead Concerning God and Creativity," in *Explorations in Whitehead's Philosophy*, ed. Lewis S. Ford and George L. Kline (New York: Fordham University Press, 1983), 3–13; cf. FABER, *Prozeßtheologie*, 524–48.

183 AI 236; ROLAND FABER, "De-Ontologizing God: Levinas, Deleuze, and Whitehead," in *Process and Difference: Between Cosmological and Poststructuralist Postmodernism*, ed.

system is programmatically an "ultimate which is actual in virtue of its accidents,"[184] then his "immanent creativity"[185] radically anticipates the immanence of Heidegger's ontological difference and, even more, that of Derrida's *différance*—and their respective dissolution of onto-theology.[186] Indeed, Whitehead's "creativity" is the *immanent self-activity* of actualities that can never be conceived devoid of its embodiments as activity of actualities, except for reasons of conceptual abstraction.[187] In this regard, Whitehead's system deconstructs Leibniz's transcendent ground[188] that has prevented "harmony" from being conceived as non-preformed "process."[189] By this deconstruction, Whitehead's system dramatically resembled Deleuze's "plane of immanence."[190]

C. ***Necessity / contingency***: If the ultimate principle of the complex rhythm of becoming and perishing is creativity, then process never mirrors an eternal order.[191] When unity is disentangled from ground and ground immanently activates the multiplicity of entities, then creative reality is (groundless) *history*. As is the case with Heidegger's history of being (*Seinsgeschichte*), "[n]o reason, internal to history, can be assigned why that flux of forms, rather than another flux, should have been illustrated."[192] History neither executes fixed orders,[193] nor does it exhibit metaphysical necessities other than the conditions allowing history to remain history.[194] From the early "formative ele-

---

Catherine Keller and Anne Daniell (Albany: SUNY Press, 2002), 209–34.

184 PR 7; cf. ROMBACH, *Substanz, System, Struktur*, vol. 1, 16–17.

185 AI 236.

186 PR 7; MG 106; cf. BRACKEN, *The Divine Matrix*, 48–51; cf. WELSCH, *Vernunft*, 259–74.

187 PR 7, 21, 25; cf. JAMES BRADLEY, "Transcendentalism and Speculative Realism in Whitehead," *Process Studies* 23, no. 3 (1994): 155–91; cf. FETZ, *Whitehead*, 225; cf. SMW 107 with PR 7 and AI 236; cf. FABER, *Prozeßtheologie*, 144–54.

188 PR 190; AI 133–4.

189 MG 106; cf. ABRAHAM LICHTIGFELD, "Leibniz und Whitehead: Eine vergleichende Untersuchung ihrer metaphysischen Grundbegriffe und deren Weiterentwicklung durch Jaspers," in *Studia Leibniziana Supplementa* 5 (1971): 169–220.

190 Cf. FABER, "Evanescence," 192–8; cf. KELLER, "Process and Chaosmos," 58–61; cf. ROMBACH, *Substanz, System, Struktur*, vol. 1, 222; cf. FABER, "De-Ontologizing God," 218–22 and IDEM, " 'Gottesmeer'—Versuch über die Ununterschiedenheit Gottes," in *"Leben in Fülle": Skizzen zur christlichen Spiritualität*, ed. Thomas Dienberg and Michael Plattig, Theologie der Spiritualität, no. 5 (Münster: Lit Verlag, 2001), 64–95.

191 PR 95–6; cf. DELEUZE, *The Fold*, 81.

192 PR 46.

193 PR 95–6.

194 Cf. MARKS, *Gilles Deleuze*, 78–90.

ments" of *Religion in the Making*[195] to the seven Platonic notions of *Adventures of Ideas*,[196] Whitehead excludes metaphysical necessity, or better, he translates it into *assemblages* of problematic, paradoxical, or even antinomic "interrelations" of pluralities of metaphysical principles, realities, or ultimates.[197] At this point, Whitehead alludes to Plato's *Timaeus* in order to tell the "hypothetical history" of ultimate principles with metaphysical contingency rather than teaching a structure based on foundational necessity.[198] In contrast to Leibniz's necessary harmony,[199] Whitehead's history of the "all-inclusive universe"[200] obeys only contingent creativity. With Deleuze, the universe has no pre-formed order other than infinite virtuality.[201] Nevertheless, contingency for Whitehead is no metaphysical chaos but, by synthetical and transitory activity, it actualizes "chaosmos."[202]

D. *Description / construction*: This is probably the most problematic issue. Although Whitehead is convinced that his philosophy is a "metaphysical description" of the ultimate generalities of all experience,[203] he takes his systematization essentially to be an "interpretation."[204] Hence, when Whitehead speaks of "imaginative generalization," "rational interpretation," "imaginative rationalization," and "imaginative construction" within the same context,[205] he *fuses* description with construction.[206] The key to this fusion is his conception of "speculative reason" that, other than "practical reason" with its pragmatic functionalism,[207] proceeds with "disinterested curiosity."[208] Hence, "speculation" is free to inform a methodical procedure pro-

---

195 RM 90.
196 AI 275.
197 MT 2–3.
198 Cf. Kann, *Fußnoten*, 42–3, 118–24.
199 PR 46.
200 RM 90.
201 Cf. Hardt, *Gilles Deleuze*, 18.
202 PR 21; cf. Deleuze, *The Fold*, 81; cf. Keller, "Process and Difference," 58–61, 64–7.
203 PR 13.
204 PR 3; cf. Rolf Lachmann, *Ethik und Identität: Der ethische Ansatz der Prozeßphilosophie A. N. Whiteheads und seine Bedeutung für die gegenwärtige Ethik* (Freiburg: Alber, 1990), 74–95; cf. John W. Lango, *Whitehead's Ontology* (Albany: SUNY Press, 1972), 5–7, and Lucas, *The Rehabilitation of Whitehead*, 44.
205 PR 5.
206 Cf. Poser, "Whiteheads Kosmologie," 105–24.
207 FR 37.
208 FR 38; cf. Lachmann, *Ethik und Identität*, 130–5.

ducing systems that comprise the plurality of individual facts as an "exemplification of general principles."[209] In naming speculative reason "Platonic,"[210] however, Whitehead discovers that its real importance rests not in the systems it produces—these are in a certain sense arbitrary.[211] Rather, since the principle of speculation is *creativity*, its importance lies in *imagination* itself. In contrast to Kant's pure reason, Whitehead's speculative reason, although interested in "ultimate generalities," is *creative*; it *constructs* generalities that, as abstractions, *produce possibilities for the intensity of the process.*[212] Whitehead's "description" only works by way of "imaginative *construction*."[213] Hence, in deep agreement with Deleuze, for whom philosophy is the creation of concepts, Whitehead can neither be described a rationalist seeking coherence nor a realist seeking correspondence.[214] On the contrary, he is a *constructivist* who is well aware of the fact that, although reality is always infinite beyond description, it requires *rationality* to be creative.[215] Whitehead's constructivism is not "anti-intellectualism,"[216] but "rationalizing mysticism."[217]

VI. Whitehead's Speculative System as Event

When viewed from a historical perspective, Whitehead's "fluent system," as I have suggested, exhibits Deleuzian characteristics—protecting plurality without losing unity, integrating ground into a structural plane, deconstructing necessity within universal relativity, and constructing a system rather than describing reality—we should be able to detect these same

209 FR 38, 41; Kann, *Fußnoten*, 74.
210 FR 11–12.
211 PR 8; AI 145.
212 PR 259; cf. Poser, "Whiteheads Kosmologie," 121; cf. Wiehl, "Whiteheads Kant-Kritik," 198–239.
213 PR 5.
214 Deleuze and Guattari, *What is Philosophy?*, 3; cf. Welsch, *Vernunft*, 247–53.
215 MG 103–6; Imm. 82; cf. Faber, "Evanescence," 189; cf. Reiner Wiehl, "Whiteheads Kosmologie der Gefühle zwischen Ontologie und Anthropologie," in *Whiteheads Metaphysik der Kreativität: Internationales Whitehead-Symposium Bad Homburg 1983*, ed. Friedrich Rapp and Reiner Wiehl (Freiburg: Alber, 1986), 141–68; Rohmer, *Whiteheads Synthese von Kreativität und Rationalität*, 154–9.
216 SMW 148; PR xii.
217 MT 174.

characteristics in Whitehead's genuine "speculative systems."[218] Whitehead repeatedly outlined such "schemes": from the "three attributes" of "substantial activity" in *Science and the Modern World*[219] and the "formative elements" of *Religion in the Making*[220] to the seven Platonic notions in *Adventures of Ideas*[221] of which "all philosophy is in fact an endeavor to obtain a coherent system."[222] Only the "categoreal scheme" of *Process and Reality*, however, seems to be detailed enough to demonstrate Whitehead's historical importance in reading "system" as "event," that is, as an *actual assemblage of togetherness.*[223]

The ultimate approach to Whitehead's "categoreal scheme" is its first category, the premise on which the entire system is built.[224] This "category of the ultimate" exhibits a *threefold assemblage* as basic to the whole system: "one," "many," and "creativity."[225] We note that Whitehead does not ground his system in a way that reduces all plurality to an ultimate "one." Rather, his system is grounded on the process of three moments that cannot be reduced to one another.[226] Later in the "final summary" of *Process and Reality*,[227] when Whitehead explicates his "cosmological construction,"[228] this threefold structure appears again—there as the "final interpretation of the cosmological problem"[229]: the creative togetherness of "God" and "World" by which "Creativity achieves its supreme task."[230] Whitehead's systematic construction is an explication of *this* "ultimate triad"—many / World, one / God, and Creativity—that, in turn, by the actual interaction of its elements, creates the system as *its* process (see Table Three).

---

218 KELLER, "Introduction," 11; cf. FABER, *Prozeßtheologie*, 297–448 (part III).
219 SMW 177–8.
220 RM 90.
221 AI 147.
222 AI 275.
223 MT 3–4; cf. KLAUS OEHLER, *Aristoteles*, 2nd ed. (Berlin: Akademie-Verlag, 1986), 41–65.
224 Cf. WILLIAM GARLAND, "The Ultimacy of Creativity," in *Explorations in Whitehead's Philosophy*, ed. Lewis S. Ford and George L. Kline (New York: Fordham University Press, 1983), 212–38.
225 PR 20–2.
226 LOTTER, "Subject-Superject," 169–97; cf. DITTMER, *Schleiermachers Wissenschaftslehre*, 464.
227 PR 348.
228 PR 341.
229 Ibid.
230 PR 348; MARJORIE HEWITT SUCHOCKI, *The End of Evil: Process Eschatology in Historical Context* (Albany: SUNY Press, 1988), 168, n. 9.

Whitehead's ultimate threefold structure represents a particular interaction between three basic features of every system[231]: a system rests on a "ground" or principle; it encompasses a "whole" to which the system is applied; and it strives toward an "aim" or satisfaction.[232] In Whitehead's "cosmological construction," "ground" is creativity,[233] the "whole" is the multiplicity of the universe as the process itself,[234] and its "aim" is synthesis, although self-transcendent synthesis.[235] Consequently, this triadic structure unfolds into the ontology, cosmology, and eschatology of Whitehead's system and, in turn, governs the interpretation of its categories (see Table Three).[236]

Table Three: Threefold Structure of Speculative Systems

	Plato (later)	Descartes	Spinoza	Hegel	Nietzsche	Derrida
aim	Good (one)	certainty	necessity	Absolute	eternal return	deconstruction
whole	Khora (mixture)	substances	modes	process	becoming	differentiation
ground	ideas (infinite)	ego cogito	substance	dialectic	will to power	différance

	Leibniz	Whitehead		Deleuze
aim	harmony	ESCHATOLOGY	*principle of process*	orgiastic freedom
	freedom	one (self-transcending)	novelty/harmony	infinite speed
	best world	*God (perfected system)*	intensity	immanence
whole	monads	COSMOLOGY	*principle of relativity*	nomads
	mirror	many (communication)	organism	events
	holism	*World (khora)*	universality	rhizome
ground	God	ONTOLOGY	*ontological principle*	plane/immanence
	Why?	self-creativity	actuality	virtuality
	creation	*Creativity (novelty)*	fluency	difference

To begin with the "ground" of Whitehead's system: for the *ontology* of Whitehead's system, ground is neither God, nor a transcendent principle,

231 PR 21.

232 General "systematology" began with Christian Wolff (1729) and Johann Heinrich Lambert (1767): cf. Ritschel, *System und systematische Methode*, 60–8.

233 PR 7.

234 AI 150.

235 AI 295; cf. Dittmer, *Schleiermachers Wissenschaftslehre*, 136; cf. Faber, "Evanescence," 188–94.

236 Cf. Rescher, "Leibniz," 119; cf. Wiehl, "Whiteheads Kant-Kritik," 234.

nor any unity of the system's wholeness,[237] but *creativity*.[238] The entire system is founded on a principle of novelty, activity, and contingency that govern existence. Since creativity is the *immanent* principle of the system, it has no *fundamentum inconcussum* and, therefore, actualizes itself in a *groundless distribution* of interrelated entities (whether the entities are occasions or concepts) into a creative assemblage of "self-differenciaton" that rests on actuality, history, and fluency.[239] Within the categoreal scheme, the "ontological principle" systematically expresses this thesis.[240] This principle proposes "that actual entities are the only reasons; so that to search for a reason is to search for one or more actual entities."[241] With creativity as the radically immanent ground of the system, the "ontological principle" has three important implications. Firstly, no transcendent reason for the system can be given beyond its actualization.[242] Secondly, universal structures and ultimate principles are mere abstractions as long as they are not actualized in the event of the system.[243] And thirdly, a system *as* event functions like an actuality, a nexus, or a society.

Turing to the "whole" of the system: Whitehead's *cosmology* exhibits its wholeness as *process* of the world's manifold.[244] "Process" is a *universal singular* that envisions "The Universe" as a moving whole.[245] Whitehead, like Bergson and Deleuze, does not understand "the whole" as a structure or set that can be spatialized and hence closed, but as "living being" that is "itself the Open."[246] Deleuze notes: "If one had to define the whole, it would be defined by Relation."[247] Hence, a system is not characterized by its set of categories or structural corset, but by its relational process. In

237 PR 7; MG 106.

238 PR 21. In FABER, "Gottesmeer," 83–95, I argue that Whitehead represents the beginning of a profound criticism of onto-theology.

239 Cf. DELEUZE and GUATTARI, *What is Philosophy?*, 15–113; cf. BOUNDAS, "Deleuze—Bergson," 82–3; cf. HARDT, *Gilles Deleuze*, 1–25.

240 PR 24–5: Category of Explanation XVIII.

241 PR 24; cf. LEWIS S. FORD, "Perfecting the Ontological Principle," in *Metaphysics as Foundation: Essays in Honor of Ivor Leclerc*, ed. Paul Bogaard and Gordon Treash (Albany: SUNY Press, 1993), 122–49.

242 Cf. HARDT, *Gilles Deleuze*, 18, and FABER, *Prozeßtheologie*, 389–437.

243 Cf. FELIX KRÄMER, *Der Zusammenhang der Wirklichkeit: Problem und Verbindlichkeitsgrund philosophischer Theorien: Eine Untersuchung in Auseinandersetzung mit Fichte und Whitehead*, Fichte-Studien-Supplementa, no. 5 (Amsterdam: Edition Rodopi, 1994), 10–11, 66–73, 244–8.

244 Cf. KANN, *Fußnoten*, 86–94.

245 AI 150.

246 DELEUZE, *Cinema*, 10; cf. SMW 148; PR 209.

247 DELEUZE, *Cinema*, 10.

Whitehead's categoreal scheme, the "principle of relativity" explains this stance.[248] This principle proposes that "it belongs to the nature of 'being' to be a potential of every 'becoming.'"[249] With creativity functioning as "the universal of universals,"[250] the "principle of relativity" has three important implications. Firstly, actualized systems cannot "be adequately analyzed purely in terms of universals,"[251] structures, or sets of categories. Rather, they *move* as a "whole" of productive relations, which Whitehead terms "prehensions."[252] Secondly, the "necessity in universality" of the system is not the completeness of its structural integrity, but rests on the system's universal relatedness that, by its movement, is "defined by 'communication'."[253] Thirdly, if creativity is the "universal of universals," then there can be no universal that can possibly serve as the universal of all universals.[254] No "abstractive hierarchy" can unite all abstractions into one "vertex"[255]; there is only the "vortex" of creativity in which all relations are actualized in living communication.[256] The "whole" of this system is "one" only in the *khora*, the "real communication"[257] that comprises a "necessary community."[258] The *khora* signifies the "general interconnectedness of things, which transforms the manifoldness of the many into the unity of the one."[259] As a result, the "coherent, logical, necessary sys-

248 PR 22: Category of Explanation IV.

249 PR 22; cf. Jorge Luis Nobo, *Whitehead's Metaphysics of Extension and Solidarity* (Albany: SUNY Press, 1986), 61–106.

250 PR 21.

251 PR 157.

252 PR 22: Category of Existence II; Systems can only exist by movement because structures are self-transcendent different/ciations in the process of the production of actuality; cf. Michael Hampe, *Die Wahrnehmungen der Organismen: Über die Voraussetzung einer naturalistischen Theorie der Erfahrung in der Metaphysik Whiteheads* (Göttingen: Vandenhoeck & Ruprecht, 1990), 104–5.

253 PR 4; cf. Roland Faber, "Trinity, Analogy and Coherence," in *Trinity in Process: A Relational Theology of God*, ed. Joseph Bracken and Marjorie Hewitt Suchocki (New York: Continuum, 1997), 147–71.

254 Cf. AI 257 where Whitehead proposes creativity to be the ground of all abstractions; cf. Garland, "The Ultimacy of Creativity," 227.

255 SMW 168.

256 Cf. Lewis S. Ford, "Neville on the One and the Many," *Southern Journal of Philosophy* 10 (1972): 79–84.

257 AI 134.

258 AI 150.

259 Ibid.; cf. Garland, "The Ultimacy of Creativity," 227; cf. Helmut Maassen, *Gott, das Gute und das Böse in der Philosophie A. N. Whiteheads* (Frankfurt a. M.: Peter Lang, 1988), 146; Bracken, *The Divine Matrix*, 56–7.

tem"[260] consists of the fluent, open interconnectedness of categories, principles, and concepts, rather than its "systematicity."[261] Hence, Whitehead's "system" or "scheme" does not positively propose a closed, hierarchical structure or complete set of rules that necessarily governs the universe,[262] but negatively insists on life, embodied in the strategy that "no entity can be conceived in complete abstraction from the system."[263]

Regarding the "aim" of Whitehead's system, *eschatology* names its satisfaction and its meaning.[264] Whitehead's bifurcation of "ground" and "unity" leads to the shift of the systematic place of "unity" from "ground" and "whole," to the "aim" of the system.[265] This is an important move because it protects "ground" *as* difference and the "whole" *as* process and, by the same token, reintroduces "unity"—similar to Derrida—as the eschatological satisfaction of difference and process.[266] Since, on this new basis, satisfaction must not freeze the "differenciation" of the ground in the process of the whole, "unity" must promote difference and process. Hence, unity must be *actual, finite*, and *perishable*, otherwise it would oppress the plurality and fluency of the process.[267] This is precisely what the "category of the ultimate" accomplishes: it introduces "unity" as the *finite unification of infinite multiplicity* that, in its satisfaction, perishes and *transcends itself* into a new plurality.[268] Within the categoreal scheme, the "principle of process"[269] explains this deconstructive eschatological move further. This principle proposes: "how an actuality becomes constitutes what that actual entity is."[270] "Oneness" is never "given"; it is the aim of a process—but as self-transcendent satisfaction. Hence, the aim deconstructs itself, flowing back into the creative differentiation of the whole.[271] This "oneness"

---

260 AI 222.
261 Cf. ROMBACH, *Substanz, System, Struktur*, vol. 2, 12; ibid., vol. 1, 222; FABER, "Gottesmeer," 78–83, 88–90; cf. DITTMER, *Schleiermachers Wissenschaftslehre*, 416–7.
262 AI 117.
263 PR 3; cf. KELLER, "Introduction," 11; cf. FR 8.
264 On Whitehead's eschatology cf. FABER, "De-Ontologizing God," 222–4.
265 Cf. FABER, "Evanescence"; cf. IDEM, "De-Ontologizing God."
266 Cf. IAN ALMOND, "How *Not* to Deconstruct a Dominican: Derrida on God and 'Hypertruth'," *Journal of the American Academy of Religion* 68, no. 2 (2000): 329–44; cf. FABER, *Poet*, 230–44.
267 Cf. GILLES DELEUZE, *The Logic of Sense* (New York: Columbia University Press, 1990), 262–3.
268 PR 21; cf. DELEUZE and GUATTARI, *A Thousand Plateaus*, 351–423.
269 PR 23: Category of Explanation ix.
270 PR 23; cf. FETZ, *Whitehead*, 125–30.
271 PR 25–6: Category of Explanation xxv.

critically reinterprets the "presence" of system in its process.[272] Although concrescence is self-creative, immediate, and self-significant,[273] the process upon reaching its aim in the moment of satisfaction loses immediacy; its "perfection" is its self-transcendence whereby it becomes "perfect."[274] Consequently, the "aim" of Whitehead's categoreal scheme presents itself as an infinite process of "self-transcendent unity" with its ultimate satisfaction, to which all its categories, principles, and concepts gather, being the *production of difference.*[275]

With Derrida's *différance*, the process is never really present.[276] In a way diverging from Derrida's *différance*, however, Whitehead reinterprets this infinite eschatological "deference" of a final aim with his eschatological concept of God. God is the *final eschatological unity* that protects process with "harmony."[277] For Whitehead, God is the "essential unification of the Universe"[278] in a dual sense: as *the* "perfected system,"[279] but only *for* the process of differentiation.[280] By virtue of God's dipolar function, God is the eschatological vortex of all structures as potentialities for novelty and harmony in the process *and* the eschatological peace of all actualities.[281] In both of God's functions—as primordial and as consequent nature[282]—God is *self-transcendent unity* in the highest sense that flows back into the process.[283] With God as the primordial exemplification of the creative pro-

---

272 As long as "ground" and "unity" are not differentiated, "ground" represents the presence of unity in multiplicity. When differentiated—as is the case with Whitehead, Deleuze, and Derrida—"ground" means the presence of difference as the real generator of structure in the process.

273 PR 25: Categories of Explanation xxi–xxiii.

274 For Whitehead, only the past is perfect; the process of becoming, however, is "Imperfection" (AI 257).

275 Cf. Roland Faber, "Zeitumkehr: Versuch über einen eschatologischen Schöpfungsbegriff," *Theologie und Philosophie* 75 (2000): 180–205.

276 PR 82; AI 237–8; cf. Lotter, "Subject—Superject," 192–17.

277 RM 154; cf. Faber, *Poet*, 230–44.

278 Imm. 90; cf. Faber, "De-Ontologizing God," 220.

279 PR 346.

280 PR 350–1; cf. Roland Faber, "Apocalypse in God: On the Power of God in Process Eschatology," *Process Studies* 31, no. 2 (2002): 64–96.

281 PR 246; AI 285; cf. Faber, *Poet*, 219, 240; Marjorie Hewitt Suchocki, *God, Christ, Church: A Practical Guide to Process Theology*, 2nd rev. ed. (New York: Continuum, 1995).

282 Cf. Thomas E. Hosinski, *Stubborn Fact and Creative Advance: An Introduction to the Metaphysics of Alfred North Whitehead* (Lanham, Md.: Rowman & Littlefield, 1993), 155–206.

283 PR 350–1.

cess,[284] the aim of self-transcendent unity is manifested as "novelty," "harmony,"[285] and its ultimate satisfaction lies in the intensity of the process itself[286] of which God is the infinite measure.[287] God's process seeks intensity, not preservation or self-presence.[288] Hence, the satisfaction of the "system" is the intensity of the event of its actualization.

VII. Conclusion: System and Life

With creativity as "ground" and relational multiplicity as "whole," the aim of Whitehead's speculative system, exemplified by the categoreal scheme of *Process and Reality*, is not what it seems to be at first glance, namely a "coherent, logical, necessary system" of all possible experience,[289] that is, a structure that comprises life. The aim of system is process itself that, as life, includes structure.[290] But if mysticism and the pre-rationality of life are ultimate,[291] what is the meaning of the structures, schemes, and patterns we normally link to "systems"? Their meaning in a "fluent system" is none other than their function to further the intensity of the actual event of the process that *is* the system. A system as event is only real in *actualization*, thereby—as Lyotard has noted[292]—creating its own rules. Structure, abstracted from this actualization, is mere potentiality. On the other hand, a fluent system demands *structure* to protect the process from the oppression of irrational power.[293] Structural definiteness generates rational permanence over against a chaos of irrational powers. Ultimately, the categories of Whitehead's categoreal scheme might be seen to "organically intercohere" enough to be recognized as "Immanent Law,"[294] but open

284 PR 7.
285 AI 295–6.
286 PR 27: Categoreal Obligation VIII.
287 PR 47; cf. Maassen, *Gott, das Gute und das Böse*, 148.
288 PR 105; cf. Gregg Lambert, "On Whitehead's proposition 'Life is robbery,'" in *Secular Theology: American Radical Theological Thought*, ed. Clayton Crockett (London: Routledge, 2001), 92–103.
289 PR 3.
290 Wiehl, "Whiteheads Kosmologie," 141–68.
291 MT 174.
292 Jean François Lyotard, "Beantwortung der Frage: Was ist postmodern?," in *Wege aus der Moderne: Schlüsseltexte der Postmoderne-Diskussion*, ed. Wolfgang Welsch, 2nd ed. (Berlin: VCH-Verlags-Gesellschaft, 1994), 202–3.
293 Cf. PR XII.
294 AI 113.

enough to prevent "the closure by which the system contains and sustains itself."[295] The true satisfaction of Whitehead's system is the intensity of life that, as event, is immanently rational by being *structurally liberated* from the imposition of controlling power.[296] In this sense, structure guarantees the system's aim: the depth of intensity[297] that frees from oppression.

In this chapter, I have argued that Whitehead's speculative system must be read as self-creative event.[298] In a reversion of Whitehead's famous dictum that every actuality is a "system of all things,"[299] we can say: every system is an actuality or better, an "organism" comparable to an "entire living nexus" of a "living society" or "person."[300] Hence system is an infinite process of "interpretation."[301] Continuing with another of Whitehead's famous formulations, that "life is robbery" of structures,[302] we can say: the system's event harbors structures to protect the process from oppression of despotizing unity. Hence, in fluent systems, structures only further the process when they generate difference. A fluent structure establishes the dignity of the multiplicity of the living whole.

In view of Whitehead's invocation of the "Mind of Leibniz" ranging "from divinity to political philosophy,"[303] the ultimate concern of Whitehead's "fluent system" might be the "Harmony of Harmonies"[304] that as the "*Final* Fact" is "*initial* Eros," endlessly initiating the process of "differenciation."[305] But its political importance might be—as with Plato,[306] Adorno, Deleuze, and Derrida—to generate a *hermeneutic of difference* for a civilization of persuasion[307] over against a barbarism of power and oppression.[308]

---

295 Keller, "Introduction," 6–11.
296 AI 113.
297 RM 113.
298 Cf. Roland Faber, "On the Unique Origin of Revelation, Religious Intuition, and Theology," *Process Studies* 28, nos. 3–4 (1999): 273–89.
299 PR 36.
300 Cf. PR 104; cf. Roland Faber, "Personsein am Ort der Leere," *Neue Zeitschrift für systematische Theologie* 44 (2002): 189–98.
301 PR 3.
302 PR 105–6.
303 MT 3.
304 AI 296.
305 AI 295.
306 AI 166.
307 AI 296.
308 AI 51.

Schleiermacher between Kant and Leibniz

Predication and Ontology

JACQUELINE MARIÑA, Purdue University

In some of his most recent work, Manfred Frank has suggested that Schleiermacher's *Dialektik* can be interpreted more fruitfully by taking stock of Leibnizian influences on Schleiermacher's thought. After all, Schleiermacher was a pupil of the Leibnizian Eberhard in Halle; Frank notes that in his correspondence Schleiermacher took sides with Eberhard against Kant in the famous Kant-Eberhard debate.[1] In this paper I take stock of this claim by examining and comparing the larger systematic thought contexts of Leibniz, Kant, and Schleiermacher. My focus will be on the ontological and epistemological commitments of each of these thinkers and their implications for their understanding of predication. More specifically, the question to be explored is the following: does Schleiermacher's adoption of Leibniz's complete concept and the theory of predication it entails conflict with his adoption of Kant's two-source theory of knowledge? I conclude that it does, and that in fact it is this clash between elements borrowed from two very different systematic thinkers that makes Schleiermacher's *Dialektik* the cipher that it is. The paper will be divided into two parts. First I explore the larger systematic interrelations in Leibniz's thought regarding his theories of substance and the complete concept. This analysis will show that the theory of the complete concept cannot be so easily divorced from Leibniz's ontology of independent substance. Second, I will examine the Kant-Eberhard debate in order to make explicit the interconnections

---

1 In the Introduction to his edition of Schleiermacher's *Dialektik*, Manfred Frank notes, "Der Briefwechsel zeigt, daß Schleiermacher sich ganz entschieden auf die Seite Eberhards und gegen Kant und Reinhold stellt (vgl. etwa Brief 128, Z. 262 ff.)." MANFRED FRANK, "Einleitung," in Dial2001, 67.

between a two-source theory of knowledge (presupposing physical influx) and Kant's understanding of the nature of judgments. This controversy highlights the differences between the systems of the followers of Wolff and Leibniz, and that of Kant. The analysis develops further the intrinsic relations between the theory of the complete concept and the idea of independent substances on the one hand, and the two-source theory of knowledge and Kant's extensional logic, on the other.

I. Leibniz and Schleiermacher on Substance and the Complete Concept

An important point of similarity in the systems of Leibniz and Schleiermacher is that of the complete concept and the theory of predication implied by it. Schleiermacher, following Leibniz, often speaks of the complete concept of a thing containing marks for all its powers and attributes. In the *Dialektik*, Schleiermacher notes that "The particular, the lowest level of concepts, has being; however general concepts are not things, but rather empty marks having their source in abstraction from a mass of judgments."[2] Two things stand out in this sentence. First, Schleiermacher's identification of a particular with the lowest level of a concept implies that the individual, with all its particularities, can be given in a *concept*. As such for Schleiermacher, as for Leibniz, there is an *infima* species. Second, the sentence is notable for the fact that it fails to distinguish between concepts and things. Since the complete concept contains all the marks for an individual, everything that is true about that individual is already contained in its concept. What ontological implications, if any, should Schleiermacher's adoption of the idea of a complete concept commit him to? In order to answer this question, we must first take a look at Leibniz's development of the theory of the complete concept and its role in Leibniz's larger philosophical system.

According to Leibniz, "The concept of a singular substance is something complete, which already contains potentially whatever can be understood of it .... A complete concept is the mark [*nota*] of a singular substance."[3] In the *Discourse on Metaphysics*, Leibniz writes: "the nature of an indi-

2 DialO 225. (All translations from Schleiermacher's *Dialektik* are my own.)
3 The English translation of this passage is from Donald Rutherford's *Leibniz and the Rational Order of Nature* (Cambridge: Cambridge University Press, 1995), 110; the passage is originally found in Eduard Bodeman, *Die Leibniz-Handschriften*

vidual substance or a complete being is to have a notion so complete that it is sufficient to contain and to allow us to deduce from it all the predicates of the subject to which this notion is attributed."[4] Leibniz's theory of truth is based on concept containment, and this implies that all true propositions are analytic. The hallmark of such judgments is that they are true in virtue of the principle of contradiction; truth is based on the logical relation between *concepts*. For Leibniz the predicate in a true judgment merely reiterates that which is already included in the subject concept or that which is implied by it; to negate such a true judgment would imply a contradiction. This understanding of concepts is principally true of God's ideas on which possibility and truth are founded, but is true of human concepts as well.

What are the systematic interrelations between Leibniz's theory of predication and his ontological commitments? In order to make these clear, we need to explore the interrelations between the idea of the complete concept and Leibniz's understanding of substance.[5] In fact, I will argue that Leibniz's theory of substance is the driving force behind his whole ontology and theory of predication. Both Leibniz's nominalism and his theory of the complete concept are intrinsically bound up with his theory of substance.

Some of the most important characterizations of substance in Leibniz's theory are the following. First, the being of a substance does not depend on anything outside of itself; hence following Spinoza's characterization, substance is that which can be thought through itself. Second, if substance is truly independent, it must itself be the source of all its modifications. As such, it contains within itself its own entelechy. This principle, along with the first is what leads Leibniz to conceive of his monads as "windowless," that is, they cannot be affected from without. This view of substance is so important that it is worth quoting Leibniz at length on this issue:

---

königlichen öffentlichen Bibliothek zu Hannover (Hannover: Hahn, 1889; reprint, Hildesheim: Georg Olms, 1966), LH IV 7C, Bl. 111–4.

4 Gottfried Wilhelm Freiherr von Leibniz, *Philosophical Essays*, ed. and trans. Roger Ariew and Daniel Garber (Indianapolis: Hackett, 1989), 41.

5 This is a major issue in Leibniz scholarship. The question is whether Leibniz's theory of the complete concept drives his theory of substance, or whether it is his theory of substance that leads him to adopt the theory of the complete concept. Current Leibniz scholarship tends to the latter view, which seems to me to be the correct one. My discussion of Leibniz is indebted to Rutherford's excellent study, *Leibniz and the Rational Order of Nature*, cited above.

> [From] the notion of an individual substance it also follows in metaphysical rigor that all the operations of substances, both actions and passions, are spontaneous, and that with the exception of the dependence of creatures on God, no real influx from one to the other is intelligible. For whatever happens to each one of them would flow from its nature and its notion even if the rest were supposed to be absent.[6]

Third, a substance persists throughout its changes. Fourth, substance is a true or *per se* unity, that is, all of its attributes should be derivable from a fundamental principle that articulates its essence. The third and fourth conditions of substance are intrinsically interwoven, since insofar as substance is a *per se* unity, it can be said to persist through all its modifications. Lastly, substances are uniquely identifiable. Given this characterization of substance, Leibniz's understanding of the complete concept is uniquely suited to articulate the nature of the intrinsic connection of a substance's attributes to its essence. If substance is truly independent (the monads are windowless), all of its changes must flow from its own spontaneous action. The complete concept expresses the unique essence of individual substance from which all of its modifications flow, since God creates through it. As Rutherford notes, "a complete concept is an appropriate way to conceive of God's knowledge of a being, which is, by its nature, a spontaneous source of change."[7] In Leibniz's system, the notion of a complete concept is inherently bound up with his notion of substance. It is Leibniz's commitment to independent substance that leads him to the notion of the complete concept.

The Leibnizian understanding of substance has ontological implications that are in turn bound up with Leibniz's theory of predication. According to Leibniz's ontology, only concrete particulars, that is, substances (*substantiva*) that cannot themselves be predicated of other substances, exist. Abstract beings do not exist; they are mere "beings of reason" (*entia rationis*), the product of thought which abstracts them from concrete particulars. Leibniz tells us that the thorniest difficulties puzzled over by the Scholastics disappear "at once if we agree to banish abstract beings and resolve to speak ordinarily only in terms of concrete beings, admitting no

---

6 Gottfried Wilhelm Freiherr von Leibniz, *Philosophical Writings*, ed. and trans. Mary Morris and George Henry Radcliffe Parkinson (London: Dent, 1973), 79.

7 Rutherford, *Leibniz and the Rational Order of Nature*, 139.

other terms in scientific demonstrations than those which represent substantial subjects."[8] He therefore offers a reinterpretation of the scholastic dictum *praedicatum inest subjecto*. This was traditionally understood to refer to the actual inherence of attributes in real subjects; a true proposition reflects this real relation. Since for Leibniz there are no *abstracta*, a proposition cannot be true in virtue of its reflecting the actual inherence of attributes in a subject. Given these nominalist commitments, how then is Leibniz to preserve objective grounds for truth? The answer lies in his theory of the complete concept. Since the divine mind creates individuals through complete concepts, essences of actual particulars have as their counterparts perfect ideas, making it possible to transfer the metaphysical basis of predication from things to concepts. *Praedicatum inest subjecto* no longer refers to attributes inhering in things, but rather to the fact that a complete concept contains simpler concepts within it.[9] Hence Leibniz notes that "when we say that Alexander is strong, we mean nothing else than that strong is contained within the notion of Alexander, and likewise for the rest of Alexander's predicates."[10]

Leibniz's brand of nominalism follows from his understanding of substance. An individual substance is what it is in virtue of all the attributes that are bound up with it. These attributes are inherently tied up with the essence of a substance, that is, the substantial essence implies them. This must be the case if a substance is to qualify as a true or *per se* unity, that is, a unifying principle (its essence) must be the ground of everything that can be truly predicated of it. Hence were it the case that any one of a substance's attributes were different, one would have a different individual altogether. Leibniz notes that "what determines a certain Adam must absolutely contain all his predicates, and it is this complete concept that determines generality in such a way that the individual is reached."[11] This means that there are no free-floating *abstracta*, since attributes are always tied to particular substances.

---

8 Gottfried Wilhelm Freiherr von Leibniz, *New Essays on Human Understanding*, ed. and trans. Peter Remnant and Jonathan Bennet (Cambridge: Cambridge University Press, 1981), 217.

9 See Rutherford's treatment of Leibniz's nominalism in *Leibniz and the Rational Order of Nature*, 115–9.

10 The English translation of the passage is from ibid., 124. The passage is originally found in Gottfried Wilhelm Freiherr von Leibniz, *Fragmente zur Logik*, ed. and trans. Franz Schmidt (Berlin: Akademie-Verlag, 1960).

11 Gottfried Wilhelm Freiherr von Leibniz, *The Leibniz-Arnauld Correspondence*, trans. Haydn Trevor Mason (Manchester: Manchester University Press, 1967), 60–1.

As noted above, Schleiermacher often speaks of a complete concept in the *Dialektik*. Given the systematic interrelations between Leibniz's theory of substance and that of the complete concept, two points bear discussion. First, we need to see just how much of Leibniz's ontology Schleiermacher takes over and at what significant junctures he diverges from him. Second, we need to ask whether, given these divergences, Schleiermacher's systematic thought can remain coherent.

The most important point of divergence between Leibniz and Schleiermacher is the following: while Leibniz's monads are windowless, Schleiermacher clearly conceives of self-consciousness as involving both spontaneity and *receptivity*. In a sketch for an 1831 lecture, Schleiermacher notes that in the opposition between subject and object

> lies the opposition between doing and suffering in both its members in a twofold way. Namely, when something in the confused manifold effects the split between subject and object, so is what is outside us the active agent and the organism is receptive, what suffers. On the other hand when something is distinguished out of the confused manifold at the behest of the faculty of knowledge, then the subject becomes active and the separated something is posited as that which is thought and as passive.[12]

In the Introduction to *The Christian Faith*, Schleiermacher repeatedly mentions the opposition between our receptivity and our spontaneity: the point of consciousness through which we are conscious of the feeling of absolute dependence is the ground of both.[13] In the *Dialektik*, Schleiermacher views the difference between being and thinking as reflected in the differences between the organic and intellectual functions. Through the organic function we are *affected* by that which lies outside us; it thereby corresponds to our receptivity. On the other hand, the intellectual function, through which objects are thought, corresponds to spontaneity. This division of the sources of human knowledge closely mirrors Kant's. In fact already in the lectures on *Dialektik* of 1814/15, Schleiermacher notes that "without unity and multiplicity the manifold is undermined; without the manifold the determinate unity and multiplicity is empty."[14] The statement

12 DialJ 496.
13 CF §§ 4–9 (12–44).
14 Dial(1814/15) 23.

is a striking echo of Kant's famous dictum that "thoughts without content are empty, intuitions without concepts are blind."[15]

Manfred Frank has pointed to Schleiermacher's claim that "We can only think of the antithesis between receptivity and spontaneity as a relative, not as an absolute antithesis. A minimum of power must be posited in receptivity, and both come together in a higher power."[16] In other words, neither source of knowledge can be found in its purity. Frank traces this idea back to Schleiermacher's Leibnizian teacher Eberhard, for whom sensation and thinking were restrictions of one primal power. As such, Frank understands Schleiermacher's project as a kind of re-Leibnizianization of Kant.[17] Against such an understanding it must be remembered that for Leibniz, as for Eberhard, the *content* of perception and thought is the same. The difference between them is merely logical, that is, sensible cognitions lack the clarity and distinctness of intellectual cognitions; they are confused representations of the infinite, and this confusion is due to our finitude and limitation. Each of us perceives the infinite, and as such we are "limited deities." Insofar as we are finite beings, we cannot grasp the infinite, that is, everything that exists and the relations of these finite things to us, in a clear and distinct manner. It must be remembered that for Leibniz, substance cannot be affected by that which is outside it. Rather, each substance "expresses" the universe due to the pre-established harmony between its perceptions and those of all other substances. The ability of a substance to "perceive" is inherently tied to the fact that it expresses the universe. But Schleiermacher, following Kant, links sensation to real receptivity: beings really do affect one another. As such, receptivity is fundamentally different from spontaneity, even though both may spring from, as Kant himself put it, "a common but to us unknown root."[18] Hence Schleiermacher's insistence on the *relative* antithesis between receptivity and spontaneity can hardly be taken as a re-Leibnizianization of Kant, since the fundamental issue here is whether there is a real receptivity at all, and on this point Schleiermacher sides with Kant.

The importance of this crucial point cannot be overestimated. It is, I suggest, the key to understanding whether Schleiermacher can consistently

---

15 IMMANUEL KANT, *Critique of Pure Reason*, trans. Norman Kemp Smith (New York: St. Martin's Press, 1965), A 51 / B 75 (93). (Page numbers are given in parentheses.)

16 DialO 255–6; see FRANK's discussion of this point in the Introduction to his edition of Schleiermacher's *Dialektik*, 55 ff.

17 See ibid., 55 ff.

18 KANT, *Critique of Pure Reason*, A 15 / B 29 (61).

adopt Leibniz's complete concept theory while at the same time adopting a Kantian two-source theory of knowledge and a correspondence theory of truth. In order to look into the matter, we must once again probe the interrelations between Leibniz's theory of substance and his theory of the complete concept. Does the theory of the complete concept imply Leibniz's theory of substance, such that to affirm real receptivity implies that the theory of the complete concept is also false? At first blush, this implication cannot be affirmed, since it seems perfectly plausible that, given determinism, God could have such a concept of a subject without this implying that there can be no interaction among substances. What has been shown above, however, is that it was Leibniz's theory of substance that drove him to adopt the theory of the complete concept. If substances do not interact with one another, the complete concept through which God creates provides a useful way of articulating a substance's essence. This essence is the source of both its entelechy and its expression of the whole universe.

While it does not seem that there is a strict implication between the complete concept theory and the rejection of physical influx, there are other implications that the theory of the complete concept carries with it regarding the role of perception in our ability to make judgments. In order to explore these questions, let me now turn to the Kant-Eberhard controversy, which reveals the connections between Kant's theory of the two sources of human knowledge and his rejection of the idea of the complete concept. Taking a look at these interconnections will reveal whether Schleiermacher could consistently adopt Kant's two-source theory of knowledge while retaining the Leibnizian theory of the complete concept.

II. The Kant-Eberhard Controversy and the Metaphysical Ground of Predication

Johann August Eberhard was the leader of the extensive criticism of Kant's philosophy coming from the right, that is, the defenders of the official Wolffian-Leibnizian philosophy. As is well known, Schleiermacher was his pupil in Halle and was influenced by him. Along with J. G. Maaß and J. E. Schwab, as well as other Wolffians, Eberhard founded the journal, *Philosophisches Magazin* in 1788, whose first volume appeared in 1788-1789. The purpose of the journal was to provide an avenue through which the influential Kantian philosophy could be countered from the Wolffian standpoint. Kant's reply, *On a Discovery*, was a reply to the objections in the first volume of the journal, and appeared in 1790. The text is important in that it answers many objections that arose as a result of misconceptions of the critical philosophy. Many of these misconceptions were due to Kant's not having fully and properly expounded his views in the first *Critique* and in the more popular work, *Prolegomena to any Future Metaphysics*. The text's importance also lies in the fact that it sets in clear relief the differences between the rationalism of Leibniz and Wolff, and Kant's critical philosophy.

Kant's theoretical philosophy is an attempt to answer the question how synthetic a priori judgments are possible. In his attempt to account for these kinds of judgments, Kant's Copernican revolution in philosophy provided a fundamental reassessment of the relationship between knowing and being. As he notes in the Preface to the second edition of the first *Critique*:

> Hitherto it has been assumed that all our knowledge must conform to objects. But all attempt to extend our knowledge of objects by establishing something in regard to them a priori, by means of concepts, have on this assumption, ended in failure. We must therefore make trail whether we may not have more success in the tasks of metaphysics, if we suppose that objects must conform to our knowledge. This would agree better with what is desired, that it should be possible to have knowledge of objects a priori, determining something in regard to them prior to their being given.[19]

It is important to keep in mind that as he developed his critical philosophy, Kant was in conversation with two separate camps. It was Hume who awoke him from his dogmatic slumbers through his analysis of causation as mere constant conjunction. Hume was, of course, an empiricist; the source of all true knowledge is experience. Hence it is the things themselves and how they affect us, and not our concepts of things (as in Leibniz) that provide the metaphysical ground of predication. The Copernican revolution therefore seems to be primarily directed as an answer to the Humean problem developed in terms of empiricist assumptions. According to Hume, if things are the ground of synthetic predication, they can only be known through how they affect us and hence through experience. We therefore can

19 Ibid., B xv (22).

have no knowledge of such things a priori. We may, indeed, have knowledge of what it is that is contained in our concepts (i. e., analytic judgments, in which the predicate merely repeats what is contained in the subject). But our concepts are not things. All judgments about things must await the verdict that experience must supply. No amount of experience, however, will yield necessity and universality. Regardless of how many times in all past instances x's are y, this will not yield the requisite necessity and universality, since uniformity of past instances implies nothing concerning those instances that might be encountered in the future. Hence, Kant concluded, to assume that in experience the mind must conform to things (i. e., where the mind's modifications depend upon how it is affected) is a dead end where synthetic a priori (i. e., necessary and universal) judgments are concerned. The only way that synthetic a priori knowledge is possible is if *things* must conform to the mind. Now things, according to Kant, are given in intuition, but intuition itself contains its a priori forms conditioning the way that objects must be given. While this empiricist context is certainly important in interpreting Kant's Copernican revolution, the Copernican revolution also brought Kant into significant disagreement with the followers of Leibniz and Wolff.

One of the driving principles of Eberhard's attack on the Kantian system was the claim that insofar as it was true, it was a mere repetition of Leibniz, and insofar as it diverged from Leibniz's views, it was simply wrong. Eberhard notes that

> The Leibnizian philosophy contains just as much of a critique of reason as the new philosophy, while at the same time it still introduces a dogmatism based on a precise analysis of the faculties of knowledge. It therefore contains all that is true in the new philosophy and, in addition, a well-grounded extension of the sphere of the understanding.[20]

Eberhard's critique rested on several key interrelated points. First, Eberhard claimed that we could have real knowledge through pure concepts without the need for intuition or experience; as such he was extremely critical of Kant's limitation of knowledge to appearances. Because the cate-

---

20 Johann Eberhard, *Philosophisches Magazin* I, 289; quoted from Henry E. Allison, *The Kant-Eberhard Controversy: An English translation together with supplementary materials and a historical-analytic Introduction of Immanuel Kant's On a Discovery* (Baltimore: The Johns Hopkins University Press, 1973), 16.

gories apply only to appearances and not to things in themselves, Kant has failed to show that they were really applicable to things (i. e., possessed objective validity). Eberhard claims:

> in the theory of critical idealism synthetic judgments can have no objects outside of the representations; for they cannot relate to anything that is actual apart from the representations, and their logical truth thus consists merely in the agreement of a representation in us with the same representation in us. Thus critical idealism presupposes a theory which reduces the entire discussion to a play with words.[21]

Second, since such genuine knowledge can be had through pure concepts, we can have knowledge of things in themselves. Third, he eschewed the essential role Kant had assigned sensibility in human knowledge. The issues are inherently related with one another. Their significance and interconnections can best be gauged through an analysis of Kant's analytic / synthetic distinction, a key feature allowing Kant to frame the question driving the critical philosophy in the first place: How are synthetic a priori judgments possible? In fact, Kant's analytic / synthetic distinction was a significant point of focus of the Eberhardian attack. In order to grasp the controversy, it is useful to take a look at Eberhard's critique of Kant's understanding of the distinction and how it was grounded in the way he himself understood it.

In the Introduction to the first *Critique*, Kant frames the analytic / synthetic distinction in terms of the relation of the subject to the predicate in a judgment. Hence he notes:

> Either the predicate B belongs to the subject A, as something which is (covertly) contained in this concept A; or B lies outside the concept A, although it does indeed stand in connection with it. In the one case I entitle the judgment analytic, in the other synthetic. Analytic judgments are therefore those in which the connection of the predicate with the subject is thought through identity; those in which this connection is thought without identity should be entitled synthetic.[22]

---

21 EBERHARD, *Philosophisches Magazin* I, 321–2; quoted from ALLISON, *Kant-Eberhard Controversy*, 41.

22 KANT, *Critique of Pure Reason*, A 6–7 / B 10–11 (48).

Analytic judgments merely break up the subject concept into its constituent elements, and repeat one of those elements in the predicate. Synthetic judgments, on the other hand, are what Kant here calls ampliative, that is, the predicate contains new information not contained in the subject and which cannot be derived from it. Such judgments are thereby informative.

In his essay "On the Distinction of Judgments into Analytic and Synthetic" in the *Philosophisches Magazin*, Eberhard criticized Kant's distinction as unclear, and proposed his own analysis, one which he believed would expose the ground of its lack of clarity. Moreover, he argued, a proper development of the distinction would show that we do, indeed, have knowledge of things in themselves. His own development of the distinction revolved around the difference between the essence of a thing and its attributes. When the predicate in a judgment is contained in the concept of the essence of the subject, we have an identical judgment that may be partial or complete. In a complete identical judgment, "the predicate contains all of the determinations of the subject by means of which it can always be distinguished from all other things."[23] In a partial identical judgment, the predicate expresses only part of the essence, that is, "it is one with only one or some, but not all the determinations of the subject."[24] Both these kinds of judgments are analytic. Synthetic judgments, on the other hand, are ones in which the predicate of the judgment is connected to the subject in virtue of the subject's attributes. These attributes are different from the essence of the subject, but are nevertheless implied by it. Since, however, they are not themselves part of the essence of the subject, the predicate in the judgment does not merely repeat what is already expressed by the subject. Eberhard labels these judgments synthetic and notes that "There are a priori judgments or necessary truths in which the predicates are attributes of the subject, that is, determinations which do not belong to the essence of the subject, but have their sufficient reason in this essence."[25] Eberhard believes he has thereby accounted for synthetic a priori judgments (i. e., judgments of metaphysics and mathematics) in a way consistent with the Leibnizian-Wolffian philosophy. As such, he claims that the tools for answering Hume's quandary are already present in the philosophy of Leibniz and Wolff. Kant's distinction, on the other hand, is flawed insofar as it fails

23 Eberhard, *Philosophisches Magazin* I, 312–3; quoted from Allison, *The Kant-Eberhard Controversy*, 37.

24 Ibid.

25 Eberhard, *Philosophisches Magazin* I, 314; quoted from Allison, *The Kant-Eberhard Controversy*, 38.

to note the difference between the essence of a thing and its attributes. It is this distinction that is the true ground of the difference between analytic and synthetic judgments. Furthermore, charges Eberhard, Wolff and Baumgarten had already identified two classes of judgment, identical and non-identical, basically corresponding to those he describes. Whereas identical judgments are based on the principle of contradiction, non-identical judgments are based on the principle of sufficient reason. Given this fact, Kant's "originality" lay merely in his having changed the name of the two kinds of judgments; moreover, his own formulation of the distinction was plagued with confusion.

In regard to this issue an important criticism of Kant's distinction, first brought up by Maaß, and then raised again by Eberhard, comes up. Maaß develops the criticism in an essay that appeared in the second volume of the *Philosophisches Magazin* entitled, "On the Highest Principle of Synthetic Judgments in Relation to the Theory of Mathematical Certainty." Whereas Eberhard's original criticism of Kant's analytic / synthetic distinction rested on his alleged failure to note the difference between essence and attribute, Maaß questions whether the distinction is at all viable on the grounds that Kant assigns to it. Just how do we identify whether a judgment is analytic or synthetic? The problem, as Maaß sees it, is the difficulty in fixing the subject concept in a judgment. Just how do we know what is contained in that concept, and what is the basis for this determination? This is key, since what is contained in the subject concept determines whether the predicate merely repeats that content or not. When I think of gold, to use Locke's famous example, I may think of something that is yellow and malleable. Not being a chemist, its solubility in *aqua regia* may be news to me. Hence the judgment "Gold is soluble in *aqua regia*," would be a synthetic judgment since it provides me with new information. On the other hand, the solubility of gold in *aqua regia* may be taken for granted as part of the definition of gold by another individual. It thereby seems that a judgment that is synthetic for me may be analytic for another individual. The problem, as Maaß acutely notes, is the psychological basis of the distinction as formulated by Kant: the difference between analytic and synthetic judgments depends upon the relation between the content of the concepts *thought* in a judgment. Given this psychological basis, what is thought in a concept can vary from individual to individual, and the consequence is that the same judgment may be considered both analytic and synthetic. Maaß proceeds to argue that stipulating what is contained in the subject through a *nominal* definition will not work either. On the basis of such a nominal definition, a triangle can be defined as a figure, the sum of the angles of which are equivalent to the sum of two right angles. However, if a triangle is defined in this way, the judgment that the sum of the angles of a triangle is 180 degrees will be analytic and not synthetic. In fact, using the nominal definition to determine what is contained in the concept of the subject can make it such that most of the judgments that Kant counted as synthetic a priori could be understood as analytic. Maaß then notes that the only alternative is to stipulate what is thought in the subject concept through the essence, but doing this will land one squarely back in the Leibnizian camp. The only proper way to distinguish between analytic and synthetic judgments is by distinguishing between the essence and the attributes of a subject, as Eberhard had done. Only in this way can the inherent vagueness in Kant's formulation of the distinction be overcome. Echoing Eberhard, Maaß argued that this showed that everything that Kant wanted to accomplish in the critical philosophy was already contained in the Wolffian-Leibnizian philosophy, only conceived of and articulated there much more clearly.

Schleiermacher voices a somewhat similar critique of the analytic / synthetic distinction in the 1814 / 15 version of the lectures on *Dialektik*. There he notes that:

> In a complete concept the accidental [elements] of its possibility and its range must be contained. For when a thing can suffer or do something, the possibility of which is not contained in its concept, then the concept is not complete. Therefore in regard to complete concepts, there are no pure synthetic judgments.
>
> In regard to incomplete concepts that are still being formed, there are no pure analytic judgments. For anything can be left out at a given moment in a concept's history. There can be a concept of a human being in which the predicate mortal has not yet been added.
>
> The difference between analytic and synthetic judgments can thereby not be fixed, and is in general nothing, since identical judgments are nothing but empty formulas when one does not lay the complete concept as a basis, in which each difference is alone grounded.[26]

Here Schleiermacher, like Maaß and Eberhard, notes the difficulty in fixing the subject concept in analytic judgments. Just what is it that is being

26 Dial(1814 / 15) 33.

thought in the subject concept? We only know the answer to this question, argues Schleiermacher, when we are dealing with a complete concept, and here we are firmly in the Leibnizian camp. Judgments having a complete concept for the subject cannot be pure synthetic judgments, since everything that is true about the subject is already contained in its concept. On the other hand, if we admit that there are incomplete concepts, we will have a hard time fixing what is thought in them, for just what it is that is thought in a concept varies with its history. This is true of both the psychological history of a concept peculiar to an individual and its nominal intersubjectively agreed upon definition. Because of the question of the fixing of the content of an incomplete concept, Schleiermacher argues that in regard to them there can be no analytic judgments. Rather, if the analytic / synthetic distinction is to make any sense at all, it must be grounded in the idea of the complete concept. Judgments containing complete concepts are analytic; *only* judgments containing incomplete concepts can be synthetic. Notice that here Schleiermacher, like Eberhard, is working with what seem to be merely logical criteria for the distinction between analytic / synthetic judgments offered by Kant in the Introduction of the first *Critique*: the basis of the distinction has to do with the content of concepts and their interrelations. Because Kant had not there clarified his distinction further, he left himself wide open to attacks by Eberhard and Maaß that his distinction was vague. Schleiermacher echoes these worries in the *Dialektik* and provides his own account of how the distinction might be more clearly grounded.

Kant's answer to Eberhard's attack in *On a Discovery* exposes many of the misconceptions regarding the analytic / synthetic distinction that lead to the Wolffian attack. Moreover his discussion reveals some of the inherent interconnections between his (Kant's) rejection of the Leibnizian doctrine of the complete concept and the two-source theory of knowledge. In *On a Discovery* Kant argues that the key to the significance of the analytic / synthetic distinction is not the merely logical question of the relation between the subject and predicate in a judgment. The key issue is rather transcendental. As such, it concerns the question of whether the predicate in a judgment stands in a *real* relation to the object. We can begin to understand what Kant means by this through Kant's analysis of the way that Eberhard develops the analytic / synthetic distinction. We then contrast this with what Kant means by a real relation to the object.

For Eberhard (as for Leibniz) a judgment is true in virtue of the relation between *concepts*. In fact, Kant notes, the basis for Eberhard's attempt to distinguish between analytic and synthetic judgments rests on a distinction between two kinds of concept containment: direct containment and mediate containment, or, implication. Eberhard's analytic judgments are true in virtue of the first kind of concept containment. In these cases a mark contained in the subject concept (presumably expressing an essence) is merely repeated in the predicate of the judgment. In Eberhard's synthetic judgments, the predicate does not merely repeat what is already contained in the subject, but rather affirms something that is *logically implied* by the subject concept. Key here, as Kant is quick to point out, is that the truth of both kinds of judgments depends upon the principle of contradiction alone. Hence even Eberhard's synthetic judgments are true in virtue of concept containment, albeit of a mediate kind. As such, this is a purely intensional logic: the basis of the truth of a judgment lies in logical interrelations among concepts. Even Eberhard seemed to have recognized the inherent similarities between analytic and synthetic judgments as he himself understood them. He noted that "the dispute as to whether a proposition is analytic or synthetic, is in regard to its logical truth, trivial."[27]

As Kant notes in his 1789 letter to Reinhold, the problem with Eberhard's analysis is that "the logical relation of ground and consequent is mistaken for a real relation .... Now this distinction of ground and consequent is either merely *logical* (having to do with the manner of representation) or *real*, that is, in the object itself .... Now it is real distinctness that one requires for a synthetic judgment."[28] While an analytic judgment is true in virtue of logical interrelations between concepts, a synthetic judgment is one that is true in virtue of some third thing, distinct from both subject and predicate, to which both the subject and the predicate apply. In a synthetic judgment one predicates something about an object x that falls under the subject concept. For instance, when I say "This hat is red," what I am saying is that there is some thing, a particular hat, that stands under the concept hat and which also stands under the concept red. The key question in a synthetic judgment is therefore the material or *transcendental* question of whether the predicate in a judgment stands in a *real* relation to some object. Such real relations have to do with the relation of concepts to *things* or the relation of *things* to one another.[29] In *On a Discovery*, Kant notes:

27 EBERHARD, *Philosophisches Magazin* I, 361; quoted from ALLISON, *The Kant-Eberhard Controversy*, 150.

28 IMMANUEL KANT, *Philosophical Correspondence*, 1759–1799, ed. and trans. Arnulf Zweig (Chicago: University of Chicago Press, 1967), 138–9.

29 Henry Allison puts the matter quite clearly and is worth quoting: "Thus a logical relation is one which holds between concepts, a logical predicate is simply any concept

> It was therefore not merely a verbal quibble, but a step in the advance of knowledge when the first *Critique* made known the distinction between judgments which rest entirely on the principle of identity or contradiction, from those which require another principle through the label "analytic" in contradistinction to "synthetic" judgments. For the notion of synthesis clearly indicates that something outside of the given concept must be added as a substrate which makes it possible to go beyond the concept with my predicate. Thus, the investigation is directed to the possibility of a synthesis of representations with regard to knowledge in general, which must soon lead to the recognition of intuition as an indispensable condition for knowledge, and pure intuition for a priori knowledge.[30]

In his 1789 letter to Reinhold, Kant affirms that the principle of synthetic judgments has been unambiguously presented in the first *Critique*. It is:

> *All synthetic judgments of theoretical cognition are possible only by the relating of a given concept to an intuition.* If the synthetic judgment is an experiential judgment, the intuition must be empirical; if the judgment is a priori synthetic, there must be a pure intuition to ground it.[31]

Of key significance is Kant's affirmation that the synthetic judgment is true in virtue of some third thing, and his identification of this third something with intuition. Two things are important in this regard. First, it is only through intuition that the mind is *directly* related to objects. The key feature of an intuition is that it is a singular representation (*repraesentatio singularis*), that is, *a representation of an individual.* Concepts, on the other

---

predicated of another, and a logical essence is merely the sum total of the partial concepts making up a given concept. Rules governing logical relations are purely formal, as they abstract from the content of thought. Real relations, on the other hand, concern precisely the content of thought. They hold between things thought about, or between concepts and things. Moreover, just as a real ground is the ground of a thing, and a real essence the essence of a thing, so a real predicate is a determination of a thing. From this we can clearly see that the defining characteristic of synthetic judgments is that they relate concepts or predicates to objects, while analytic judgments are merely concerned with logical relations between concepts (the predicate and the concept of the subject)" (Allison, *Kant-Eberhard Controversy*, 54–5).

30 Kant, *On a Discovery*; quoted from Allison, *The Kant-Eberhard Controversy*, 154–5.

31 Kant, *Philosophical Correspondence*, 141.

hand, are always representations of representations; they can only stand in a mediate relation to an object. This means that it is always possible that more than one object stand under any given concept, no matter how specific. As such, for Kant, as opposed to Leibniz and Schleiermacher, there can be no *infima* species and the individual can only be given in intuition. Second, in human beings intuition is always sensible. While a divine understanding might have intellectual intuitions and thereby stand in an immediate relation to its object without having to be affected by it, for human beings this is not possible. The only way that we can stand in an immediate relation to an object is by its affecting us. Hence while through the understanding objects are thought, it is through receptivity that objects are *given*.

This discussion has shown the systematic interrelations between a two-source theory of knowledge and Kant's analytic / synthetic distinction. The latter distinction can only be developed on the supposition that our concepts can stand in a real relation to objects. But such a real relation presupposes that *objects* are given to us, that is, that we can stand in an immediate relation to them through our intuition, and thereby through our receptivity. In fact such a real relation is presupposed by a correspondence theory of truth, the theory that Schleiermacher adopts in the *Dialektik*. Schleiermacher notes in the lectures of 1811 that "Knowledge is the correspondence of thinking with being as that which is thought."[32] What is significant about Schleiermacher is that he adopts the two-source theory of knowledge while at the same time adhering to a Leibnizian view of predication. The discussion above has shown that the analytic / synthetic distinction implies a two-source theory of knowledge. The question still remains, however, whether a two-source theory of knowledge implies Kant's extensional logic (and thereby the analytic / synthetic distinction). This certainly seems to be the case. If the understanding merely provides functions of unity for our representations, whatever content there is for thought must be *given* in intuition. Hence any real extension of knowledge depends upon intuition, and this implies an extensional logic. Schleiermacher's adoption of a Leibnizian view of predication does not mix well with his adoption of a two-source theory of knowledge and a correspondence theory of truth. Rather, Leibniz's theory of the complete concept is much better suited to an ontology wherein substances do not interact with one another, that is, in which they are completely independent, windowless monads. Much of the

---

32 Dial(1811) 13.

opacity of Schleiermacher's *Dialektik* derives from Schleiermacher's having borrowed elements from two very different philosophical systems. These elements, I have suggested, are impossible to square with one another.

Relative Freedoms:

The Influence of Spinoza on the Systems of Whitehead and Schleiermacher

KRISTA DUTTENHAVER, University of Notre Dame

One of the most intractable issues plaguing the history of Western philosophy and theology has been the question of what to do with indeterminacy. Indeterminacy—something that, in its numerous guises is a very plainly present reality in the universe—poses a particularly troubling hurdle for systematicians to clear, for, in the past, the whole point of system-building has been to banish—or at least tame—indeterminacy in order to create the most unitive and comprehensive system possible. This problem of indeterminacy extends back into Platonism, in which the existing universe is derived from two fundamental absolutes: an absolutely determinate reality that limits (and that grounds the forms), and an absolutely indeterminate reality that receives limitation. The principle of absolute indeterminacy is for Plato something that requires the absolutely determinate in order to be "bounded" or "tamed"; without determinate limitation, absolute indeterminacy is a dangerous breeding ground for chaos, error, and irrationality. Within this dual set of ultimates, the determinate clearly succeeds in taming the indeterminate, but Platonic philosophy cannot simply do away with the latter, for it is the interaction between the two principles that accounts for all there is in the universe.

It is with the Continental Rationalists that we see this drive to eliminate indeterminacy operating at its most decisive and systematic pitch. At this point in history, the Copernican revolution and the subsequent development of Newtonian physics offer for the first time in Western civilization the prospect of a perfectly ordered universe that conforms to a series of laws that are, with the appropriate and persistent application of human reason, fully intelligible. With the advent of the new scientific methodologies and commitments, crafting a philosophical system that does not require an ultimate principle of indeterminacy begins to seem a plausible and desirable endeavor. For Descartes, Leibniz, and—most prominently—Spinoza, the inexorable and mechanistic necessity of these laws furnishes precisely what is needed to ground a system that eradicates all traces of indeterminacy. The elimination of indeterminacy and, by implication, its concomitant surd elements, reaches a high-water mark in Spinoza's system. In applying a geometric method to the fundamental Cartesian insights, Spinoza strives to demolish all vestiges of these surd elements, thereby constructing a system that is entirely determinate.

While banishing indeterminacy arguably makes for a system that is extraordinarily unified, such unity comes at the obvious cost of human freedom and any meaningful sense of growth. Spinoza does away with Descartes' dual-substance ontology in order to achieve the kind of thoroughgoing mechanism that Descartes sought to avoid, but this drive to overcome or eliminate indeterminacy results finally in the attendant drive to do away with all forms of genuine freedom and genuine self-realization. Spinoza's mechanistic system eradicates indeterminacy by means of the ultimate absorption of the finite into the infinite, and hence authentic but finite human freedom disappears completely. For Whitehead and Schleiermacher, the elegance of Spinoza's system possesses considerable appeal. However, neither Schleiermacher nor Whitehead wishes to subscribe to the comprehensively deterministic mechanism that the Spinozistic system entails.

This paper, then, will be an attempt to examine in greater detail the extent to which Spinoza's influence is visible in the systems of both Whitehead and Schleiermacher, and the extent to which each thinker appropriates and transforms certain Spinozistic elements in order to accommodate the need for genuine freedom and authentic self-formation. Using the categories of determinacy and indeterminacy as lenses through which to focus both upon this issue of self-formation as it unfolds in Whitehead's thought, and upon the related concepts of activity and receptivity in Schleiermacher's thought, I intend to draw out important connections between the two thinkers. It is my contention that these connections become clearest in an examination of their accounts of self-formation. By looking carefully at these connections, I hope to be able to draw some conclusions regarding system and history, as well as position myself to respond briefly to Dr. Mariña's arguments regarding Schleiermacher, Leibniz, and Kant.

I. Whitehead and Spinoza

While the hostile intellectual atmosphere surrounding Spinoza's thought at the turn of the nineteenth century poses challenges for the historian pursuing clear interpretations of the philosopher's influence on Schleiermacher, in the case of Whitehead, the evidence needed for an accurate assessment is far more plentiful. Indeed, Whitehead frequently draws explicit connections between his own philosophy of organism and Spinoza's monism, translating Spinoza's concepts into the terminology of his own system and providing detailed criticism of the monistic philosophy's central flaws. In contrast to Schleiermacher's reading of Spinoza, which comes early in his career and arguably remains a lifelong influence, Whitehead's philosophical meditations on Spinoza unfold relatively late in his career, and it is evident from these that Spinoza is simply one philosopher among many who figured in Whitehead's intellectual formation. However, to say that Spinoza exerts a minor influence on Whitehead's philosophy is not to minimize the importance of Spinoza's system in comprehending Whitehead's; on the contrary, Whitehead often illustrates how the basic structure of his own organic philosophy can be intuited by analogy with Spinoza's conception of the one substance and its modes. Although Whitehead is patently unsympathetic to the specific content of Spinozism, it is difficult to overlook the frequent indications that Whitehead's thinking often orients (and reorients) itself in accordance with the overall "impression" that his predecessor's system inspires in him. For this reason, the philosophies of Spinoza and Whitehead display what an evolutionary biologist might describe as *analogous* rather than *homologous* growth.

Although Whitehead is unequivocal in his repudiation of Spinozism's monistic doctrine, he never hesitates to express his philosophy's deep affinities with what he perceives to be the underlying framework of Spinoza's thought. For instance, Whitehead readily acknowledges the fundamental structural similarity between the one substance of Spinoza's system and his own ultimate—creativity—as well as the further connection between the Whiteheadian actual occasions and Spinoza's modes. In spite of these significant parallels, Whitehead frequently reveals such similarities only in order to emphasize the profound differences between the two philosophies. On the one hand, creativity as the Whiteheadian ultimate is only actual by virtue of the facts of God and the actual entities that are its accidents; in Spinoza's monistic scheme, on the other hand, "the ultimate is illegitimately allowed a final, 'eminent' reality, beyond that ascribed to any of its accidents .... One side makes process ultimate; the other side makes fact ultimate."[1] With this analysis, Whitehead announces his own departure from the main strand of Western philosophy of which Spinoza's thought is among the foremost exemplifications. Here, Spinoza is implicated as a main offender in Whitehead's larger polemic against what he variously calls "the subject-predicate form of expression,"[2] "the 'substance-quality' metaphysics"[3] or the Aristotelian "primary substance."[4] In his clearest and most succinct assessment of Spinoza, Whitehead expresses the ambivalence with which he views his predecessor:

> The philosophy of organism is closely allied to Spinoza's scheme of thought. But it differs by the abandonment of the subject-predicate forms of thought, so far as concerns the presupposition that this form is a direct embodiment of the most ultimate characterization of fact. The result is that the substance-quality is avoided; and that morphological description is replaced by description of dynamic process. Also Spinoza's 'modes' now become the sheer actualities; so that, though analysis of them increases our understanding, it does not lead us to the discovery of any higher grade of reality.[5]

Although this passage captures the basic nature of Whitehead's criticism and appropriation of Spinoza, most of the special claims made in it cannot be elucidated without a more detailed exploration of Whitehead's general interpretation of Platonism, his critique of Aristotelian logic, and his description of that logic's rise to complete dominance in Spinoza's thought.

Throughout *Process and Reality*, Whitehead suggests that his essay in cosmology is an attempt to redirect Western metaphysics to some important strands of Greek thought that either never received sufficient development by the Greeks themselves or withered from neglect during the medieval period. Occasionally, Whitehead states this project boldly: he identifies his construction of the philosophy of organism as an attempt "to render Plato's general point of view with the least changes made necessary by the intervening two thousand years of human experience."[6] Elsewhere,

---

1 PR 7.
2 PR xiii.
3 PR 137.
4 PR 157.
5 PR 7.
6 PR 39.

Whitehead describes his system as a synthesis of the Platonic cosmology of the *Timaeus* and the pioneering scientific cosmology of the seventeenth century, expressing his belief that "perhaps the best solution consists in a fusion of the two schemes."[7] Although *Process and Reality* endeavors to construct a bridge between the two cosmologies, the work is also highly significant in the present context of system and history for its analysis of the historical trajectory by which the mechanistic cosmology emerges from its Platonic antecedents.

Buried underneath Whitehead's technical discussions of seemingly isolated historical controversies and complex scientific quandaries is actually a consistent and dramatic account of the manner by which early modernity frees Western cosmology from the fundamental deficiencies of the older mode of thought, but only at the price of suppressing those conceptually pregnant elements in Greek thought that could serve as resources for its own correction. Through the course of his exposition, Whitehead argues that the decisive breakthrough of modernity comes through a gradual recognition of individual matters of fact as the loci of determinacy and value. Through the incremental Scholastic transformation of the Aristotelian metaphysics and its further revision at the hands of early modern science, Western philosophy is able to forge an account of concrete fact that liberates it from its Platonic status as imperfection, chaos, irrationality, and indeterminate unintelligibility. In Whitehead's view, however, the great irony of this achievement is that the thinkers of the seventeenth century, especially Spinoza, accomplish it by means of the very same conceptual instrument that problematizes individuality and particularity in the first place.

The complex pathways by which modern mechanistic cosmology both overturns and perpetuates the unsavory habits of Greek thought can only be retraced by a work the size of Whitehead's magnum opus. It is sufficient to note here, though, that Whitehead perceives the common flaw of the two cosmologies to be manifested in their shared desire to identify the activity and actuality of ultimate reality with the static being of a principle of absolute determinacy. For Whitehead, the only abiding difference between the two cosmologies is that the mechanistic worldview transmogrifies the static, eternal, and necessary reality of Plato's transcendent realm into the ordered perfection of Newton's world-machine. Whitehead identifies the great villain of this philosophical tragedy in the Aristotelian guise

---

[7] PR xiv.

through which it achieves exclusive dominance in the Western tradition: primary substance. Although Whitehead admits that both Plato and Aristotle "included the strains of thought which issued in this doctrine," he emphasizes repeatedly that the substance-quality metaphysics, as an ultimate logical and ontological framework in Greek thought, is always "inconsistently mingled with other notions."[8] Unlike the totalizing systems of modern mechanism, in which the substance-quality metaphysics is made fully explicit and all-encompassing, "Greek philosophy was subtle and multiform, also it was not inflexibly consistent."[9] Minority counter-currents are present in the philosophies of Plato and Aristotle, and these are, in fact, the very elements that Whitehead appropriates and extends through his philosophy of organism.

Whitehead attributes to both of these thinkers the first germs of a thoroughly "evolutionary" philosophy, in which ultimate reality is characterized, not by the imposition of determinate form upon indeterminate matter, but by the gradual emergence of determinate order from an indeterminate chaos. It is in this respect, asserts Whitehead, that "the organic philosophy only repeats Plato. In the *Timaeus* the origin of the present cosmic epoch is traced back to an aboriginal disorder, chaotic according to our ideals. This is the evolutionary doctrine of the philosophy of organism."[10] Whitehead speculates that "the full sweep of the modern doctrine of evolution would have confused the Newton of the *Scholium*, but would have enlightened the Plato of the *Timaeus*."[11] While it is clear that a genuinely evolutionary cosmology is ultimately inaccessible to the mechanistic worldview, the freshness of insight and flexibility of mind possessed by the Greeks expose their development of the substance-quality metaphysics to its potential overthrow by alternative systems. In spite of Whitehead's clear admiration for Plato, his exoneration of Platonism is qualified by an awareness that a latent form of the substance-quality metaphysics is the dominant feature of Plato's thought.

Aristotle, by contrast, though he bears such explicit responsibility for the formulation of the "primary substance" concept, is a far more ambiguous thinker in terms of his relation to the substance-quality metaphysics. Although the substance idea lies at the very core of Aristotle's logic, White-

8 PR 137.
9 PR 158.
10 PR 95.
11 PR 93.

head insists that the Aristotelian logic "does not seem to be characteristic of Aristotle's own metaphysical speculations."[12] Indeed, in comparison with the implicit Aristotelianism of the modern period, in which Aristotle's logic becomes thoroughly "impressed on the European mind," Whitehead quips that "probably Aristotle was not an Aristotelian."[13] The total impression of Aristotle that Whitehead's commentary presents is of a thinker whose philosophical temperament remains deeply divided. According to Whitehead:

> Aristotle corrected his Platonism into a somewhat different balance. He was the apostle of 'substance and attribute,' and of the classificatory logic which this notion suggests. But, on the other side, he makes a masterly analysis of the notion of 'generation.' Aristotle in his own person expressed a useful protest against the Platonic tendency to separate a static spiritual world from a fluent world of superficial experience.[14]

It is important to note that what Whitehead unflinchingly affirms here in Aristotle's thought is the truly evolutionary side of the philosopher's account of generation. Alongside the negatively Platonic tendencies that can easily be found in Aristotle's exposition of "substance," Whitehead interprets the "genuine" Aristotelian understanding of substance to be that of an embryonic kind of being—such as the being a tree possesses in its seed or the being of a future event—a being that can only be characterized by a free and contingent process of development. Aristotle calls this embryonic kind of being matter, which he believes to be identical in all substances, and which takes on form only in the course of its development.

According to Whitehead, it is not until "medieval Aristotelian thought allowed the static notions of Aristotle's logic to formulate some of the main metaphysical problems"[15] that Aristotelianism reverses the order of Aristotle's original evolution. Yet even Aristotle's most static characterization of substance for example, in his logical formulae, betrays these evolutionary undertones. For instance, by construing substance as the permanence that underlies change and the basic subject of predication, Aristotle in effect sees it as a bare substratum, a completely indeterminate, neutral "stuff" whose

---

12 PR 30.
13 PR 51.
14 PR 209.
15 Ibid.

only determinacy derives from the attributes that qualify it. In fact, Whitehead's own construction of creativity as "another rendering of the Aristotelian 'matter,'"[16] and a category that "replaces Aristotle's category of 'primary substance'"[17] may be viewed as an endeavor to reorient Western philosophy to the evolutionary aspect of Aristotle's account and achieve a coherent logical and ontological framework on the basis of this theme.

At some points, Whitehead suggests that the philosophies of Continental Rationalism are themselves aware of this profound ambiguity in the Aristotelian notion of substance, and he often portrays their radical revisions of the concept of substance as attempts to purge it of its original evolutionary character for the sake of a greater coherence. Arguing that "the substance-quality metaphysics triumphed with exclusive dominance in Descartes' doctrines,"[18] Whitehead presents Descartes' modification of substance as a selective overemphasis on the static aspects of Aristotle's concept, which undergoes further truncation at the hand of Spinoza. Indeed, "the attraction of Spinoza's philosophy lies in its modification of Descartes' position into greater coherence."[19] Although Descartes' definition of a substance as a "thing which requires nothing but itself in order to exist"[20] is generally thought to be the major step in modernity's break from Aristotelianism, Whitehead frames it as a "true derivative of Aristotle's definition."[21] Where Descartes is decidedly in continuity with Aristotle, though, is in his traditional view of substance as the subject of predication. As with Aristotle, this idea also implicates Cartesianism in a conception of substance in which substance acts as an indeterminate substratum, a "naked" subject that is only intelligible by virtue of the qualities that it possesses. Substance, by definition, does not depend upon its accidents, but because the Cartesian (and basically Platonic) view is that the reality of a substance is a function of its static determinacy, Descartes' formulation opens itself up to embarrassing charges of logical incoherence.

Spinoza, in an attempt to exclude a reading of substance that divorces it ontologically from the universal predicates that determine it, makes a subtle but crucial modification of Descartes' doctrine of substance. For Spinoza,

---

16 PR 31.
17 PR 21.
18 PR 137.
19 PR 6.
20 René Descartes, *Principles of Philosophy*, trans. Valentine Rodger Miller and Reese P. Miller (Dordrecht: Kluwer, 1991), part 1, principle 51.
21 PR 50.

not only is it necessary for a substance to exist in itself, but it must also be *conceived* through itself. Because of this, the locus of ontological determinacy (the substance) and the locus of conceptual determinacy (the universal concepts that qualify it) are no longer only contingently conjoined; rather, they are essentially related and become virtually identical. In this way, Spinoza retains the definition of the independence of substance, but the price is an abolition of any significant ontological difference between substance and its attributes. Consequently, attributes become only "what the intellect perceives of a substance as constituting its essence."[22] As such, they are merely different perspectives from which an eternal, independent substance can be viewed. Ironically, then, the substance that Aristotle posits originally as a principle of permanence underlying *real* change becomes for Spinoza the means by which apparent change is explained away as only so many renderings of the essential, eternal, and necessary order of infinite substance.

Whitehead's move, on the other hand, is to reject what he considers to be the "vacuous actuality" of the Cartesian substance, conceived as a fully determined actuality capable of existing independently from those very attributes needed to determine it. In his attempt to rehabilitate the *res vera* of Cartesianism, Whitehead rejects the move Spinoza makes in sacrificing the ontologically meaningful distinction between substance and attribute for the sake of preserving independent substance. Whitehead argues instead that the concept of independent substance itself must be abandoned in favor of an account that takes seriously both the distinction between a substance and its attributes as well as the constitutive role that attributes play in the actuality of substances. In respect to his own approach to Descartes' logical incoherence, Whitehead sees his philosophy as a total reversal of Spinoza's direction:

> Spinoza is practically a logical systemization of Descartes, purging him of inconsistencies. But this attainment of logical coherence is obtained by emphasizing just those elements in Descartes which the philosophy of organism rejects ... The philosophy of organism may be conceived as a recurrence to Descartes and to Locke, in respect to just those elements in

22 Benedict de Spinoza, *Ethics*, ed. and trans. George Henry Radcliffe Parkinson (Oxford: Oxford University Press, 2000), part I, def. 4 (75). (Page numbers given in parentheses.)

> their philosophies which are usually rejected by reason of their inconsistency with the elements which their successors developed. Thus the philosophy of organism is pluralistic in contrast to Spinoza's monism.[23]

By substituting a relational account for the substance-quality metaphysics, Whitehead effectively rescues the pluralism (not to be read as dualism) of Descartes' metaphysics from a reduction to Spinozistic monism. Whitehead diverges from the Cartesian account by reconceiving Descartes' attributes as "really the forms of internal relationships *between* actual occasions, and *within* actual occasions."[24] Whitehead's shift here can be seen, likewise, as a reformation of the traditional understanding of universals and particulars. "Every so-called 'universal' is particular in the sense of being just what it is, diverse from everything else; and every so-called 'particular' is universal in the sense of entering into the constitutions of other actual entities. The contrary opinion led to the collapse of Descartes' many substances into Spinoza's one substance."[25] Fundamentally, Whitehead's "universals"—the eternal objects—are those principles of definition by means of which actual occasions grow into order, as well as the patterns by which such order is passed on to other occasions. Thus, Whitehead continues to distinguish the creative indeterminacy that composes the "substance" of the actual occasion from the eternal objects that define it by their attribution, but he insists that the *actuality* of the occasion can only be the *process* of evolutionary development by which indeterminate matter grows freely into determinate form.

It is only in this evolutionary, relational understanding of actuality that Whitehead refers to the actual occasion as Spinoza's *causa sui*. According to Whitehead, "an actual entity feels as it does feel in order to be the actual entity which it is. In this way an actual entity satisfies Spinoza's notion of substance: it is *causa sui*."[26] Where Spinoza finds actuality in those substances that possess a timeless essence, Whitehead locates an entity's actuality in its freedom to decide what it will be. "To be *causa sui* means that the process of concrescence is its own reason for the decision in respect to the qualitative clothing of feelings. It is finally responsible for the decision by which any lure for feeling is admitted to efficiency. The freedom in-

23 PR 73.
24 PR 309.
25 PR 48.
26 PR 222.

herent in the universe is constituted by this element of self-causation."[27] The freedom experienced by an entity is not a function of its participation in a realm of eternal act; on the contrary, Whitehead establishes freedom as the oscillation between the activity and the receptivity of the entity in the world, forging its agency in response to its influence from the world and the influence it simultaneously exerts upon the world. As will soon be evident, Schleiermacher shares with Whitehead this account of freedom and self-formation as the order that emerges through the mutual play of activity and receptivity in the finite sphere.

II. Schleiermacher and Spinoza

In his early essay *On Freedom*, written between 1790 and 1792, Schleiermacher celebrates the necessary interconnectedness, or *Zusammenhang*, of all knowledge and action. He writes that human freedom, or "our feeling of self-activity and personality[,] is greater the more we are able to see into the mutual interconnection of the various individual workings of the soul in relation to an action, and into the interconnection of an entire state with preceding states according to the rule of necessity."[28] Free self-formation is at least in part a function of our ability to discern the way that our actions and ideas are situated within the larger, intricately interconnected framework of all actions and ideas. Schleiermacher observes that the greater our understanding of the causes and circumstances behind a particular occurrence in our lives, the better suited are our intellectual and emotional faculties for making sense of and accepting that occurrence in light of our own self-activity and in light of the nexus of experiences and ideas within which we at every moment exist. This notion of an interrelated causal nexus, which he describes as "the necessary interconnection of actions with the universal chain of causes and effects,"[29] is a significant element of Schleiermacher's thought that remains with him throughout his career.

While the essay *On Freedom* is meant primarily as a response to and critique of Kant, it is likely that, at the time of its composition, Schleiermacher had already been exposed to the thought of Spinoza via F. H. Jacobi's text,

---

27 PR 88.

28 Friedrich Schleiermacher, *On Freedom*, trans. Albert L. Blackwell (Lewiston: The Edwin Mellen Press, 1992), 81.

29 Ibid., 49.

Über die Lehre des Spinoza.[30] Indeed, Schleiermacher's notion of the interconnected nexus of actions and ideas seems to be a near cousin of Spinoza's famous deduction in proposition seven of the second part of the *Ethics* that ideas and things both exist in an ordered and connected system. For Spinoza, human freedom is located in our capacity to see the world in light of the necessary orderings and interconnections that characterize it. Our desires and emotions are obstacles to this capacity, and insofar as we can align our reason with the necessary chain of causes and effects in reality, we will be able to understand and then gain a knowing control over our desires and emotions.

A seemingly similar conception of the world is at work, not only in Schleiermacher's essay *On Freedom*, but also in the second of his *Speeches*, in which he exhorts the cultured despisers of religion to recognize the beauty and the order of the laws operative in the world. These governing laws "embrace everything, the largest and the smallest, the world systems and the small mote of dust that flutters about restlessly in the air."[31] This all-encompassing quality reflects the unity of the world, which reveals itself in the great and small alike. Most importantly, explains Schleiermacher, "only a person who in fact sees [this unity] everywhere, who, not only in all alterations but in all existence, finds nothing else but a production of this spirit and a representation and execution of these laws, only to him is everything visible really a world, formed and permeated by divinity."[32] Only such a person—one who has developed her capacity to see and understand the underlying order and interconnectedness of the multiform elements of existence—truly can be said to have grasped her own place in the world of causes and effects. Developing this understanding is an important part of free self-formation.

Given this description, it would seem as though Schleiermacher merely takes up into his system the Spinozistic ideas of interconnection and self-formation exactly as they appear in Spinoza's system. If we are to comprehend the substantial way in which Schleiermacher in fact appropriates and transforms these Spinozistic ideas, however, a brief return to Spinoza's metaphysics is in order. According to Spinoza, there is only one infinite substance, God, which is indivisible and is characterized primarily by ab-

---

30 Julia Lamm, *The Living God: Schleiermacher's Theological Appropriation of Spinoza* (University Park, Pa.: Pennsylvania State University Press, 1996), 13.

31 *On Religion*, 35 (speech two).

32 *On Religion*, 37 (speech two).

solute activity. This infinite substance is present in the world in the guise of finite modes. These modes are *only* modes, and are not independently existing substances or entities. As Spinoza explains,

> Whatever is, is in God, and nothing can be or be conceived without God. Except for God, there neither is, nor can be conceived, any substance that is, thing that is in itself and is conceived through itself. But modes can neither be nor be conceived without substance. So they can be in the divine nature alone, and can be conceived through it alone. But except for substance and modes there is nothing. Therefore, everything is in God and nothing can be or be conceived without God.[33]

Clearly, then, Spinoza asserts that all finite modes are in reality manifestations of the attributes of the one, indivisible, infinite substance. Only God is *causa sui*, or absolute activity. The finite modes, as aspects of God's infinite activity, are necessarily interconnected within a framework of mutual causing and conditioning precisely because they are all merely modes of divine, infinite activity. The underlying, necessary, infinite activity of God's essence—of which everything finite is a mode—accounts for the fact that every finite thing is related causally and mechanistically to every other finite thing. All finite modes are finally and at the highest level of intelligibility interconnected, because they are ultimately the same thing: a modification of the absolute divine activity.

As evidenced by the citations from *On Freedom* and the *Speeches*, Schleiermacher wishes to preserve in his own system a sense of this mutual interconnectedness, but with at least one very significant difference. Schleiermacher wants an interrelated causal nexus without the absolute, mechanistic determinacy necessitated by Spinoza's system. Because Spinoza crafts a wholly determinate system in which everything is coincident with absolute divine activity, every interaction within the causal nexus that occurs between finite modes comes about in strict conformity to the laws of nature. There is no sense in which the experience, history, or environment of a finite mode significantly impacts its trajectory or contributes to its way of being in the world. Thus, the freedom about which Spinoza speaks is ultimately reducible to a kind of *fait accompli* acceptance of the necessity

---

33 Spinoza, *Ethics*, part I, prop. 15 dem. (86).

of things. Schleiermacher, on the contrary, wants a non-mechanistic interconnectivity—one in which the freedom that comes with seeing and understanding one's place within the causal nexus is genuine and not illusory or merely reducible to a species of Stoicism. Schleiermacher attempts to leave some room for the sort of indeterminacy that is always present when real interaction—in terms of both receptivity and activity—is part of the process of free self-formation.

Thus, Schleiermacher's brand of interconnectedness is not identical to a pure Spinozistic mechanism operating at the expense of individuality and authentic freedom. This is so because the interconnectedness in Schleiermacher's system carries greater weight and greater responsibility for the direction in which an individual's self-formation moves than it does in the thought of Spinoza. Schleiermacher intends for the interactions within the causal nexus to play a real part in the actualization of the individual, rather than operating merely as necessary byproducts of the laws of nature. Unlike Spinoza, whose hard and fast rationalism renders all things wholly intelligible and determinate, Schleiermacher construes his causal nexus to allow for the existence of some level of indeterminacy and unintelligibility—at least from the perspective of finitude, which is the perspective all humans necessarily possess. An individual is self-actualizing precisely in the context of the nexus of mutually conditioning causes in which she plays a part, both receptively and actively.

Schleiermacher is able to appropriate and reconfigure these fundamental elements of Spinoza's system for himself by retaining the Spinozistic notion of an underlying unity of some sort that accounts for the interconnection of everything in the finite world. The significant difference—which is precisely what permits Schleiermacher to allow for a genuine and profound individuality and a concomitant indeterminacy within his system—is that Schleiermacher does not make this underlying unity reducible to the absolute divine activity as Spinoza does. This move on the part of Spinoza brings God squarely into the world—indeed, makes God's infinite and absolute activity entirely coextensive with the world—and hence closes and determines the system. Rather than claiming that finitude's interconnectedness stems from the infinite divine activity of which all finite modes are a modification, Schleiermacher contends that the interconnectedness of finitude in fact comes about as a result of the unifying characteristic shared by all finite entities: absolute receptivity. In order to understand the steps Schleiermacher takes in making this not insignificant shift, we must now look more carefully at Schleiermacher's systematic theology, focusing particularly on his psychology as it relates to the categories of activity and receptivity.

1. Schleiermacher's Psychology

In paragraph four of *The Christian Faith*, Schleiermacher discusses at some length the two primary experiential elements that shape human life within the context of finitude: receptivity (*Empfänglichkeit*) and activity (*Selbsttätigkeit*).[34] We experience receptivity whenever we are conditioned by another entity, such that we depend upon or react to that entity in some fashion. Contrarily, we experience activity whenever we condition another entity, such that we are at least partially free in relation to that entity and can exercise some sort of determinative function with regard to that entity. For Schleiermacher, receptivity and activity must be conceived of as two endpoints on a minimum-maximum continuum. This means that, rather than casting them as mutually exclusive opposites, they should be thought of as two limiting factors along a continuous spectrum. Considering them in this way allows us to see that, within the causal nexus, both activity and receptivity—which in Schleiermacher's system correspond respectively to relative freedom and relative dependence—are present at all times in life in proportionate quantities. At any given moment in which a human being is thinking or doing anything whatsoever, that moment will be characterized partially by freedom and partially by dependence. No moment within the horizon of life can be wholly and completely active without a single element of receptivity present as well. There can be no such thing as an either-or situation in which a moment consists either entirely of receptivity or entirely of activity. Even if the proportion of one element is almost infinitesimally present, its presence is nonetheless assured. As Schleiermacher observes:

> If we consider our relations to Nature, or to those which exist in human society, there we shall find a large number of objects in regard to which freedom and dependence maintain very much of an equipoise: these constitute the field of equal reciprocity. There are other objects which exercise a far greater influence upon our receptivity than our activity exercises upon them, and also vice versa, so that one of the two may diminish until it is imperceptible. But neither of these two members will

34 CG § 4,1 (24). (Page numbers given in parentheses.)

> ever completely disappear ... Our whole self-consciousness in relation to the World or its individual parts remains enclosed within these limits.[35]

This nexus of relatively free and dependent relations comprises all of what is within finitude, from babies and emperors to the heavenly bodies and the ground on which we walk. The finite causal nexus can be thought of as an horizon of antitheses and oppositions, in which nothing is wholly unconditioned, and in which every finite entity is relatively free and relatively dependent with regard to every other finite entity. This holds true as much for the mutually conditioned interconnection between, for example, a planet or a dust mote and a human being, as it does for the mutually conditioned interconnection between two human beings.

While every entity within the sphere of finitude oscillates continually between relative dependence (or receptivity) and relative freedom (or activity), the fundamental interconnectivity of finite reality is nonetheless held together, not in virtue of its existence as modes of the absolute divine activity, but rather in virtue of its underlying unity of absolute dependence, or absolute receptivity. Schleiermacher derives this ontological notion of the absolute dependence or receptivity of the whole of finite reality from his psychological account of human self-consciousness—specifically from the interaction between its two higher manifestations, the sensible self-consciousness and the immediate self-consciousness.[36] The relative freedom and dependence that characterize an individual's existence in the causal nexus pertain to the workings of the sensible self-consciousness. This is because the sensible self-consciousness is precisely that stage of consciousness that "belongs to the realm of reciprocal action ... For every moment which is made up of a partial feeling of freedom and a partial feeling of dependence places us in a position of co-ordinate antithesis to a similar Other."[37] The sensible self-consciousness rests upon the antithesis between subject and object, self and other, receptivity and activity. The immediate self-consciousness, on the other hand, does away with this antithesis: it "is not the consciousness of ourselves as individuals of a particular

---

35 CF § 4,2 (15).

36 There is an initial, lowest stage in Schleiermacher's three-stage exposition of human consciousness. This stage, which he calls the "confused animal grade" (CF § 5,1 [19]), exhibits itself primarily in pre-verbal children and in people in transition between sleeping and waking states.

37 CF § 5,1–2 (19–20).

description, but simply of ourselves as individual finite existence in general; so that we do not set ourselves over against any other individual being, but, on the contrary, all antithesis between one individual and another is in this case done away."[38] In the immediate self-consciousness, self and world come together in light of their underlying unity in absolute dependence, and subject becomes—in some sense—indistinguishable from object.

This coming together of self and world—or, psychologically speaking, this subsuming of the antithetical categories of relative freedom and dependence into the absolute dependence of all of finite reality—should not be understood as the substitution of the immediate self-consciousness for the sensible self-consciousness. Indeed, Schleiermacher is careful to observe that the immediate self-consciousness exists always and only in relation to and yet in distinction from the sensible self-consciousness. He writes:

> It is impossible to claim a constancy for the [immediate self-consciousness], except on the supposition that the sensible self-consciousness is always conjoined with it. Of course, this conjunction cannot be regarded as a fusion of the two ... It means rather a co-existence of the two in the same moment ... It is impossible for anyone to be in some moments exclusively conscious of his relations within the realm of antithesis, and in other moments of his absolute dependence in itself and in a general way; for it is as a person determined for this moment in a particular manner within the realm of the antithesis that he is conscious of his absolute dependence.[39]

This necessary relation between the immediate self-consciousness and the sensible self-consciousness means that, although individuals continually experience themselves in opposition to a world and to other individuals in that world, these experiences of the sensible self-consciousness are always coupled with the immediate self-consciousness, in which self and world fuse together into an original identity that acquires its unity in virtue of the feeling of absolute dependence. This notion points to an obvious question: upon *what* is the individual and the whole of finite reality absolutely dependent? To answer this question, we must turn to Schleiermacher's presentation of divine causality.

38 Ibid. (19).
39 CF § 5,3 (21).

2. Schleiermacher and Divine Causality

The doctrine of creation is the first doctrine Schleiermacher addresses in the body of part one of his dogmatic theology. He gives it this place of consequence because it expresses the all-important notion that the "totality of finite being exists only in dependence upon the Infinite."[40] It is precisely this absolute dependence that characterizes not only the individual entity, but also the whole of finite reality. This dependence is not partial or limited in any way. Rather, it is an "unconditional, all-inclusive dependence"[41] that pertains to all that is, and that points to a God who is absolutely, unconditionally *in*dependent of all that is (insofar as such human categories can even be applied to the divine). According to Schleiermacher, this dependence of the created order extends even to space and time, neither of which can be exempted from the Divine Sovereignty.[42] Only God can be described as the "sole original activity (*Gott allein Tätigkeit ursprünglich zugeschrieben wird*)"[43]; God does not derive this characteristic from any source outside Godself, and nothing in the created order can be described in the same fashion. Because God creates everything that exists, everything that exists is absolutely dependent upon God.[44]

Everything that exists—the totality of finite reality—is characterized by the reciprocal relations between activity and receptivity. The finite sphere is the "realm of antitheses" in which every entity experiences both limiting and being limited in relation to other entities. God, as the free creator of this finite order and the cause of the whole causal nexus of relative freedom and dependence, transcends the horizon in which this arrangement of activity and receptivity plays out. For this reason, Schleiermacher opposes any view of God which amounts to "bringing the Supreme Being within the sphere of limitation."[45] God causes this sphere of limitation to exist, and therefore cannot even be conceived of in terms of limitation or opposition. In his discussion of God as absolute causality in paragraph 51 of the *Chris-*

40 CF § 36,1 (142).
41 Ibid. (143).
42 CF § 37,1 (144).
43 Ibid.; CG § 37,1 (188).
44 Schleiermacher emphasizes the fact that the creation of the world is the result of a single divine act that is not conditioned by anything either before or after that act. Because we are constrained by our finitude, we have no way of truly comprehending how such a singular act could have occurred; the mechanics of it lie beyond the human capacity for knowing, and for this reason, Schleiermacher refuses to affirm the traditional Christian formulation of the doctrine of creation in terms of *creatio ex nihilo*.
45 CF § 47,1 (179).

tian Faith, Schleiermacher carefully explains that speaking of this causality in terms of divine omnipotence "is always an inexactitude."[46] He first describes the antipodal interrelationship of relative freedom and dependence and the causal nexus of reciprocal action engendered by this interrelationship. He then notes that the divine causality, which causes all that exists in the sphere of reciprocal action to be, is not to be opposed directly to the finite causality that is at work in this sphere. To put divine and finite causality in direct opposition would be to conceive of them as like in kind, for such a directly contrastive opposition only makes sense when the two opposed entities are comparable.[47] According to Schleiermacher, this notion is an anthropomorphizing one, and it artificially makes God "a part of the totality of the natural order."[48] As the cause of the natural order, God transcends the natural order. God cannot be contained within the nexus of antitheses that constitutes the created realm. Because these antitheses and oppositions all exist on account of the original divine positing of the finite sphere, they are—together as one unit—absolutely dependent upon God.

Pursuing a doctrine of God as Schleiermacher does—namely, along the lines of a *via causalitatis*—can undoubtedly raise red flags with regard to the question of human freedom. In light of Schleiermacher's understanding of God as absolute causality, how are we to make sense of the claim that Schleiermacher's system in fact permits the authentic development of self-formation? To answer that question, let us examine the way in which Schleiermacher understands his doctrine of God in terms of a guarantee of genuine individuality.

3. Schleiermacher and the Freedom of the Individual

If God is the cause of all that is, how can human freedom and human agency be accounted for? According to Schleiermacher, the absolute dependence of the world on God does not countermand the actual efficacy of human activity. Indeed, Schleiermacher adamantly rebuts the "not infrequent misunderstanding"[49] that human free will runs in opposition to the divine will. He explains that, while the finite sphere is characterized by

46 CF § 51,1 (201).

47 Elsewhere, Schleiermacher explains that "in this sphere a comparison of more or less can only be made between similars; but the divine causality" stands outside this sphere and thus outside of comparison. CF § 52,1 (204).

48 Ibid. (202).

49 CF § 49,1 (189).

the mutual reciprocity of activity and receptivity, human and divine agency cannot be placed on such a reciprocal spectrum. There is no maximum-minimum continuum with regard to divine and human causality, such that, the more divine causality is in effect, the less human causality is in effect (and vice versa). Divine agency causes the continuum of human activity and reciprocity to be; it does not compete with the human agency that is operative along this continuum in the created order. As Schleiermacher explains, "God co-operates (*mitwirkt*) in every case with activities which are appropriate to the nature of the active thing, but only and always according to His own causality, which is entirely different (*gänzlich verschieden*) from that which belongs to the sphere of reciprocal action."[50] Because God's causality is qualitatively different from human causality, it does not threaten human agency.

This does not mean, however, that our consciousness of being free to act in the world and to discern our place within it is merely an illusion. Schleiermacher warns against holding such an idea—which he clearly associates with Spinozistic and other mechanistic philosophies—in strong language. If we espouse a mechanistic position, then we necessarily will be reduced to saying:

> that men should regard themselves too as simply part of this mechanism, and should treat consciousness of self-activity as only an unavoidable illusion. Fortunately, few have ever been capable of this self-annihilating renunciation by which, after robbing the whole world of life, they sacrifice their own selves also to the completeness of their theory. By this method all causality of the finite is transformed into appearance, and consequently there is no ground for regarding individual finite being as existing for itself.[51]

Schleiermacher sees precisely such an eliding of genuine finite causality occurring in Spinoza's system, for, in making over all of finitude into mere modes of infinite causality, Spinoza is unable to grant the finite modes authentic self-determination and growth in light of experience. According to Schleiermacher, this error can quite easily be avoided so long as one holds that divine causality is neither mechanistic nor in direct competition with

---

50 CF § 49,2 (192); CG § 49,2 (252–3).
51 CF § 49,1 (190–1).

human causality. In stating that free causes are themselves objects of divine preservation,[52] Schleiermacher is not relegating human freedom to the realm of illusion; rather, he is taking the notion of the absolute dependence of the finite sphere upon God to its logical conclusion. If God is the cause of all that is (including space, time, and the system of nature), then God is necessarily the encompassing cause of finite causing. Free causes are not thereby chimerical, but are in fact quite real. As humans, we experience and are aware of our status as free actors throughout life. Free causality is included in all that is real, and all that is real depends absolutely upon God for its very being. For Schleiermacher, the absolute dependence of the created order upon God and God's status as the absolute (and absolutely Other) cause of the created order are thus mutually co-implying tenets within his theological system, and both tenets together serve to ground the specific type of relative but very real freedom possessed by human beings.

4. Schleiermacher, Self-Formation, and Receptivity

In the second of his *Speeches*, Schleiermacher criticizes Kant's presentation of religion qua ethics for several reasons, chief among them the way in which the Kantian exposition of the moral law tends to deprivilege the idiosyncratic and free self-formation of individuals. Schleiermacher asks, "How does praxis arrive at an impoverished uniformity that knows only a single ideal and lays this as the basis everywhere? It is because you lack the basic feeling for infinite and living nature, whose symbol is multiplicity and individuality. Everything finite exists only through the determination of its limits, which must, as it were, 'be cut out of' the infinite. Only thus can a thing be self-formed within these limits; otherwise you lose everything in the uniformity of a universal concept."[53] These plainly are not the words of a mechanist or a fatalist, for they protest vehemently against a worldview that would purchase systematic uniformity at the cost of authentic self-formation—a purchase that Spinoza unapologetically makes in his system on account of his one underlying infinite substance of divine activity. In Schleiermacher's system, however, the principle of systematic uniformity—the underlying, unitive, and absolute dependence of the finite sphere upon the divine causality—is in point of fact precisely that which permits the process of individual self-formation to take place within the interconnected limits of the causal nexus. This process of self-formation,

52 CF § 47,1 (179).
53 *On Religion*, 23–4 (speech two).

to use the language of the *Speeches*, is a process of exploring and discerning one's possibilities and boundaries—of finding and understanding one's place as an individual within the whole. This self-location occurs in and through the ever-present oscillation between moments of relative receptivity and activity—moments which always exist in conjunction with the accompanying absolute receptivity and dependence of the immediate self-consciousness.

It is important to note that this category of absolute dependence cannot be construed as bald passivity. Indeed, the immediate self-consciousness, which is the seat of the feeling of absolute dependence, "is not to be thought of as something either confused or inactive, since ... [it] either directly or indirectly lies at the root of every expression of our wills and ... on the other hand, it can be grasped by thought and conceived of in its own nature."[54] The immediate self-consciousness—that state of human self-consciousness in which inward feeling emerges—is thus always closely linked to knowing and doing, and it can be fixed in knowing and expressed in doing. Although the feeling of absolute dependence remains self-identical at every moment,[55] it cannot on that account be excluded from the process of self-formation occurring in and through the external interactions and alterations of the sensible self-consciousness, for the immediate self-consciousness never ends with a merely inward awareness of absolute dependence upon God. Because humans must live at every moment within the sensible and concrete realm of reciprocal conditioning, there is no sense in which the immediate self-consciousness can exist in and for itself without any connection to the sensible self-consciousness. The immediate self-consciousness is "an essential element of human nature,"[56] and as such, necessarily occurs in conjunction with the sensible self-consciousness. It is present in and alongside every human activity, whether it involves thinking or doing.[57] For this reason, it is integrally related to the self-activity and self-actualization of the individual, and cannot be dismissed as a simplistic or impotent passivity that has no real connection with the flux and reciprocity that characterize self-formation.

54 CF § 3,5 (11–12).
55 CF § 5,3 (21).
56 CF § 6,1 (26).
57 This does not mean that the immediate self-consciousness derives from or is grounded in the sensible self-consciousness. The immediate self-consciousness and the feeling of absolute dependence have their source—as Schleiermacher repeatedly notes—in that which is wholly other: God.

The fact that Schleiermacher conceives of the absolute receptivity of the immediate self-consciousness as a constitutive element of the human being and of the human being's self-formation leads to a consideration of the most significant characteristic of receptivity: its indeterminacy. In a discussion of relative dependence—an exemplification of partial receptivity—Schleiermacher describes receptivity as coming into contact with "another factor besides the Ego, a factor which is the source of ... particular determination (*Bestimmtheit*), and without which the self-consciousness would not be precisely what it is."[58] To be receptive is by definition to be open to the possibilities that present themselves in the context of self-formation. As the individual engages with the world and goes about influencing and being influenced by others, the antithetical forces of partial activity and receptivity—"Being and Having-by-some-means-come-to-be"—provide the tools by which the individual begins to discern her limits and her location in the causal nexus. Perhaps most important of all, the unity which underlies the back-and-forth of reciprocal causality is an absolute receptivity that is experienced as a feeling of absolute dependence upon God, who is the cause of all that exists within this nexus of reciprocal causality. What this means for Schleiermacher's system is that underlying the entire finite order of things is a profound and penetrating openness toward God. This unconditional openness—or absolute receptivity—leaves space for the very real freedom we have to discern, to react to, and to appropriate the manifold possibilities that present themselves to us as we move through our lives. This real freedom occurs within the limits of the finite causal nexus originally posited in creation—an intricate and only incompletely knowable web of interrelated causes and effects—and hence it is a freedom that is relative in the truest sense of the word. In light of this conclusion, I'd like to turn to Dr. Mariña's paper in order to propose a tentative response to what I see to be the thrust of her argument.

5. Response to Dr. Mariña

I think that Dr. Mariña reminds us of an important historical truth in pointing out the difficulties Schleiermacher has in systematically reconciling the influence of Leibniz with that of Kant. I would like to suggest, however, that Schleiermacher's all-important category of absolute dependence (or receptivity), which lies at the very heart of his systematic theology, ultimately

58 CF § 4,1 (13).

furnishes him with the tool he needs to synthesize the particular Leibnizian elements with the particular Kantian elements that find their way into his own system. As Dr. Mariña notes, Schleiermacher in the various versions of the *Dialectic* appears to subscribe simultaneously to a theory of the complete concept and to a theory that "contentless" concepts receive their content in and through the sense-related workings of the organic function.[59] To begin with, such a discrepancy might plausibly and charitably be explained quite simply in virtue of Schleiermacher's inability to render the *Dialectic* into a systematic and properly organized whole before his rather abrupt death. Given Schleiermacher's criteria for the scientific value of propositions—definiteness of concept and coherent interconnection[60]—I am not unwilling to grant him the benefit of the doubt as I attempt to make sense of these unfortunate ambiguities and inconsistencies. However, beyond this very basic appeal for the most irenic reading possible in light of the circumstances surrounding the *Dialectic*, I also would like to look at the way in which my discussion of Schleiermacher 's category of receptivity might at least sketch the barest outlines of what I see to be Schleiermacher's approach to synthesizing the elements of Leibniz and Kant which Dr. Mariña mentions.

The individual's feeling of absolute dependence upon God is the expression of the essential, pre-reflective consciousness of the uniting of all subject-object antitheses in a collective dependence upon the creator. It signifies the pervasive character of receptivity and dependence in relation to God that is the hallmark of the entire finite sphere. It points to the imprint of bare existence—the sheer, open receptivity of the immediate self-consciousness—that, in continual conjunction with the sensible self-consciousness, allows for the development and self-formation of the individual in the context of the world. The initial creation of the world not only calls forth the feeling of absolute dependence, but also imbues the world with reason, which works itself out in and through the interconnected nexus of finite causality. Inasmuch as the human individual is the locus of reason, she (and every other human) possesses a system of what ought more properly to be called innate concepts (rather than wholesale Leibnizian complete concepts). For Schleiermacher, this means that, in looking at the process of knowing from the side of the intellectual func-

---

59 See JACQUELINE MARIÑA, "Schleiermacher Between Kant and Leibniz: Predication and Ontology," section I in this volume.

60 CF § 17,2 (84).

tion, "we must have concepts before we have notions of individual things as such."[61] It is this innate concept-tree that permits us to form true judgments and correct predications in processing sense input from the organic function.

Considered from the side of organic function, however—which would be to proceed historically—Schleiermacher explains that "we would have to take judgment first, because obviously actions rather than things are posited in the consciousness of children."[62] The process of growth and self-formation is in part the process of holding the organic and the intellectual functions together as the individual engages in thinking. What is more, he adds, despite the fact that these innate concepts—as universals—in some sense precede the process of judgment formation, it is nonetheless not a bad idea when examining instances of real knowing "to set aside the existing formation of a concept and to view everything independently from the viewpoint of action."[63] This illustrates the extent to which Schleiermacher understands the receptivity of the organic function and the activity (or spontaneity) of the intellectual function to be interdependent and—at different moments—more or less in ascendancy in relation to one another. Schleiermacher seems to be making the point here that knowing—correct concept and judgment formation—is the joint outcome of innate concepts and an embeddedness in history, both the history of an individual life and history considered as the interrelated causal nexus.

What underlies this entire, ongoing process of knowing in its interconnected moments of partial activity and receptivity is the absolute dependence or receptivity of all of history. It is this unifying experience of the absolute dependence of finitude upon God that allows Schleiermacher to grant individuals a freedom that is both entirely relative and nonetheless very real. For both Schleiermacher and Whitehead, Continental Rationalism, despite its appeal and influence, does not hold the answer—the geometric method provides coherence and determinacy at a price that is too high: a real, relative freedom.

---

61 *Dialectic(1811)* 30.
62 *Dialectic(1811)* 51.
63 Ibid.

System without Certainty

Marjorie Hewitt Suchocki, Claremont School of Theology

> There remains the final reflection, how shallow, puny, and imperfect are efforts to sound the depths in the nature of things. In philosophical discussion, the merest hint of dogmatic certainty as to finality of statement is an exhibition of folly.[1]

Certainly systems of philosophical theology have become problematic—particularly insofar as systems tend to explain God and the universe within the compass of one's own small systematized perspective. Indeed, the twentieth century as a whole increasingly turned the philosophical quest away from the study of being to the study of language, and the deconstruction movement pressed the study of language to its radical conclusion in the empty spaces between words themselves. Systems per se fell by the wayside.

And yet the most recent studies in physics from the result of the Wilkinson Microwave Anisotropy Probe (or W-MAP) surely challenge scholars in the area of philosophical theology to look anew at the way our theological understanding of the universe coincides with the intriguing data emerging from increasingly sophisticated scientific instruments. That is, while philosophers immerse themselves in the intricacies of language, physicists explore the outer reaches of space. But surely one of the tasks of philosophy and theology is to interpret as best we can the nature of this universe of which we are a part, particularly as we receive increasingly reliable data from the sciences. Insofar as the data uncovered by physicists enhance our understanding of the universe, we are bound to integrate this data into our own work, and to ponder the hard questions concerning possible meanings relative to human experience. To do so pushes us in the direction of cos-

---

1 PR XIV.

mology, and cosmology—the study of the whole—requires us to develop systems.

One of the problems that drove philosophers and theologians away from system was the pretension to absoluteness, and therefore certainty, attendant upon a theory that intended to speak about the whole. A philosophical legacy bequeathed us by the twentieth century is the knowledge of the perspectival limitations that beleaguer all our thought. Our attempts to discuss the whole are necessarily hedged in by our own finitude and fallibility. But what of that? Rather than drop systematic thought altogether, why not simply build provisionality into our systems? Nearly one hundred years ago, Werner Heisenberg proclaimed an uncertainty principle governing the world of particle physics. By taking that same uncertainty principle into our ways of developing systems, we develop systems without certainty. With the relinquishment of absoluteness, our systems conform more closely to the world about which we speak.

I recognize here the caveat that could be interjected by linguistically oriented philosophers of religion. "What do you mean by 'conform'?" they might ask, tempting us to dance around the issue of a correspondence theory of truth. Betraying my Whiteheadian perspective, I respond that the intention of a "system without certainty" approach is to correspond with some degree of adequacy to the world as we experience it. By definition, such a system is necessarily open to data that force the system into continuous evaluation of itself. In a sense, a system without certainty is radically open to its own creative transformation in light of new data. Hence a Whiteheadian system, which portends to be a cosmology, must be peculiarly open to information such as that received from the W-MAP, and ready to revise itself if its basic model is called into question by new data.

So then, with this introduction, I must first discuss Whitehead's fundamental model, the actual entity. To discuss his reasons for an open system requires recourse to the system itself, and the system itself depends upon his fundamental model. Following this, I show how an open system also follows from the epistemology he uses to develop his model. Finally, I test his model insofar as I am able against the data of the Wilkinson probe, the W-MAP.

I. Whitehead's Basic Model

I summarize the model briefly for the purpose of showing how the model not only presumes openness, but suggests openness. Models themselves, in

Ian Barbour's words, are "limited and inadequate ways of imagining what is not observable. They remain hypothetical."[2] Whitehead, then, on the basis of his knowledgeable experience in physics, imagines what might be the case in order to account for this swirling, dynamic, interactive universe.

To be is the result of a process of becoming, and becoming itself emerges from the influences of a particular past. Each process emerges from a past that, like itself, was an emergent process in its own time, so that the model Whitehead developed presupposes innumerable instances prior to and following the model described. Each and every one of those instances conforms to the model, and there is no actuality that does not conform to the model.

To be is to have an effect. The emergence of a new instance of actuality (the "actual entity," "actual occasion," or "occasion of experience" are Whitehead's terms for this actuality) follows from the effects of the many actualities preceding the becoming occasion of experience. These effects are like tangible influences that force the emergence of new accountings of those influences. In order to respond to the past, however, the becoming actuality needs some sense of how it might respond. In a sense, the becoming actuality emerges through the power of an accomplished past and the power of the possible future. Its own power is its ability selectively to contrast these powers in light of its aim for itself. It accepts some data, rejecting others—but clearly, its rejection of data is as significant as its acceptance of data, deeply affecting its becoming. The entity emerges through its selectivity. Once it accomplishes this, it joins the past in calling for a new present to emerge. Whitehead, with Plato, calls each emergent actuality that "which is always in a process of becoming and perishing and never really is."[3]

Clearly, a single paragraph attempting to summarize a model that took the 351 pages of *Process and Reality* to delineate is hardly sufficient. Whitehead expands upon the internal and external dynamics of the actual entity. Internally, he focuses on the process of the instinctive contrasting and comparing of all the received data. This he called the "physical pole" of an entity, which is itself contrasted with the decisive "mental pole," whereby the emergent entity values "this" rather than "that" data, "this" rather than "that" becoming. The physical pole is essentially oriented toward its feel-

---

2 Ian Barbour, *Myths, Models and Paradigms* (New York: Harper & Row, 1974), 38.

3 PR 82. He is quoting from Section 28A of Plato's *Timaeus*.

ings of its past, whereas the mental pole is essentially oriented toward its feeling of a possible way of selectively organizing and integrating that past into its own subjective becoming.

Externally, Whitehead focuses upon the instrumental effectiveness of the actual entity as each combines with an infinite number of others continuously to produce the world of our experience. Because the relations between entities are internal, there is a necessary connectedness among entities, like a web that conjointly and dynamically yields an ever-successively-new universe. The forms of connectedness yield the extensive continuum that underlies our experience of space-time; within this continuum are the various combinations of entities that yield an unending variety of organisms, including ourselves. Internally, every actual entity is a confluence of creativity as it unifies the discrete manyness of its past into the unique unity of itself. Externally, the universe is the pulsating, dynamic, interwoven result of these myriad internal becomings.

Whitehead's philosophy of organism, then, unifies what previous philosophies had considered disjunctive. Internal and external are interrelated phases of a dynamic existence, so that the internal yields the external and the external yields the internal. Whitehead's memorable phrase is that "The many become one, and are increased by one."[4] Likewise, "physical" and "mental" are not two discrete substances such as were posited in a Cartesian world, but are necessary components of every actuality whatsoever. If we push "physical" and "mental" into their usual human terms of feelings and thought, the unification still holds. In a Whiteheadian world, feeling is the substratum of thought, or, thought depends absolutely on feelings. Thought emerges through the intensity of contrast gained through the ability to hold divergent feelings from the past together in light of a contrasting future. Also, it should be clear that the usual Western dichotomy of "subject / object" is also overturned, for every subject of its own becoming is an object for the becoming of its successors. There is no subject that is not objectified, and no object that was not a subject to itself.

There is surely a role for God in Whitehead's remarkable model, and again, the usual disjunction of "God" and "world" is meliorated in Whitehead's system. Whitehead's systematic principle was that if something exists, an actual entity is the reason for that existence; or, to put it another

---

4 PR 21.

way, if the actual entity is the basic model for existence, then any existent thing must be rooted in the actual entity. But possibilities exist. Each emergent actuality deals with possibilities for its own becoming. While some process thinkers insist that all possibilities emerge naturally from the past, Whitehead thought differently. His consideration of dramatic breaks from the past—what would be called "saltations" in evolutionary theory—demanded that there be a realm of possibilities beyond that which the past alone could provide—for indeed, that very past was once searching for its own possible future. If the becoming entity feels not only its past actual world, but also its possible future, those possibilities must come from somewhere, and that somewhere must be an actual entity. Thus Whitehead found himself forced to posit a unique entity—God—to account for the availability of possibilities. The power of the past, rooted in the world, had to be conjoined with the power of a future rooted in God in order fully to explain the dynamism of the actual entity.

But just as Whitehead joins the seeming opposites of internal / external, physical / mental, feeling / thought, he also joins the traditional opposites of God and the world. By the stratagem of reversing the mental / physical polar structure of the entity he calls God, he not only creates an actual source of all possibilities, but also an actual destiny for the world. For God, like all other entities, must experience the world and have an effect upon the world. Through God's experience of the world, God takes the newly become world into the divine self, moment by moment, in each completed instance of the universe's becoming.

This, then, is an outline of Whitehead's system. Its foundation is in the unique model of the actual entity, but in order to explain not only the entity, but also the world and indeed the universe of our experience, Whitehead so constructs his model that its implications radically offer a unified understanding of our world. Its usefulness extends to varied disciplines in both the humanities and the sciences. Most of my own work has focused on implications of the Whiteheadian model for the development of a comprehensive theology. But given the essential interrelatedness of the model, there is no responsible way to isolate theology from other realms, such as the various disciplines of the humanities and the sciences. Indeed, one of the requirements of the model is that it be continuously tested against all the data given in our experience, and adjusted accordingly. This is to say that Whitehead's system is necessarily an open system—a system without certainty—and to this we now turn.

II. Whitehead on Openness

Whitehead put it this way:

> Plato and Hume illustrate that system is essential for rational thought. But they also illustrate that the closed system is the death of living understanding. In their explanations they wander beyond all system. Thus they illustrate in their own procedures that our primary insight is a mixture of clarity and vagueness. The finite focus of clarity fades into an environment of vagueness stretching into the darkness of what is merely beyond. The partly comprehended forms of succession dimly illuminate this environment within experience.[5]

Whitehead's primary argument for the necessary openness of a system is because of our essential inability to incorporate clearly all that we experience. This follows from the model just described. The actual entity is created in and through its cutting away of much of the data of its experience. Becoming is a process of unification that depends upon selectivity. If we apply this to our own thought processes, the clarity of our thinking is purchased by the vast background of data that we have rejected, whether consciously or nonconsciously. Furthermore, in Whitehead's model, consciousness does not pervade human existence, but is a late phase of each momentary entity successively defining itself. Thus we must say that thought not only depends upon the cutting away of vast amounts of data external to ourselves, but also depends upon the depths of feeling that precede our final phase of conscious thinking. Positively, thought is a focusing upon some detail in relation to a whole, but that whole extends beyond our power fully to comprehend it.

But if this is the case, then all systems necessarily exclude some data of our external and internal experience. By definition, thought—and therefore system—is a cutting away of some things in order to focus on other things. Even as Whitehead strove to develop a model that would account for all facets of our experience, he recognized the anomaly that models are hypothetical constructs needing a constant testing against further data of our experience.

He framed this in one of his few but famous illustrations, that of the aeroplane.

---

5 MT 83.

> The true method of discovery is like the flight of an aeroplane. It starts from the ground of particular observation; it makes a flight in the thin air of imaginative generalization; and it again lands for renewed observation rendered acute by rational interpretation.[6]

Thus Whitehead's imaginative model of the actual entity is not a closed model against which all data must be interpreted, but an open model that interprets data, and is tested and itself interpreted by new data.

The danger attendant upon any model, of course, is that its adequacy in some fields can tempt us to assume that it is adequate in all fields, without need for further testing. And the more successful a model is in a variety of fields, the more susceptible are those enamored of the model to illusions of absoluteness. We at the Center for Process Studies have, in fact, been accused of this very illusion given our penchant for interpreting all things according to the measure of Whitehead's model. We must rather live with the paradox that we cannot be "true believers" in the system, but "testing believers" in the system, eager to revise it where we find it inadequate to the data at hand.

We have, of course, severely revised and extended Whitehead's basic model, most notably in the direction established by Hartshorne toward revision of the Whiteheadian notion of God. But internal tinkering with the model based on internal criteria is not the force of an open system. By definition, a system must not only be open to revision according to the criterion of internal consistency, but it must also be open to revision based upon neglected or previously unknown data from any source whatsoever. Whitehead himself cited two forms of critical testing, the one rational (coherence and logical consistency) and the other empirical (adequacy to the data and applicability).[7] To date, revisions to his system have been made primarily on the basis of the first, or rational criteria, and these not without controversy. I am not aware of revisions made on the basis of the second, or empirical criteria. The force particularly of the empirical criteria is that we must always build a certain tentativeness into our way of utilizing the system, always ready to test it and if need be revise it in light of new data.

---

6 PR 5.
7 PR 3.

In addition to speaking about the necessary vagueness attendant upon thought, Whitehead also cautions system-builders because of the very finitude of our ability to grasp data. Hear him again from *Modes of Thought*:

> Systematic thought has clarified insights and has directed attention to aspects of experience which exemplify special systems. But the universe stretches beyond our finite powers of understanding.[8]

If my first quotation from *Modes of Thought* stressed the necessity of openness because of internal reasons concerning the selective nature of thought, this quotation stresses the external reason of the simple vastness of data throughout the universe. Shakespeare may have said it best: "There are more things in heaven and earth ... than are dreamt of in your philosophy."[9] Whitehead's system intends to explain / interpret all of the data of experience, and his understanding of that data is that it extends throughout the universe of our experience. In the interconnectedness of things, there is nothing that does not have an effect upon all its successors. We know this through data from the realm of physics as we learn of chaos theory on the one hand, and of the interpenetration of things on the other. Our very bodies are constituted through their incorporation of material that once was generated on distant stars. Yet despite this vast network of interconnectedness—or perhaps because of this network—there is no way by which Whitehead could incorporate all the data of experience into his imaginative construction of his model. The universe stretches infinitely beyond each individual, even though the universe as a whole has an impact on each individual. Given this vastness, there is no way to test a model against each and every element of existence. Thus there is a necessary provisionality built into the model. Insofar as we can test it, this construction of experience seems to fit as much of the data as we can find and / or use. But given the vastness of the universe and the limits of our perspective, we can never be certain that we have adequately accounted for all that must be taken into account in order to produce a workable system.

This leads to yet a further caveat against a closed system, once again from *Modes of Thought*:

---

8 MT 82.

9 William Shakespeare, "Hamlet," in *Complete Works* (New York: Oxford University Press, 1998), 679. The quotation is from Hamlet's speech to Horatio in Act I, Scene V.

> The character of the lecturer arises from the moulding it receives from the social circumstances of his whole life. These social circumstances depend on the historic epoch, and this epoch is derivative from the evolution of life on this planet.[10]

If the intricacies of the internal dynamics of thought processes, and the vastness of the external world about which we wish to think, challenge the impulse to close any system of thought, it is also the case that the necessary social conditioning that we bring to our thinking also challenges us. In the above quotation, Whitehead puts social conditioning into the nesting boxes of one's particular history, and indeed, the evolution of life on this whole planet. In his own day, scholars were barely conscious of the social conditioning that they brought to their scholarship; in our day, marginalized groups from all directions have raised this issue with force. We know that we see and interpret existence from a framework that is inescapably conditioned by the data of our personal and social histories.

While Whitehead was aware of women's issues, the extent of his awareness was congruent with the social situation of his day. But Whitehead, like Nietzsche before him, assumed that thought takes place within a vast network of pervasive but unexamined presuppositions. A task of philosophy is to probe for those unexamined suppositions and, once any are uncovered, to investigate their relevance for how one thinks. But in the nature of the case, for every one uncovered there will be a myriad of others yet to be discovered—if, indeed, they are ever captured by the meshlike nature of thought. Their very pervasiveness accounts for their elusiveness. As Whitehead notes, we do not usually notice that elephants are not in the room unless someone first calls attention to the fact of elephants at all.

Feminists are witnesses to the truth of this suppression of pervasive suppositions. We live in a patriarchal world that shapes our consciousness and our conscience, so that the imposed subservience of women to men is made to appear as the natural rather than the social ordering of things. But feminists themselves, having raised the suppressed subordination of women to consciousness, soon learned that suppression—or oppression—is a hydra-headed monster. Oppression of women was but one social construction woven into the fabric of society. It was like the proverbial opening of Pandora's box, for we no longer noted the oppressive values imposed upon women than we discovered value distinctions based upon things such

10 MT 87.

as race, color of skin, poverty, able-bodiedness, sexual orientation. Nor could feminists maintain the posture of innocent victim over against guilty oppressors, for feminists—like everyone else—think within a system that buries its social constructions into unexamined suppositions. Women of color quickly called white feminists to account over issues of classism and racism. In a world where we necessarily think within social systems, no one stands outside the net; everyone participates to some degree or other, whether consciously or unconsciously, in the social structures that assume as normative the ill-being of designated others.

When we attempt to construct our systems of thought, we do not leave these suppositions behind. To the contrary, they exercise their silent influence on the way thought takes shape. Whitehead created an imaginative model that applied fundamentally to a subatomic world that is neither male nor female, and hence to strata much deeper than the divisions of our societies. In astonishing ways, his model undercuts the binary oppositions of a patriarchal world, primarily because of the thoroughgoing relationality and interdependence that is at the heart of his system. Yet even in developing this model, he consistently uses the masculine as normative language. His only reference to a woman is to a mother, harboring thoughts in her heart like Mary of the Christmas story. The saving grace is that he seldom refers generically to men, either—only to the steady stream of male philosophers with whom he is in dialogue.

Nancy Howell has shown that the patriarchal suppositions that nonetheless underlie Whitehead's thought have shaped the application of his thinking to hierarchical structures of being that reinforce patriarchal orders of society.[11] The system itself, she contends, is not necessarily hierarchical, for its value of intensity (which could lead to hierarchical orders) is balanced by its value on diversity (which stresses the importance of multiplicity for the very achievement of intensity). Uncovering the patriarchal suppositions that tend toward hierarchical use of the model leads to needed adjustment. In a closed system, unquestioned suppositions quietly take their influential places, nudging thought along paths so well-worn that one never notices they are there. In an open system, one recognizes that unquestioned suppositions invariably affect the way one thinks, and therefore one seeks to raise questions for the sake of greater adequacy not only in the model, but in its social implications.

---

11 NANCY R. HOWELL, *A Feminist Cosmology* (New York: Prometheus Books, 2000), 53–61.

Yet a fourth reason for the necessity of an open as opposed to a closed system lies in Whitehead's understanding of symbolic reference, or the inevitable errors of judgment that fall to human thinking. "In derivative judgments," says Whitehead, "there can be error."[12] Briefly, Whitehead's epistemology entails an interpretative union between sensory experience and a felt proposition relevant to that experience. This union, which can take place at preconscious levels, yields conscious feelings that entail a judgment about that which is felt. Knowledge is not a simple one-to-one correlation of thinking to observed fact, but is itself a process involving stages of interpretation and judgment. The final result, which in our case is a system of thought, is an imaginative construction that is based in one's feelings of the world, and yet separated from those direct feelings through intricate processes of judgment. There is no such thing as pure sensory experience; sensory experience is always interpreted.

Whitehead illustrates one facet of his epistemology through exegesis of a simple statement, "Caesar has crossed the Rubicon," showing that the meaning of the statement varies according to the perspective of the speaker; hence any one statement yields a multiplicity of meanings.[13] This, combined with the complex nature of the process of thinking itself, yields an inexactitude to our statements and, indeed, our systems, for in the very process of communicating, additional data relevant to the hearer enters into the interpretative process. Thus there is a certain infinality awaiting infinite completions attendant upon our statements concerning the structure of reality, whether as process or any other construal.

These, then, are some of the reasons internal to Whitehead's own system that force his system to an essential incompleteness. The paradox is that one strives for a more perfect, a more adequate, a more holistic way of thinking about existence, but such perfection of necessity eludes us. A closed system is, by virtue of its closedness, inadequate. The more open a system is, the more adequate it is—and yet its adequacy is necessarily of a provisional kind. It always must be tested against data, and even in its revisions it remains incomplete.

So then, let us test Whitehead's model against the world of physics that seems originally to have suggested the model to Whitehead.

---

12 PR 192.
13 PR 195–7.

III. LISTENING TO W-MAP

Recently the National Aeronautic Space Agency announced information from their interpretation of the data returned to earth from the Wilkinson Microwave Anisotropy Probe, which began its exploratory journey into the universe on June 30, 2001.[14] The probe is capable of taking photographs of the full universe with unprecedented clarity, allowing physicists to determine properties of the universe hitherto only guessed at. Among these properties are the now fixed age of the universe at 13.7 billion years, following a "Big Bang" of rapid expansion. The analysis of components of the universe yields 4% atoms of ordinary matter, 23% dark matter, and 73% dark energy. Dark energy is thought to be the repulsive force responsible for the accelerating expansion of the universe. Another finding indicated by W-MAP is that the universe appears to be unending; its expansion will be infinite. Because of these and other findings, cosmologists now have the opportunity to test various theories, such as an inflation theory that has to do with the first nanoseconds of the Big Bang that resulted in our universe. A more coherent cosmic theory is made possible by the evidence compiled through W-MAP.

And what has this to do with Whitehead and his "system without certainty"? Whitehead's model purports to describe the fundamental structure of existence. As mentioned above, he employs four criteria to test his theories, the first two rational criteria of coherence and logic, and the second two empirical criteria of applicability and adequacy. In his words, "'applicable' means that some items of experience are thus interpretable, and 'adequate' means that there are no items incapable of such interpretation."[15] And Whitehead defines his system as being a cosmology. Surely this new data about our universe ought to provide a critical testing ground for an open system. Is Whitehead's model of the actual entity consistent with a universe that is infinite but not eternal? That is, the universe appears to have begun, but it does not appear that it will end. And while Whitehead's model might well apply to the 4% of the universe that is composed of atomic matter, is it the case that the model can apply as well to the 23% of dark matter and 73% of dark energy—whatever that may be? And if we can define it no better than that, how do we deal with its presence in a

---

14 Information on this probe is available through NASA's internet address, Spacelink.

15 PR 3.

cosmology that hopes for some adequacy relative to the way the universe is?

In the nature of the case, a mere theologian is hardly the person to answer these questions. Here the well-established road between process philosophers and physicists needs to be traveled again. But a theologian needs not only to be an eavesdropper on conversations among the journeyers, but she also needs to raise the questions relative to the theology implied by the process model.

In Whitehead's model, God is the prime exemplar of relationality. To exist is to be in relation, and existence is not possible apart from relation. Is the relational God of process thought consistent and compatible with a universe that starts with a Big Bang and, once begun, is infinite in its expansion? An expanding and contracting universe is very consistent with the process model of God, for—as Whitehead notes—we exist in a cosmic epoch which pertains to this epoch alone. There is a certain degree of arbitrariness in the parameters defining existence in this universe, and other parameters might well define some other universe. And since God is infinitely relational, God presumably related to some other universe prior to this one's beginning, and presumably will relate to a successive universe when this one collapses into some dark hole. But if this universe is infinitely expanding, then clearly there was no infinitely expanding universe prior to our own. What was there then? Our relational model cannot survive a nonrelational period with a God existing independently of a universe. Does the data from W-MAP suggest that there was a time when the universe was not? Or was there eternally a "something" that finally exploded in its own transformation from "something" to "universe"?

Perhaps it is the case that the data from W-MAP suggests that Whitehead's system be adapted to a somewhat parallel system developed by Samuel Alexander in his magnum opus, *Space, Time, and Deity.*[16] Alexander's God emerges along with the universe. While Whitehead's God expands with the universe through incorporation of the universe into the consequent nature, the primordial aspect of Whitehead's God is eternal. Does data suggesting that the universe once began, but is now infinite, require adjustment to this system? I am not physicist enough—which is to say, not at all!—to answer my questions. But I am theologian enough to raise them. And an open system that recognizes the final inadequacy of all forms of thought, including

---

16 SAMUEL ALEXANDER, *Space, Time, and Deity*, 2 vols. (London: Macmillan, 1920).

the most finely-wrought system, must dare the questions and any revisions their answers suggest.

There are other questions raised to Whitehead's system by the discoveries of W-MAP, not the least of which is the applicability of his model to dark energy and dark matter as well as to atomic bits. If it is the case that the model is not fundamental to all existence, but only to the 4% that conforms to matter, then what kind of a model must be constructed to deal more fully with the complexity of the universe? In a sense, Whitehead already provides for more than one kind of existence in his opening statement concerning the "Categories of Existence" in *Process and Reality*.[17] There he contrasts the two major forms of existence, actuality and potentiality, each of which requires the other for its existence. His other six categories are subcategories of these two, and concern private and public matters of fact: prehensions, nexus, subjective forms, propositions, multiplicities, and contrasts.

But the closest corollary to dark energy in Whitehead's system is his understanding of creativity, which only exists in its instantiations. Dark energy appears to be a force driving the expansion of the universe; in Whitehead's system, creativity drives the expansion of the universe. Are the two correlated in any way? If yes, there is a problem with creativity seemingly existing apart from its instantiations; if no, how are creativity and dark energy related? For the coherence of any system means that all data should either be incorporated within the existing system, or provide for a revision of the existing system in order that it shall be more adequate to the data. This is what the openness of the system means.

Again, I am a theologian, not a physicist. But the questions I raise cannot be dismissed, and they illustrate the challenges that inevitably address any system. To the extent that a system is closed against such challenges, the system fails. To the extent that a system is open to such challenges, it faces its own transformation. But that is what an open system is all about. Paradoxically, a system provides a certain amount of certainty, but its certainty is always provisional, restless, seeking the further questions, seeking its own supersession. We are left with system without certainty.

---

17 PR 22.

II

System and Categorization

System and Dynamism in Whitehead's Thought

The Category of the Ultimate and the Concept of God

Anna Case-Winters, McCormick Theological Seminary

I. Introduction: Personal Perspective

I write this paper, not as a philosopher but as a Christian theologian seeking a more "adequate" concept of God, and one doing a better job of pointing toward the God who is incomparably greater than all our concepts. I seek a vision like the "brief Galilean vision"[1] that has flickered uncertainly through the ages, a concept of God that is more credible, more religiously viable, more morally adequate than what many popular and traditional notions have offered. In this dawning of the twenty-first century, it seems that talk of God has been eclipsed, or where public talk of God occurs, it is discredited by its invocation in connection with national interests, narrow-mindedness, various fanaticisms, and quests for personal blessing. Those who would continue to affirm and inquire into this much-abused, yet holy name of God, need to give an account of the nature of God. A good beginning might lie in articulating a concept of God that is more worthy of its subject, in Hartshorne's terms more "worshipful."[2] Can there be progress in theology? As Victor Lowe has observed, "Theology, like metaphysics, is dead when it ceases to be a continuing business."[3]

The tools of philosophy have been an ever-present help to generations of theologians for illuminating, clarifying, and testing theology. So I join

---

1 PR 342.

2 Charles Hartshorne, *Divine Relativity: A Social Conception of God* (New Haven: Yale University Press, 1948), 41.

3 Victor Lowe, *Understanding Whitehead* (Baltimore: Johns Hopkins University Press, 1962), 92.

a great cloud of witnesses in seeking out the best philosophical thought to aid in the theological endeavor. Fruitful resources are to be found in the creative imagination of Alfred North Whitehead and the good company of process thinkers.

II. Can "System" be Justified?

1. "What cannot be."

In *Process and Reality*, Whitehead comments that, "The concept of 'God' is the way in which we understand this incredible fact—that what cannot be, yet is."[4] It may be in this era that "system" falls under the category of, playing upon Whitehead's words, "what cannot be, yet is." It would seem that "system" cannot be justified. Metaphysics has fallen upon hard times. Many treat such work with suspicion as to its usefulness or even its possibility. The anti-metaphysical bias of our postmodern era presents a challenge to those who would continue to entertain the big questions or embrace broad generalities and comprehensive systems of thought.

In *Society and Spirit*, process theologian, Joseph Bracken, has argued that however troubling are the critiques of contemporary thinkers such as Jacques Derrida and Richard Rorty, the charges that they level at classical metaphysics do not really strike the target when it comes to Whitehead's metaphysics.[5] Bracken has insisted—and I concur—that Whitehead's system, because of its distinctive character, is not finally vulnerable to the criticisms of Rorty and Derrida.

According to Bracken, both Rorty and Derrida owe some of the impetus of their critical reflection to Martin Heidegger, who raised fundamental issues concerning the limits of human knowledge of reality or being and the particular problems of subjectivism and anthropocentrism that characterize modern thought.[6] Heidegger's reservations have taken root in our era. The postmodern mind assumes that what human beings perceive as objects of experience are not necessarily the "things in themselves" but are products of the synthesizing, constructing activity of human thought. As Heidegger suggests in *The End of Philosophy and the Task of Thinking*, metaphysics is

---

4 PR 350.

5 Joseph Bracken, *Society and Spirit: A Trinitarian Cosmology* (London: Associated University Presses, 1991), 24–35, esp. 31.

6 Ibid., 29.

at an end. The task of thinking remains, but "the development of philosophy into the independent sciences which, however independently communicate among themselves ever more markedly, is the legitimate completion of philosophy. Philosophy is ending in the present age."[7] Philosophy and philosophical reflection may continue, but not as metaphysics.

According to Bracken, Richard Rorty, much influenced by Heidegger, describes and rejects the "ocular metaphor" for human knowledge, knowledge as seeing things as they really are.[8] Such an understanding, Rorty claims, leads to "foundationalism." According to Bracken, foundationalism means "an understanding of reality as grounded in unchanging concepts of principles grasped by the mind intuitively insofar as the latter is the mirror of nature."[9] Philosophy may continue as an "edifying" conversation, but not as a "systematic" inquiry into the nature of things.

Jacques Derrida, on the other hand, according to Bracken, directs his criticism of metaphysics toward its inevitable "logocentrism." In metaphysics, as in the spoken word and in writing, all we have is "a self-contained system of signs that have no intrinsic meaning apart from their necessary connection with one another."[10] For both Rorty and Derrida, then, the starting point and *terminus* of metaphysics is and must be the experience, thought, and the language of the human subject.

2. "Yet is."

It would seem that system cannot be justified. Yet, in Whitehead's case, it is. Whitehead urges that there can and should be system. He is willing to defend speculative philosophy as a "method productive of important knowledge."[11] He offers a formidable defense of metaphysics, and a proposal of a way of doing metaphysics that is eminently defensible against charges common to the present context. Whitehead understands the purpose of speculative philosophy to be "the endeavor to frame a coherent, logical, necessary system of general ideas in terms of which every element of our experience can be interpreted."[12] By "coherent," he means that "the

7 Martin Heidegger, "The End of Philosophy and the Task of Thinking," trans. Joan Stambaugh, in *Martin Heidegger: Basic Writings*, ed. David Farrell Krell (New York: Harper & Row, 1977), 376.
8 Bracken, *Society and Spirit*, 31.
9 Ibid.
10 Ibid., 32.
11 PR 3.
12 Ibid.

fundamental ideas, in terms of which the scheme is developed, presuppose each other."[13] He uses the term "logical" in "its ordinary meaning, including 'logical' consistency, or lack of contradiction, the definition of constructs in logical terms, the exemplification of general logical notions in specific instances and the principles of inference."[14] The philosophical scheme should be "necessary" in the sense of "bearing in itself its own warrant of universality throughout all experience."[15] Whitehead holds that such a system of general ideas offers an interpretation that may be both applicable and adequate to experience. "Here 'applicable' means that some items of experience are thus interpretable, and 'adequate' means that there are no items incapable of such interpretation."[16] In this formulation of the work of speculative philosophy, Whitehead is making a strong claim for the relevance of such systems to reality as such.

This is an ambitious (even audacious) claim to postmodern ears, but Whitehead is very clear that this is an "endeavor," one that is humble, tentative, imaginative, and readily and necessarily correctable. The humility with which he approaches the endeavor delivers it from the assumption that reality can simply be read off the face of things. As he himself writes, "Progress is always a transcendence of what is obvious."[17] Reality cannot be fully and finally described. Here Whitehead shows himself a sympathizer with Heidegger's conviction concerning the limitations of human knowledge of reality and being.

Whitehead recognizes that the "chief error in philosophy is overstatement. The aim at generalization is sound, but the estimate of success is exaggerated."[18] We readily fall into the "Fallacy of Misplaced Concreteness."[19] That is, we regularly mistake our abstract logical constructions for concrete reality. We make a "false estimate of logical procedure in respect to certainty and in respect to premises."[20] Whitehead attributes this tendency, in part, to our being misled by mathematics and the habit of thinking that certainty can by found by establishing clear premises and then erecting upon them a deductive system of thought. Mathematics may indeed

13 Ibid.
14 Ibid.
15 PR 4.
16 PR 3.
17 PR 9.
18 PR 7.
19 SMW 51.
20 Ibid.

deduce, but philosophy describes. For philosophy, verification does not depend upon clarity and certainty of first principles, but upon general descriptive success. Philosophy should not allow the methods of mathematics to be foisted upon it, except insofar as these methods may help as an auxiliary mode of verification with which to test the scope of the generalities.

Similarly, philosophy has been misled by religion into habits of dogmatism. Whitehead is as critical as anyone of dogmatism in philosophy. As he writes in *Science in the Modern World*, "Philosophy has been afflicted by the dogmatic fallacy, which is the belief that the principles of its working hypotheses are clear, obvious, and irreformable."[21] He insists instead that, "Metaphysical categories are not dogmatic statements of the obvious; they are tentative formulations of the ultimate generalities."[22] As with the method of deduction in mathematics, the dogmatism of religion is unhelpful in constructing a system. It is forgetful of the tentative nature of any formulation of ultimate generalities, and may lead to uncritical acceptance of working hypotheses that need continually to be re-evaluated and reformed.

Here I must diverge a bit from Whitehead's position to clarify that I do not believe dogmatism *belongs* to religion in the same way that deduction belongs to mathematics. Dogma belongs to religion, but dogmatism does not necessarily belong. In Christian tradition, for example, *dogma* simply means *the teaching of the church.* Such teaching may well be viewed as humble, tentative, and reformable, just as Whitehead would urge in the formulation of ultimate generalities of speculative philosophy. Dogma is transmogrified into *dogmatism* when it is imposed upon believers by church authorities in ways that do not invite further reflection, but rather oppress and constrain thought. *Dogmatism* is as destructive for theology as it is for philosophy. *Dogma*, on the other hand, as the collective wisdom of our religious forbears, may in fact prove illuminating for our reflection. We are not the first to ask theological questions concerning meaning and ultimate reality, and there may be distinctive insights that become available to us through a critical appreciation of what others have thought and said on these questions.

Whitehead urges that we should not let the problems of deductive and dogmatic habits of thought deter us from systems altogether. He in fact criticizes those who, reacting to the "dogmatic fallacy," have gone to the

---

21 AI 223.
22 PR 8.

other extreme of completely discarding system and method. These philosophers, he writes, "boast that they have no system. They are then prey to the delusive clarities of detached expressions which it is the very purpose of their science to surmount."[23] In this way, Whitehead is claiming that avoiding systems, because of the deductive and dogmatic habits of thought that often attend them, does not necessarily deliver one from misplaced certainties. "Delusive clarities" may arise in the construction of a system, but they may also arise when one avoids constructing a system.

Whitehead discovers a way out of deductive and dogmatic approaches by welcoming the dimension of imagination in the work of speculative philosophy. There is, implicit in this step, a recognition of the *constructive* character of human knowing. He admits into the endeavor of speculative philosophy, "the play of a free imagination, controlled by the requirements of coherence and logic."[24] His metaphor for the process of discovery is illuminating:

> The true method of discovery is like the flight of an aeroplane. It starts from the ground of particular observation; it makes a flight in the thin air of imaginative generalization; and it again lands for renewed observation rendered acute by rational interpretation.[25]

Imagination becomes an integral part of the process as one seeks to make generalizations, but these generalizations are, in a sense, testable[26]—they can be confirmed or falsified—in terms of whether they prove to be adequate and applicable to the interpretation of a wider range of experience.

Whitehead welcomes imagination and understands the constructive character of the endeavor to formulate generalizations, but imagination and construction, for him, have a logical limitation. Their products are testable

---

23 AI 223.

24 PR 5.

25 Ibid.

26 Ibid. "[T]he conditions for the success of imaginative construction must be rigidly adhered to. In the first place, this construction must have its origin in the generalization of particular factors discerned in particular topics of human interest; for example in physics, or in physiology, or in psychology, or in aesthetics, or in ethical beliefs, or in sociology, or in languages conceived as storehouses of human experience. In this way the prime requisite, that anyhow there shall be some important application is secured. The success of the imaginative experiment is always to be tested by the applicability of its results beyond the restricted locus from which it originated."

in terms of applicability and adequacy to experience. Postmodern deconstructionists go considerably further than Whitehead, insisting on the radically imaginative and constructive nature of *all* knowing. There is not for them, as there is for Whitehead, a clear path from imaginative construction to a testing ground in reality. Deconstructionists generally have a higher degree of skepticism than Whitehead admits regarding access to *reality*. Michel Foucault, for example, applies a radical skepticism to the area of social sciences. He insists, as Rorty summarizes, that

> there is nothing deep down inside us but what we have put there ourselves, no criterion we have not created in the course of creating a practice, no standard of rationality that is not an appeal to such a criterion, and no rigorous argumentation that is not obedience to our own conventions.[27]

Bruno Latour brings a similar skepticism to the field of natural science. In his book, *Science in Action*, he proposes that the social and institutional context of science shapes what science "discovers" and counts as true. He argues that even the notion of "fact" is "susceptible of degrees," and, to a great extent, a function of "interests."[28] It would seem that for these thinkers, in these limited areas at least, imaginative construction is the primary factor at work.

For Whitehead, who assumes a greater access to *reality*, stronger claims can be made for the work of speculative philosophy. The end result of its endeavors is not *nothing*. It would be, to borrow Hartshorne's expression, a "species of metaphysical false modesty"[29] to claim that imaginative construction is all that is at work here. At the end of every argument and every generalization, philosophy faces a confrontation with the circumstances to which it should apply. The realm of experience becomes a testing ground that interrogates. There is no "benumbing repression of common sense" in these imaginative and constructive endeavors.[30] Whitehead shows, it seems to me, a fitting modesty in not claiming finality while avoiding a false modesty that claims no progress whatsoever. At the end of the day,

---

27 RICHARD RORTY, *Consequences of Pragmatism* (Minneapolis: University of Minnesota Press, 1982), XLIII.
28 BRUNO LATOUR, *Science in Action* (Cambridge: Harvard University Press, 1987), 25.
29 HARTSHORNE, *Divine Relativity*, 35.
30 PR 9.

"[t]he proper test is not that of finality but of progress."[31] Whitehead has eluded the charge of not knowing the limits of knowing.

Whitehead's system is also invulnerable to the charge of not recognizing the limits of language and symbol systems. He pays particular attention to the problem of language and symbol systems, their inadequacies, and their inevitable and necessary connections to one another. In this, he incorporates some of the concerns later raised by Derrida and others. Whitehead himself introduces the notion of "The Fallacy of the Perfect Dictionary."[32] We assume that we already have a complete vocabulary for expressing what we experience. But our language enshrines old patterns of thought. New language is needed to express new insights. Language is inadequate for the task of describing reality; it must be continually redefined and even redesigned. (Whitehead himself does this rather freely, much to the consternation of all first readers of Whitehead!) Even when language is honed to a greater precision, it is not the case that there will be some kind of direct correspondence between reality and the linguistic representations of it. Even as he enters these reservations about language, Whitehead does not go so far as to allow that language is entirely self-referential, as Wittgenstein does when he proposes that meaning is only "use in a context" and that the representational, "picture theory" of words must be let go.[33] Whitehead assumes that it refers to reality, reality that is *out there* and not only *in here* (in human thought and imagination). He has, however, no illusions about *direct* reference. As he claims in *Process and Reality*,

> Words and phrases must be stretched toward a generality foreign to their ordinary usage; and however such elements of language be stabilized as technicalities, they remain metaphors mutely appealing for an imaginative leap.[34]

Whitehead's metaphysical system is also defensible against the charge of inattention to the experiencing subject. In this connection, the "reformed subjectivist principle" of Whiteheadian metaphysics merits attention. This principle repudiates older notions of "vacuous actuality," a *res vera* that is devoid of subjective immediacy.[35] "The whole universe consists of el-

---

31 PR 14.

32 MT, Epilogue.

33 Ludwig Wittgenstein, *Philosophical Investigations* (London: Macmillan, 1953), part 1, sect. 46.

34 PR 4.

35 PR 29.

ements disclosed in the analysis of the experiences of subjects. Process is the becoming of subjects."[36] The aim of a system should be "elucidation of immediate experience."[37] Whitehead formulates this view that "apart from the experiences of subjects there is nothing, nothing, nothing, bare nothingness."[38] By this principle, the experiencing subject is placed on center stage in a way that is responsive to contemporary concerns. For Whitehead, however, there is no assumption that only the *human being* can be an experiencing subject. Rather, a universal subjectivity is admitted. This might have potential for addressing Heidegger's concerns regarding the anthropocentrism of modern thought. Although human beings cannot stand outside themselves and experience things from these other vantage points, the admission that there are these other vantage points signals "progress."

The risk of foundationalism, as articulated by Rorty, seems to be avoided by Whitehead's categoreal system, since Whitehead intends this system to be an *imaginative* scheme. Whitehead articulates criteria for any system; the scheme must be coherent, logical, adequate, and in respect of its interpretation, applicable. He does not require that the system be able to establish a firm foundation on first principles. Whitehead admits that weakness of insight and deficiencies of language stand inexorably in the way of articulating a set of first principles. Foundationalism in a metaphysical system would be an example of the "Fallacy of Misplaced Concreteness,"[39] wherein one begins to treat abstract ideas as if they had a concrete manifestation.

Whitehead is clear concerning the role of the metaphysical system as a wider horizon for a variety of fields, and the importance of these fields being open not only to metaphysics but also to one another. He offers a helpful exploration of relations among mathematics, science, philosophy, and religion. He argues that each of these fields with their respective general truths must necessarily condition the other fields, as "the systematization of knowledge cannot be conducted in watertight compartments."[40] Philosophy's proper role is the elaboration of categoreal schemes. The work involves reconciling the differences among competing schemes each with their own merits and failures. In this way, philosophy may help the various

36 PR 166.
37 PR 4.
38 PR 167.
39 SMW 51.
40 PR 10.

fields to become more internally consistent and coherent as well as better integrated with other fields.

To summarize, I am claiming that the devastating charges leveled at classical metaphysics do not find their mark in a Whiteheadian metaphysical system. Whitehead does not fall into the particular habits of thought so fittingly criticized in others. Given that this book brings Whitehead and Schleiermacher into conversation, we might at this point borrow a page from Schleiermacher's notebook. We might say to these contemporary critics of metaphysical systems something akin to what Schleiermacher said in *On Religion: Speeches to its Cultured Despisers*, that "what you despise is not it"[41]; the thing you despise is not the thing itself. What is attacked in metaphysical systems as foundationalism, or inattention to the limits of language, or lack of awareness of the import of subjective experience is not essential to metaphysics as such. The claims articulated in these criticisms do not, in fact, characterize Whitehead's metaphysical system.

III. What Kind of System is Whitehead's System?

In this paper, I have argued that *system* as such—at least one of a Whiteheadian sort—may both be possible and useful if it avoids certain problematic habits of thought. We may now turn to consider what Whitehead in fact proposes. Here I will paint with broad strokes, looking for the most fundamental characteristics.

- It is a system of thought that aims to be coherent, logical, adequate, and applicable. These would be Whitehead's own criteria for his own system.
- It is a system of thought conversant with and open to all fields of human knowledge and experience (math, science, and even religion).
- It is a pluralistic system (beyond the monism of Spinoza). There is sociality at the heart of things.
- It is a connected pluralism (beyond the disconnected pluralism of Leibniz). There is relationality at the heart of things. Whitehead appreciated the contributions of both Spinoza and Leibniz but thought

41 *Speeches*, 15 (speech one).

that their best insights, along with those of Plato, could be reconciled in a new system—the philosophy of organism.[42]

- It is an antidualistic system (beyond dualisms of fact and value, Creator and created).

- It is a system that conveys the dynamism that is life.

Of the general attributes of Whitehead's system, the one I want to draw out more fully is the attribute of dynamism. We might at first assume that a system as such cannot really be dynamic. Talk of methods, schema, and categories evokes a sense of static fixity. Once again, it seems that "what cannot be, yet is," in Whitehead. He privileges "becoming" over being. All things are in process for Whitehead: every actual entity and even God. It might be more accurate to say that, for him, all things *are* process. Whitehead's system is a "philosophy of organism"; it is pervaded by the dynamism appropriate to organisms. His system is inherently dynamic. That this is the case can be illustrated by attention to dynamism in the "Category of the Ultimate"[43] and the Concept of God.

IV. System and Dynamism in the Category of the Ultimate and Whitehead's Concept of God

I take up the question of God, not only because I am a theologian, thinking and talking about God all day. I am that. I take up, however, the question of God here because Whitehead does so rather extensively in *Process and*

42 Whitehead's philosophy of organism assumes that relations are internal rather than external. "[E]very item of the universe ... is a constituent in the constitution of any one actual entity" (PR 148). These internal relations are essential to what the entities are in themselves. They have their own identity, and yet they are in a sense co-constitutive, interdependent, drawing completion from one another in reciprocal relation. This philosophy of organism also acknowledges subjectivity in all entities. Whitehead spoke of "physical and mental poles" present in varying degrees in all entities. His meaning for "mentality" is simply the capacity to experience. Thus, as David Griffin has proposed, "panexperientialism" might be a better way of naming Whitehead's proposal than Whitehead's own term "panpsychism."

43 The topic of category in Whitehead is a complex matter because Whitehead stipulates four sets of categories (PR 20–6): 1) The Category of the Ultimate, 2) Eight Categories of Existence, 3) Twenty-Seven Categories of Explanation, 4) Nine Categoreal Obligations.

Reality and elsewhere as well.[44] By his own admission, his reason for doing so is not because of personal religious experience or because he has an explicitly religious agenda; rather his reasons pertain to his metaphysical system. Whitehead holds that no special religious experience is required to grasp the necessity of these functions of God in relation to the world.[45] Whitehead has a place for God in his categoreal scheme. Yet in that scheme, God is not a "category" in the proper sense. The next section will show why God is not a "category" for Whitehead, and the one that follows it will undertake to articulate Whitehead's concept of God and how this concept fits in the larger system. Each in turn will provide particular instances of the dynamism found in Whitehead's system as a whole.

1. The Category of the Ultimate: The One, The Many, and Creativity

In Whitehead's description of the Category of the Ultimate, God is neither one of the elements nor the sum total of them. The Category of the Ultimate has three elements: the one, the many, and creativity.[46] A closer look at the Category of the Ultimate as Whitehead describes it reveals the fundamental dynamism among these three elements. The many and the one continually flow into one another in the creative process. Wherever a unity is achieved, it is in fact an incorporation of multiplicity, a complex unity embracing contrasts. It does not remain a closed unity, but it becomes, in its turn, a datum of experience for the multiplicity of other actual entities. Each droplet of experience entails a new integration of multiplicity into a new unity.

The dynamism between the one and the many has a shape that can be particularly appealing for our time. In our time, there are, on the one hand, new quests for unity, a hankering after wholeness. In tension with this, there is, on the other hand, an insistence that difference be respected: a blessing of diversity—our *manyness*. Whitehead's system allows for a dynamic interchange between the one and the many. There is not so much an opposition as a flow between the one and the many.

44 PR 342–52.
45 RM 115.
46 PR 20.

In Whitehead's system, God cannot simply be identified with the one, the many, or creativity as such. For the Christian tradition, the most likely candidates might have been "the one" or "creativity." Even a cursory look at the system, however, clearly reveals that neither works in this way. For Whitehead, "the one" is not a being or an agent. It is each subject considered in its private experiencing. Such a "one" arises out of the many; it is a concrescence co-constituted by a multiplicity of events. Upon its completion, it is there to be the datum of the experience of the many. The end of its private experience is the beginning of its public career.[47] In a way, the one and the many are just different ways of looking at the same actual entity.

Neither can God, according to Whitehead, be identified with creativity as such—that other place theologians might seize upon as the location of God. Creativity is simply that drive toward the endless production of new syntheses in which the many become one and are increased by one.[48] Creativity is not a being or an agency; it is what actuality exhibits in the creative advance. God might be thought of—as is Whitehead's habit—as the chief exemplification of creativity, even as the leader in the creative advance, but God does not have a monopoly on creativity, for creativity characterizes all actual entities, and not God only. Here Whitehead resists the old duality of Creator and created. God is not the *Creator* as distinct from all other things that are only *created*. "It is as true to say that God creates the World, as that the World creates God . . . . God and the World are the contrasted opposites in terms of which Creativity achieves its supreme task of transforming disjoined multiplicity with its diversities in opposition, into concrescent unity, with its diversities in contrast."[49] Creativity, then, is not to be identified with God. It is rather the dynamism inherent in all actual entities. Within each entity, becoming and transition manifest creativity. It is manifest as well in the larger scope of God's interaction with the world—a mutual dynamic creativity.[50]

47 "Experience is a relation between private centers of experience and public objects experienced." Lowe, *Understanding Whitehead*, 38. "The end of an occasion's private life—its 'perishing'—is the beginning of its public career." Ibid., 41.

48 PR 21.

49 PR 348.

50 PR 21.

2. The Concept of God

As with the Category of the Ultimate, Whitehead's concept of God is an instance of the dynamism of his system. This will become increasingly evident as the active and interactive aspects of the Primordial and Consequent Natures of God are explored.

If God is not the Category of the Ultimate as such or any element within that category, what is the place of God in the system? According to Whitehead, the place for God is as one among the actual entities.[51]

> 'Actual entities'—also termed 'actual occasions'—are the final real things of which the world is made up. There is no going behind actual entities to find anything more real. They differ among themselves: God is an actual entity and so is the most trivial puff of existence in far-off empty space.[52]

Anything that exists in itself and over time turns out to be an abstraction from those final real things, the actual entities.

God, however, is a peculiar member of the category of actual entities, a *non-temporal actual entity*. It is this non-temporality that makes possible the particular functions that God has in Whitehead's system. These functions, to be detailed in the following sections, are multiple. They are functions that cannot be performed by temporal actual entities. Actual entities require them as the conditions of the possibility of their existence. But the "ontological principle" requires that the reason for anything be located in an actual entity.[53] Thus a non-temporal actual entity, namely God, is necessary to the system.

---

51 According to Hosinski, three important concepts sum up with completeness all the other categories that Whitehead identifies and employs. "[T]hese categories are actual entities, eternal objects, and creativity. All the other categories are either elements of aspects of actual entities (such as prehensions and subjective forms), groupings of actual entities (such as nexus), interactions between actual entities and eternal objects (such as propositions), or categories explanatory of actual entities (the categories of explanation and obligation). If we divide the category of actual entities into the concepts of God and the temporal actual entities, we have four principal ideas summing up Whitehead's philosophy." THOMAS HOSINSKI, *Stubborn Fact and Creative Advance* (Savage, Md.: Rowman and Littlefield, 1993), 277.

52 PR 18.

53 PR 18–19.

a. God's Primordial Nature[54]

α. Ground of the Eternal Objects

Among the functions associated with God's Primordial Nature are aspects that make experience possible. The concept of God in Whitehead's system functions as the Ground of the Eternal Objects. God is the *ground* of the eternal objects in the sense of being their source or repository. These eternal objects might be thought of as "potentialities of definiteness."[55] What is actual in world process is so by virtue of participation (or "ingression") of potentialities of definiteness.[56] The things that are temporal, having actuality—actual entities—, arise by their participation in the things that are eternal, having potentiality—eternal objects.

One might reasonably ask why the historical environment cannot provide all that is needed in the way of potentiality. Why is there a need for this realm of eternal objects? As Victor Lowe has summarized Whitehead's position, "the answer is that if the past provided everything for the present, nothing new would appear. Novelty and adventure were too real to Whitehead to permit him to say, like the materialists, that the apparently new is a reconfiguration of the old."[57] The world is not restricted to mere repetition of what has gone before. The realm of eternal objects offers to the world of actualities the potential for genuinely new possibilities.

β. Ground of Order

Another important function of the concept of God in Whitehead's system is to provide a ground of order. There is an order in the relevance of the eternal objects to the process of creating. According to Whitehead, "God is the actual entity in virtue of which the entire multiplicity of eternal objects obtains its graded relevance to each stage of concrescence."[58] It is a necessary step to order the unlimited possibilities in terms of their relevance.

---

54 This "side of God's nature is free, complete, primordial, eternal, actually deficient, and unconscious. The other side originates with physical experience derived from the temporal world and then acquires integration with the primordial side. It is determined, incomplete, consequent, 'everlasting,' fully actual, and conscious." DONALD SHERBURNE, *A Key to Whitehead's Process and Reality* (Chicago: University of Chicago Press, 1966), 181. Whitehead's use of the image of "weaving"—of the one nature upon the other—conveys a sense that this is one "fabric viewed from two different aspects."

55 PR 44.

56 Ibid.

57 LOWE, *Understanding Whitehead*, 47.

58 PR 164.

In this ordering, God "is the lure for feeling, the eternal urge of desire."[59] God has particular relevance to each creative act and is therefore, for each, the "initial object of desire" establishing the initial phase of each subjective aim.

> God is the unconditioned actuality at the base of things; so that, by reason of this primordial actuality, there is an order in the relevance of eternal objects to the process of creation. His unity of conceptual operations is a free creative act untrammeled by reference to any particular course of things .... His conceptual activity at once exemplifies and establishes the categoreal conditions.[60]

The steps Whitehead takes here seem not unlike the traditional cosmological and teleological arguments for the existence of God such as we might find in Thomas Aquinas.[61] Whitehead's conclusions are not the same—no unmoved mover for Whitehead! Whitehead, like Thomas, works from experience to the conditions of the possibility of that experience. In Whitehead's system, God is the condition of the possibility of there being any experience whatsoever, the answer to why there is something and not nothing (the cosmological argument). God is also the answer to why the apparent order we see in world process obtains (the teleological argument).[62] Thomas argued similarly for the necessity of there being an "Orderer" to account for why there is order in the world process. Thomas's arguments expressed in his "Five Ways" may not function as "proofs" for the existence of God. As *proofs* they are probably only convincing to those who already believe on other grounds. But as *arguments* that offer a reasoned accounting of world process and its workings, these proposals are still *arguable*. Whitehead's presentation of the concept of God in his system operates similarly. Given the order that obtains in world process, he locates its ground in God as "that ultimate unity of direction in the Universe, upon which all order depends,"[63] bringing order to the possibilities in gradations of relevance. Actual entities prehend this divinely given graded relevance and "the World lives by its incarnation of God in itself."[64]

---

59 PR 344.
60 Ibid.
61 THOMAS AQUINAS, *Summa Theologiae*, I, q. 2, a. 3.
62 PR 88.
63 MT III, 3.
64 RM 149.

b. God's Consequent Nature[65]

α. Ground of Novelty

Among the functions of God associated with God's Consequent Nature are aspects that contribute to the richness and meaning of experience: novelty, value, and concreteness. As the Ground of Novelty, God is the leader of the creative advance. In this, God is not the exception to the rule. The world can also be thought of as introducing novelty into the divine experience due to the interdependence of God and the world and the final emphasis on creativity.[66] As Victor Lowe has expressed it, "Both are in the grip of the ultimate metaphysical ground, the creative advance into novelty. Either of them, God and the World, is the instrument of novelty for the other."[67] God can be the Ground of Novelty because God's Primordial Nature is also the Ground of the Eternal Objects. This repository of all unrealized potentialities offers itself to the world process. Novelty arises as actual entities participate in and thereby actualize these potentialities. God is, in this way, the source of truly novel possibilities and, in this sense, may be said to be their "Creator."

With this addition, we fill out the sense in which God may be said to be "Creator." It is not by determining what will be but by providing all that

---

65 "The final summary can only be expressed in terms of a group of antitheses, whose apparent self-contradictions depend on neglect of the diverse categories of existence. In each antithesis there is a shift of meaning which converts the opposition into a contrast.
It is as true to say that God is permanent and the World fluent, as that the World is permanent and God is fluent.
It is as true to say that God is one and the World many, as it is to say that the World is one and God many.
It is as true to say that, in comparison with the World, God is actual eminently, as that in comparison with God, the World is actual eminently.
It is as true to say that the World is immanent in God, as that God is immanent in the World.
It is as true to say that God transcends the World, as that the World transcends God.
It is as true to say that God creates the World, as that the World creates God." PR 327–48.

66 A more clearly panentheistic vision might suit the system better. As JOSEPH BRACKEN has observed in his book, *Society and Spirit*, Whitehead tends toward a pantheism, in which everything is finally absorbed into God. Some appreciators, according to Bracken (notably Meland and Loomer), have actually tended toward a pancosmism. Bracken's proposal that we think in terms of God and the world acting together in a common field, creatively forming a cosmic society is very intriguing and allows for both to stand in their integrity and alterity even while being utterly connected. BRACKEN, *Society and Spirit*, 152.

67 LOWE, *Understanding Whitehead*, 57.

actual entities need (both order and novelty) to create themselves. Beyond this, God and other actualities are co-creators of the World. God's priority is metaphysical and not temporal.[68]

β. Ground of Value[69]

Another key function of the concept of God in Whitehead's system is as Ground of Value. God has a preserving, transforming, even "redeeming" role. God assures the everlasting value of each occasion by its preservation in the life of God. Each one gains an immortality in God and therefore in the temporal world. "The objective immortality of actual occasions requires the primordial permanence of God, whereby the creative advance ever re-establishes itself endowed with initial subjective aims derived from the relevance of God to the evolving world."[70] God, as Ground of Value, provides assurance of value through preservation in God's permanence.

Two persistent problems in theological and philosophical reflection are addressed in this one step. On the one hand, God's capacity to interact with the world is a problem from the standpoint of God's perceived permanence. On the other hand, it would appear that all good things pass away.

> The problem of the fluency of God and of the everlastingness of passing experience are solved by the same factor in the universe. This factor is the temporal world perfected by its reception and its reformation, as a fulfillment of the primordial appetition which is the basis of all order. In this way God is completed by the individual, fluent satisfactions of finite fact, and the temporal occasions are completed by their everlasting union with their transformed selves, purged into conformation with the eternal order which is the final absolute "wisdom."[71]

---

68 RM 85.

69 Bracken proposes that a panentheism is the best way of understanding the God-World relation in Whitehead and that this understanding is enhanced by notions of society and spirit. According to this view, God is immanent in the world and the world is immanent in God without loss of the independent status of either God or the world. "[A] genuinely panentheistic understanding of the God-world relationship will only be achieved if one thinks of God and the world as interpenetrating fields of activity with the field proper to creation contained within the even larger field of the divine intentional activity." BRACKEN, *Society and Spirit*, 159. Bracken's field-oriented approach avoids pantheism on the one hand and pancosmism on the other.

70 PR 347.

71 Ibid.

There is a component of judgment and transformation as well. "The revolts of destructive evil, purely self-regarding, are dismissed into their triviality of merely individual facts; and yet the good they did achieve in individual joy, in individual sorrow, in the introduction of needed contrast, is yet saved by its relation to the completed whole. The image—and it is but an image—under which this operative growth of God's nature is best conceived, is that of a tender care that nothing be lost. The Consequent Nature of God is his judgment on the world."[72] God "saves the world as it passes into the immediacy of [God's] own life. It is the judgment of a tenderness which loses nothing that can be saved. It is also the judgment of a wisdom which uses what in the temporal world is mere wreckage."[73] With this addition, there is an important contribution to addressing the problem of evil in world process. Evil is "dismissed into triviality" in its passing into the divine life, whereas all that is of value is preserved and cherished everlastingly.

God's role in Whitehead's system is salvific and redeeming.

> God's role is not the combat of productive force with productive force, of destructive force with destructive force; it lies in the patient operation of the overpowering rationality of his conceptual harmonization. He does not create the world, he saves it; or more accurately he is the poet of the world, with tender patience leading it by his vision of truth, beauty, and goodness.[74]

This presents a compelling picture of God's persistent and effective care of the world. It is expressed as a continual luring toward truth, beauty, and goodness through the persuasive power of God's attractive vision.

In the divine workings, God's goodness is prior to God's power. There is a notable loss of the preoccupation with divine omnipotence conceived as power to dominate and control the world process. Whitehead seeks a concept that is in this sense "purified." As Whitehead expressed it,

> [i]n a communal religion you study the will of God in order that He may preserve you; in a purified religion, rationalized

---

72 PR 346.
73 Ibid.
74 Ibid.

> under the influence of the world-concept, you study his goodness in order to be like him.[75]

The concept of God presented in Whitehead's system articulates an understanding of God that is religiously viable and morally adequate.

γ. Principle of Concretion

The concept of God also serves, in Whitehead's system, as the Principle of Concretion. God is the eros, the lure toward satisfaction, and, as Victor Lowe suggests, the initiator of the move toward a definite outcome from an indeterminate situation.[76] The way this work of concretion is accomplished is described by Whitehead in terms of the two poles; physical and conceptual. As he says, "In each actuality there are two concrescent poles of realization—'enjoyment' and 'appetition,' that is, the 'physical' and the 'conceptual.' For God the conceptual is prior to the physical, for the World the physical poles are prior to the conceptual poles."[77] The world's objectification in God is a completion of God's nature, weaving physical feelings onto conceptual feelings. In this process, Whitehead concludes,

> What is done in the world is transformed into a reality in heaven, and the reality in heaven passes back into the world. By reason of this reciprocal relation, the love in the world passes into the love in heaven, and floods back again into the world. In this sense, God is the great companion—the fellow-sufferer who understands.[78]

Here again we have an alternative to a view that has proven problematic in Christian tradition, namely the view that in relation to suffering, God is either the direct cause or the one who permits suffering. The defense is that this is either deserved punishment or necessary learning. Whitehead steps outside this framework when he proposes that, in relation to suffering, we should think of God as the "fellow sufferer who understands."[79]

---

75 RM I, VI.
76 LOWE, *Understanding Whitehead*, 47.
77 PR 348.
78 PR 351.
79 Ibid.

V. Summary and Conclusion

The schematization laid out above makes clear the dynamism of Whitehead's system with his Category of the Ultimate and his concept of God each as a case in point. I have set forth the necessity and functionality of the concept of God in Whitehead's system. According to Hosinski, Whitehead does not introduce the concept of God into his philosophy because he has some religious agenda; he introduces it on rational grounds as necessary for the consistency and coherence of his system.[80] Nor is the concept of God brought in *ad hoc* in response to problems of Whitehead's system. His God concept is well-integrated into the system and necessary to it. This is as it should be given Whitehead's commitment to coherence and his insistence that central elements should "require one another."[81] Coherence matters. Whitehead is careful to fully integrate the concept of God systematically. As he claims, "God is not to be treated as an exception to all metaphysical principles, invoked to save their collapse. He is their chief exemplification."[82]

Though the concept is fully integrated and not introduced late to provide an *ad hoc* answer to difficulties in the system, it does in fact function to answer some central conundrums. For example, according to Sherburne, the cosmological problem is set by the ideal opposites of permanence and flux, novelty and order.[83] As Whitehead writes,

> The world is thus faced by the paradox that, at least in its higher actualities, it craves for novelty and yet is haunted by terror at the loss of the past, with its familiarities and its loved ones. It seeks escape from time in its character of "perpetual perishing."[84]

Further,

> The most general formulation of the problem is the question whether the process of the temporal world passes into the formation of other actualities bound together in an order in which novelty does not mean loss.[85]

80 Hosinski, *Stubborn Fact and Creative Advance*, 207.
81 PR 3.
82 PR 343.
83 Sherburne, *A Key to Whitehead's Process and Reality*, 171.
84 PR 340.
85 Ibid.

God is the ground of these ideal opposites and the method by which each is preserved in relation to the other. The dynamism between God and the world is central to this resolution. "In God's nature, permanence is primordial and flux is derivative from the World; in the World's nature, flux is primordial and permanence is derivative from God."[86] It is in the dynamic interchange between God and the world that the problem of God's permanence and seeming stasis, along with the world's flux and perpetual perishing, are effectively addressed.

> The theme of Cosmology, which is the basis of all religions, is the story of the dynamic effort of the World passing into everlasting unity, and of the static majesty of God's vision accomplishing its purpose of completion by absorption of the World's multiplicity of effort.[87]

Whitehead's philosophical theology provides a system that justifies and satisfies two longings of rational agents: the longing for novelty and the longing for permanence.[88] The challenge is to find "actuality with permanence requiring fluency as its completion; and actuality with fluency requiring permanence as its completion."[89] God's "Primordial Nature" is "an unchanging unity of conceptual feeling which embraces all eternal objects in its vision" and serves as the instrument of novelty for the temporal world. Conversely, God's "Consequent Nature" receives the finite occasions of the temporal world, which is the instrument of novelty for God. Both are bound into an ordered unity and then pass into the formation of other actualities in such a way that the passing is not "loss."

The fact that Whitehead's system has found a way of effectively addressing the challenges of permanence and fluency, order and novelty is, of course, not "proof" that his system is "true." But neither should the system's ability to respond to this issue be held against it! The system seems to stand up to the criteria that Whitehead set for himself in being logical, coherent, adequate, and applicable. Nor should the system's theistic character be held against it. As Victor Lowe has pointed out, "If a thinker produces

86 PR 348.

87 PR 349.

88 The temporal creature is "everlasting in the being of God .... In this way, the insistent craving is justified—the insistent craving that zest for existence be refreshed by the ever-present, unfading importance of our immediate actions, which perish and yet live forevermore." PR 351.

89 Lowe, *Understanding Whitehead*, 36.

a theistic metaphysics which, among other things, justifies an insistent religious craving, this result does not discredit the metaphysics."[90]

The real question for Whitehead is whether the system passes the coherency test. He believes it does.[91] Interestingly, it is in his work on God and the world that Whitehead locates his primary discussion of the importance of coherence.[92] From this point, he goes on to show how basic elements of his system require one another. What Whitehead demonstrates here, according to Hosinski,[93] is that temporal actual entities cannot be understood in abstraction from eternal objects (which are exhibited in them and constitute their potentiality), creativity (which is the dynamic self-creation of each temporal occasion), and God (which is the ultimate ground of the possibility of temporal occasions and the final unification, harmonization, and transformation of temporal occasions). The fundamental notions in Whitehead's system are coherent; they require one another. The concept of God is crucial to the system and God has a unique role. But God remains very much within the system and not outside or behind the system as the ultimate explanation of it all. Whitehead's system does not treat the concept of God as an exception to all the metaphysical principles, but rather as their chief exemplification. The concept of God is thoroughly integrated into the larger system. "The concept of 'God' is the way in which we understand this incredible fact—that what cannot be, yet is."[94]

Whitehead's system offers a dynamic and distinctive vision of God. He explicitly rejects prominent understandings of God as an imperial ruler, a cosmic moralist, or an unmoved mover. As he says:

> There is, however, in the Galilean origin of Christianity yet another suggestion which does not fit very well with any of the three main strands of thought. It does not emphasize the ruling Caesar, or the ruthless moralist, or the unmoved mover. It dwells upon the tender elements in the world, which slowly and in quietness operate by love; and it finds purpose in the present immediacy of a kingdom not of this world. Love neither rules nor is it unmoved; also it is a little oblivious as to

---

90 Ibid., 102.
91 PR 3–6.
92 PR 349.
93 Hosinski, *Stubborn Fact and Creative Advance*, 287.
94 PR 350.

> morals. It does not look to the future; for it finds its own reward in the immediate present.[95]

Whitehead once said that, "it is a disease of philosophy when it is neither bold nor humble, but merely a reflection of the temperamental presuppositions of exceptional personalities."[96] While Whitehead is surely an exceptional personality and one we might be tempted to imitate, I think he would not call us to be "Whiteheadians," but to progress in this work using our constructive imaginations and testing all systems in the reality of lived experience. We should be both bold and humble. In our postmodern era when it seems audacious to even propose a system, we must surely see how Whitehead was bold.[97] But when he had done his best he concluded,

> There remains the final reflection, how shallow, puny, and imperfect are efforts to sound the depths in the nature of things. In philosophical discussion, the merest hint of dogmatic certainty as to finality of statement is an exhibition of folly.[98]

Whitehead is a good example of a philosopher who was both bold and humble in his endeavors. So may we all be.

---

95 PR 343.

96 PR 17.

97 "First, that the movement of historical, and philosophical, criticism of detached questions, which on the whole has dominated the last two centuries, has done its work, and requires to be supplemented by a more sustained effort of constructive thought. Secondly, that the true method of philosophical construction is to frame a scheme of ideas, the best that one can, and unflinchingly to explore the interpretation of experience in terms of that scheme. Thirdly, that all constructive thought, on the various special topics of scientific interest, is dominated by some such scheme, unacknowledged, but no less influential in guiding the imagination. The importance of philosophy lies in its sustained effort to make such schemes explicit, and thereby capable of criticism and improvement." PR XIV.

98 Ibid.

Novelty and System in Schleiermacher's Thought

CHRISTINE HELMER, Claremont School of Theology

On the stage of contemporary theology, specifically systematic theology, experience has emerged as both a significant source and a crucial norm for reflection. Although "experience" is often used in an uncritical and epistemologically naïve sense, it remains the hallmark of the postmodern privileging of local narrative that is coupled with an implied criticism of a eurocentric, universal, onto-theological narrative imposed onto conceptually resistant experiential data. This privileging does justice, on the one hand, to a dimension of the gospel that is alleged to be lost in Western theoretical thought: the liberation in reality of those groups marginalized and oppressed by the dominant culture, gender, and socio-economic class. What challenges the totalizing effect of the concept is precisely that which Hegel attempted to explain: the necessity of the particular that is, in an un-Hegelian dialectic, to be liberated from the dominance of the universal.

The highlighting of particular experience as the stuff of contemporary theology is, on the other hand, plagued by a few methodological questions. Even if local consensus agrees that a representation of local experience is more or less accurate, the nagging question remains in view of the individual. It is individual experience that ultimately resists subsumption into a larger whole; it is the individual who continues to upset consensus by virtue of its novelty. Every conceptualization of local experience is a model, and as a model, it has by epistemological necessity abstracted from the individual in order to make transindividual claims. A second question accompanying the first is actually its converse. In order to make a claim to knowledge regarding the individual, the epistemological move of locating that individual within the whole is required. The allocation of an individual to a distinct site within the whole presupposes knowledge of the whole structuring and structured by all its constitutive parts.[1] Hence knowledge of the indi-

[1] According to Schleiermacher, a requirement of knowledge consists in allocating the

vidual requires knowledge of the system. Finally, the third question thematized by contemporary theology concerns theology's communication to those addressees beyond respective regional boundaries. By whatever differentiating criteria particular regions are bounded, such as race, class, gender, communicability beyond the local presupposes a hermeneutical unity connecting speakers and hearers. The actuality of communication and its converse, understanding, are made possible by unifying conditions without which transdiscursivity, and by implication, the communication of the mandate for liberation, is not possible.

What might possibly be a solution—at least an attempt at a solution—could be what I propose as a transdiscursive theology, a theology that relates various regional discourses with each other. Such a transdiscursive theology would rehabilitate system as a goal of theological thinking, while decisively concentrating on the individual perspective as the site of building a system in dialogue with others.[2] As such, a transdiscursive theology would account for the individual as a principle of novelty; system is, after all, articulated from an individual perspective. Yet the individual is also located within a larger whole; systematic interconnections, whether epistemological or dialogical, make possible the dialogical building of system by appealing to common structures of reason and presupposing a common horizon of understanding. By virtue of transdiscursive permeability, an individual articulation of system would be, in principle, open to revision, but would also be constituted by relatively stable features, such as elementary concepts and hermeneutical bridges. In this paper, I will take a close look at three basic issues in a proposed transdiscursive theology. And in so doing, I will address the theme of this collection of essays by showing how, on the basis of Schleiermacher's thought, the individual articulation of a system relates novelty to transdiscursive features of individuality that are

individual to a site within the whole. In *Der christliche Glaube*: "um so durch eine Einteilung des ganzen Gebietes die Örter zu bestimmen, in welche die individuellen Gestaltungen, sobald sie geschichtlich aufgefunden sind, eingestellt werden könnten" (CG § 2,2 [I, 13]). (Page numbers given in parentheses.)

2 In her political philosophy, Iris Marion Young makes a similar move. She proposes a model for inclusion of the individual in the democratic dialogical process in order to overcome what she diagnoses to be a political impasse related to special interest groups. By including the individual through various discursive strategies (i. e., greeting, narrative), the individual influences the outcome of decision-making that she will subsequently experience. See Iris Marion Young, *Inclusion and Democracy*, Oxford Political Theory (Oxford: Oxford University Press, 2000), 52–120.

inherently systematic.[3] First, and from Schleiermacher's perspective, I will look at individuality as the ontologically basic element of system, and show how it is the individual alone who introduces novelty into the system. Second, I will appeal to Schleiermacher's dialectic to show how knowledge of the individual requires system, and conversely, I will make use of Schleiermacher's hermeneutics to show how knowledge of individual novelty is acquired. Third, I will discuss the theological implications of the preceding philosophical discussion. The articulation of a system of systematic theology privileges the novelty introduced by individual experience, while also presupposing specific identity across various systems that makes transdiscursive communication and understanding possible.

I. Individual and Identity

If the individual is to be known as an individual in relation to the whole, then it must be related to structures of identity, or universality.[4] In both his philosophical ethics and dialectic, Schleiermacher isolates particular structures of identity of which the individual is a particular manifestation. In this section, I locate the novelty introduced into identity by analyzing Schleiermacher's conception of the individual in terms of the Kantian distinction between absolute positing and relative positing. This discussion clears the way to the epistemological question of knowing the individual that is the theme of the next section.

It was perhaps his intense love for Eleonore von Grunow that motivated Schleiermacher to think about the individual in relation to identity in a way

3 I have chosen novelty as a key mark of the individual in Schleiermacher's thought because it facilitates a comparison with Whitehead's metaphysics in which novelty plays a significant role. In Whitehead's account of concrescence, the lure to novelty is one of many possible aims presented to each actual occasion. The source of this particular lure is God, specifically originating in the primordial nature. As constitutive of the initial aim, the lure to novelty is envisaged by God in order to metaphysically maximize cosmological beauty. Whether this lure is adopted into a subjective aim, however, depends on the freedom of the actual occasion. See PR 343–9. John Cobb has adopted Whitehead's insights on novelty in his Christology. For Cobb, the "Logos" is God's primordial and ordered plan for the world. This plan is diffused throughout creation by lures to novelty, which, when integrated into concrescence, are evidence of "creative transformation." See John B. Cobb, Jr., *Christ in a Pluralist Age* (Philadelphia: Westminster, 1975).

4 I use the term "identity" as the English translation of *Identität*, Schleiermacher's term for features common to all individuals, yet constituting them in different combinations.

radically differing from Kant. In his text *Soliloquies* (*Monologen*), and in which the references to love's uniting effect suggest the spiritual presence of Eleonore in its composition, Schleiermacher laments the flattening of individuality under the weight of Kant's categorical imperative.[5] An appreciation for the individuality of each individual is nonetheless linked to an understanding of the individual's constituent features that are resistant to change. In order to account for the I's continuity through time, Schleiermacher looks into that place that he would take up at the center of his life's work. Introspection leads to the inner I, which does not change in spite of its outer exposure to life's ebb and flow.[6] Yet the secret of individual stability is itself understood as a distinct mixture of features identical in each individual. According to Schleiermacher, the individual is idiosyncratically constituted by antitheses common to all individuals.[7] Flexibility and stability, youth and age, are all opposites, shaping each individual who embodies them according to his or her own peculiar proportion. By this same elementary composition appearing differently in individual shapes, the principle of identity, meaning the identical elements constituting each individual, is connected to the principle of individuality. This insight will later become the metaphysical cornerstone of Schleiermacher's *Philosophische Ethik* and *Dialektik*, works explaining the relation of the individual to system.

Even in the immature version of the *Dialektik* from 1811, Schleiermacher envisions a metaphysical system relating the individual to the totality of reality.[8] In a manner similar to Plato's *Sophist*, Schleiermacher conceptualizes this system from its highest to its lowest points as the progressive disjunction of the concept into its two constitutive opposing concepts. The totality

---

5 FRIEDRICH SCHLEIERMACHER, *Monologen. Neujahrspredigt von 1792. Über den Wert des Lebens*, ed. Friedrich Michael Schiele, expanded and corrected by Hermann Mulert, 3rd ed., Philosophische Bibliothek, no. 84 (Hamburg: Felix Meiner, 1978). See ch. 4 (74–7) for explicit references to love that implicitly refer to Eleonore.

6 The psychological argument must be read together with the sermon Schleiermacher preached on New Year's Day, 1792, that is included in the German Meiner edition of the *Soliloquies*. The inner and self-same I that Schleiermacher addresses in these early works will later be explicated in the CG as immediate self-consciousness.

7 For a similar discussion of the identity of opposites that is combined differently in each individual, see *On Religion*, 5 (speech one).

8 The 1811 edition of the *Dialektik* is the only version to be as yet translated into English (see *Dialectic(1811)*); for research purposes, however, this version is less than useful. Schleiermacher based all his *Dialektik* lectures after 1814 on the foundational text of 1814/15.

of reality is constructed as the progressive bifurcation from higher to lower concepts that Schleiermacher calls "hovering" concepts.[9] At each level of bifurcation, the hovering concept is seen, together with its opposing concept, from the proximate higher concept as one of two concepts divided by the proximate concept. From the perspective of the higher, unified concept, the lower concept is one into which the higher concept bifurcates. On the other side, when a hovering concept is seen from the lower angle of bifurcation, that concept unites those two disjunct opposing concepts below it, and is thus deemed the essence of the two concepts that it unites. The concept hovers between its proximate genus and the lower opposites for which it itself functions as a proximate genus. In Schleiermacher's thought, the definition of a thing, the genus together with its specific difference, is metaphysically conceived in the context of the whole system of hovering concepts.

The complete system of concepts has its correspondence in reality by virtue of the identity transcending the two highest concepts of thought. The highest disjunction of thought is the antithesis between ethics, according to Schleiermacher the science of the principles of history, and physics, the science of nature.[10] The two concepts, ethics and physics, refer to two spheres of reality, the sphere of human spirit and the sphere of physical

---

9 Schleiermacher's German terminology is the following: "[D]er Begr*iff* ist an *sich* immer *eine* schwebende *Einheit*" (DialKGA II, 497 [*Kolleg* 1822; 32nd lecture]). (Italics in original.)

10 The question in Schleiermacher research has been whether Schleiermacher conceives his dialectic as the science of the *Ur-wissen* uniting ethics and physics in a way transcending the bounds of reason, or as a technical-critical discipline that mediates between speculative and empirical knowledge. In his groundbreaking work on Schleiermacher's philosophical ethics, Birkner understands Schleiermacher's dialectic to be the science of the *Ur-wissen*. See Hans-Joachim Birkner, *Schleiermachers christliche Sittenlehre im Zusammenhang seines philosophisch-theologischen Systems*, Theologische Bibliothek Töpelmann, no. 8 (Berlin: Verlag Alfred Töpelmann, 1964), 36. The debate between Fichte, Schelling, Steffens, and Schleiermacher, preceding the foundation of the University of Berlin in 1809, focused on the conceptualization of the relationship between an *Ur-Wissen* and its bifurcation into the two foundational academic spheres of study, ethics and physics. For a transcript of their essays, see Ernst Anrich, ed., *Die Idee der deutschen Universität: Die fünf Grundschriften aus der Zeit ihrer Neubegründung durch klassischen Idealismus und romantischen Realismus* (Darmstadt: Wissenschaftliche Buchgesellschaft, 1956). By distinguishing his dialectic between the transcendental and the formal parts, Schleiermacher aims to account for both the dialectic as the science of the identity between the ideal and real, and as the science describing the process of thinking "in the pure sense," or thinking that is oriented to knowing.

process. Yet because they are transcendentally (or transcendently[11]) united in an identity transcending thought, these two spheres are found in reality, not as independent spheres of actuality, but as mutually interrelated with each other. In the existence of an individual, the ethical and the physical are both present in a unique mixture. Ethics and physics are extended on a minimum / maximum continuum; even the slightest maximum of matter contains a minimum of reason, and conversely, the maximum of reason contains a slightest minimum of matter. The possibility for all mixtures in existence is given with the transcendental (or transcendent) unity that Schleiermacher understands together with the Absolute Idealists as the identity between reason and matter.[12] The identity is the transcendental (or transcendent) guarantee for the correspondence between thought and reality.

By virtue of the transcendent / transcendental identity of the ideal and the real, Schleiermacher's system of hovering concepts corresponds to the totality of reality as the sphere of ethics and as the sphere of physics. The Leibnizian distinction between power (*Kraft*) and appearance (*Erscheinung*) plays a crucial role in Schleiermacher's post-Kantian conceptualization of a system corresponding to reality; in fact, Manfred Frank describes the move of the post-Kantian search for a foundational philosophy (in Reinhold's and Fichte's case: *Grundsatzphilosophie*) or the Romantic longing for system (in Schlegel's and Schleiermacher's case) as a re-Leibnizianization of Kant's critical philosophy.[13] In Schleiermacher's system, the distinction in reality between powers and appearances corresponds to the distinction in thought between concepts that unify their lower opposing concepts and the bifurcated concepts; the unifying concept corresponds

---

11 Schleiermacher research has yet to determine the equivalence between these two terms that are, at least in terms of the definitions Kant assigns to them, distinguished in Kant's philosophy.

12 The principle of identity characterizing both Schleiermacher's and the Absolute Idealists' conception of system is the point of indifference (to use Schelling's term) that is not a logical but an ontological identity of two opposites. The point of indifference is the ontological site at which one pole merges with its opposite. At this point, the transition from nature to spirit and conversely, from spirit to nature, takes place. For an analysis of the principle of identity in Gottfried Wilhelm Freiherr von Leibniz, through Gottfried Plouquet, and finally in the Idealists, see MANFRED FRANK, "Identität, Korrespondenz und Urteil: Fragen an Schleiermachers *Dialektik*," in *Schleiermachers Dialektik: Die Liebe zum Wissen in Philosophie und Theologie*, ed. Christine Helmer, Christiane Kranich, and Birgit Rehme-Iffert, Religion in Philosophy and Theology, no. 6 (Tübingen: Mohr Siebeck, 2003), 3–22.

13 Ibid., 6.

to the power holding together its distinctive appearances and the bifurcated concepts correspond to the appearances of their respective underlying powers.[14] Like the concepts, powers also "hover." Each power is an essential unity holding together its appearances. Yet each power also functions as the appearance of a higher power when viewed from the perspective of that higher unity.

The *Dialektik* discusses both the ontological conditions that must be presupposed if correspondence between thought and reality is to be obtained within the limits of finite thinking, and the epistemological features of "pure thought"[15] producing knowledge. What the *Dialektik* does not thematize, but presupposes is the way in which the knowledge of reality is at all possible given a particular relation between the ideal and the real. It is in his *Philosophische Ethik* that Schleiermacher undertakes the task of showing how the relation between reason and nature in reality structures the possibility of knowing the world of physical and socio-cultural organisms. The dialectical correspondence between thought and reality is already given in reality by the discrete ways in which reason has imbued nature with rationality. According to Schleiermacher, ethics is the activity of reason in the ways that it organizes nature, and physics is nature that exhibits the rational structures resulting from ethic's activity.[16] Reason structures nature in such a way as to render reason transparent in nature to the possibility of knowing.

Ethics is further determined by two fundamental directions describing the way reason acts upon nature. In one activity, its "organizing" activity, reason acts in order that nature in general, and human nature in particular, becomes reason's "organ" (*Organ*) or tool.[17] Nature is organized in such a way as to give definite physical or socio-cultural shape to reason's activity. By virtue of reason's activity, these shapes, whether they are physical organisms or socio-cultural institutions, such as the political sphere, can be known by humans. In the second of its activities, the "symbolizing" activ-

---

14 DialKGA II, 528 (*Kolleg* 1822; 40th lecture).

15 Schleiermacher distinguishes between three types of thought: thought that aims at activity (*geschäftliches Denken*), artistic thought (*künstlerisches Denken*), and pure thought (*reines Denken*). Only the latter aims at knowledge. See the 1833 Introduction to the *Dialektik* in DialKGA I, 393 (*Einleitung* 1833, § 1,1).

16 See FRIEDRICH SCHLEIERMACHER, *Ethik (1812 / 13)*, on the basis of Otto Braun's ed., ed. and intro. Hans-Joachim Birkner, Philosophische Bibliothek, no. 335 (Hamburg: Felix Meiner, 1981), 8.

17 BIRKNER, *Schleiermachers christliche Sittenlehre*, 38–41.

ity, reason represents itself in a system of signs and symbols.[18] Through this representational activity, nature can be recognized and known by reason. Reason's twofold activity transforms nature into an entire system exhibiting reason's structuring properties and thereby facilitating knowledge of nature in the first place.

Another area of the *Dialektik* that is complemented by Schleiermacher's *Philosophische Ethik* is his description of socio-cultural organization in nature. Where the *Dialektik* concentrates primarily on physical entities, the *Ethik* focuses on the more complex ethical unities qualifying human life.[19] For Schleiermacher, the highest socio-cultural entities given in his whole system of powers are identified by the distinction between reason's organization and symbolization activity that itself is divided through the two principles of individuality and identity.[20] Individuality and identity are the two features of reason as it qualifies, on the one hand, the uniqueness of an entity that is irreplaceable by another, and on the other hand, the features of that unique individual that are also common to all members of its class. From this double division, four distinct areas result (that are nevertheless not entirely separate from each other). These are Schleiermacher's four socio-cultural shapes that are the four appearances of reason's power in human history.[21] The site of identical organization is allocated to the state, or the arena of public transaction characterized by exchange of property. Individual organization, such as *freie Geselligkeit* and friendship, is characterized by the organization of individual property as well as its exchange in more intimate spheres of human relations. The academy, the arena of knowledge, occupies the site of identical symbolization because it both symbolizes inner thought in external speech and stipulates requirements for the sameness of thinking in all.[22] Finally the site of individual

18 Ibid., 41.

19 Schleiermacher addresses the relation between nature and human history in DialKGA II, 392 (*Kolleg* 1818 / 19).

20 SCHLEIERMACHER, *Ethik (1812 / 13)*, 23–35. See also BIRKNER's discussion in *Schleiermachers christliche Sittenlehre*, 38–41.

21 Schleiermacher also refers to three of these four shapes (religion, academy, state) in his CG § 3,1 (I, 15–16). Herms focuses on the relation of ethics to knowledge in view of the relation between immediate and sensible self-consciousness in EILERT HERMS, "Philosophie und Theologie im Horizont des reflektierten Selbstbewußtseins," in *Schleiermachers Dialektik: Die Liebe zum Wissen in Philosophie und Theologie*, ed. Christine Helmer, Christiane Kranich, and Birgit Rehme-Iffert, Religion in Philosophy and Theology, no. 6 (Tübingen: Mohr Siebeck, 2003), 23–52.

22 For Schleiermacher, there are two requirements for knowledge: correspondence between thought and reality, and an intersubjective identity concerning the pathway of

symbolization is occupied by religion and art, two spheres of human activity that represent the immediacy of self-consciousness in symbols. By the mutual interplay between reason and nature as well as the division of this interplay into individuality and identity, Schleiermacher explains how the plural ways of reason's activity in nature can be known.

The site at which knowledge of reason in nature is accessible in nature is the human individual. For Schleiermacher, as for the Absolute Idealists, the human individual represents the location in creation at which reason and nature coincide; for Schleiermacher, in contrast to the Absolute Idealists, however, this place of co-incidence is not represented in reason but in immediate self-consciousness. In immediate self-consciousness, or feeling, the individual has access to the transcendent / transcendental identity of reason and nature. This identity is felt, rather than known, because in Schleiermacher's conception, the immediate ground of the self cannot be mediated through and exhausted by reason. The specific content of this feeling is a relative identity; a relative identity between the real and the ideal is felt as the ground of individuality that itself is ontologically connected to the absolute identity transcending it. What is intriguing about this insight, first articulated in the 1820 edition of *The Christian Faith* and later explicated in the 1822 lectures on *Dialektik*,[23] is that this connection between absolute and relative identity is felt precisely as the lack of being. As Christiane Kranich has shown in a detailed essay on the subject of "Selbstbewusstein—Nähe zu und Mangel an Sein," the identity constituting the individual is a relative, not an absolute identity.[24] Its relative identity is given precisely in immediate self-consciousness that signals not

---

arriving at knowledge. The latter condition eliminates fantasy or fluke from knowledge. See DialKGA I, 408–12 (*Einleitung* 1833, § 3) as well as the essay by Rehme-Iffert in which she describes in detail these two requirements in view of the idea of knowledge: BIRGIT REHME-IFFERT, "Wahrheit und Wissen in der *Dialektik* Schleiermachers," in *Schleiermachers Dialektik: Die Liebe zum Wissen in Philosophie und Theologie*, ed. Christine Helmer, Christiane Kranich, and Birgit Rehme-Iffert, Religion in Philosophy and Theology, no. 6 (Tübingen: Mohr Siebeck, 2003), 294–308.

23 Although it is Odebrecht's own prose compilation of Schleiermacher's lecture notes and student transcripts, his edition of the *Dialektik* is of scholarly value for precisely the following reason: the 1822 *Dialektik* places the discussion of immediate self-consciousness, the pillar of CG 1820 / 21, at the end of the first or transcendental part as the answer to the question concerning access to the absolute identity.

24 CHRISTIANE KRANICH, "Selbstbewusstsein—Nähe zu und Mangel an Sein," in *Schleiermachers Dialektik: Die Liebe zum Wissen in Philosophie und Theologie*, ed. Christine Helmer, Christiane Kranich, and Birgit Rehme-Iffert, Religion in Philosophy and Theology, no. 6 (Tübingen: Mohr Siebeck, 2003), 275–93.

the possession, but the "lack of being" (*Mangel an Sein*).[25] The individual's constitution as a relative identity of reason and nature is itself characterized by a feeling that the ground of its own existence fundamentally eludes the self's grasp.

It is precisely in the individual's lack of being that the condition for novelty resides. Novelty is introduced into being from the individual's constitution as an ontological lack. This occurs at two dimensions of the individual's existence. Firstly, according to Schleiermacher's general theory of a historical series—that can also be applied to individual biography—novelty first enters into the series by an absolute positing (*absolute Setzung*) of an entity at an original point. Although the terminology of absolute positing is Kant's and not Schleiermacher's, it helps to describe what Schleiermacher means by the sheer fact of an origin to a historical series.[26] It is furthermore Kant's sense of absolute positing as the existence of an entity, its *Dasein*, or its actuality—Kant uses these three terms interchangeably—that Schleiermacher probably has in mind. According to Kant, existence is not a real predicate but the sheer ascription of actuality to an entity.[27] For Kant, existence does not fall under the category of quality that determines the reality of an object as its "whatness." Rather, existence falls under the category of modality, which means that the existence of an entity is acknowledged by perceiving it.[28] If Schleiermacher's account of the positing of the individual as a lack of being can be read from the perspective of Kant's terminology, then it seems that the absolute positing of the human individual

25 The term used by Frank, "Mangel an Sein" (MANFRED FRANK, *Der unendliche Mangel an Sein: Schellings Hegelkritik und die Anfänge der Marxschen Dialektik*, 2nd expanded and rev. ed. [Munich: Fink, 1992]), is the philosophical specification that is given a religious determination in CG § 4,3 (I, 28) as "ein Bewußtsein schlechthinniger Abhängigkeit."

26 The distinction between absolute and relative positing is given in the early Kant's demonstration for the existence of God. IMMANUEL KANT, *Der einzig mögliche Beweisgrund zu einer Demonstration des Daseins Gottes* (1763), in *Kant's gesammelte Schriften*, ed. Königlich Preußischen Akademie der Wissenschaften, vol. 2 (Berlin: G. Reimer, 1900ff.), 70–81.

27 IMMANUEL KANT, *Kritik der reinen Vernunft*, newly ed. Raymund Schmidt, Philosophische Bibliothek, no. 37a (Hamburg: Felix Meiner, 1956), A598–9 / B626–7. For a discussion of these warrants in Kant, see MANFRED FRANK, *"Unendliche Annäherung": Die Anfänge der philosophischen Frühromantik*, 6th ed., Suhrkamp Taschenbuch Wissenschaft, no. 1328 (Frankfurt a. M.: Suhrkamp, 1998), 83–4; also more recently: IDEM, *Selbstgefühl: Eine historisch-systematische Erkundung*, Suhrkamp Taschenbuch Wissenschaft, no. 1611 (Frankfurt a. M.: Suhrkamp, 2002), 44–51.

28 KANT, *Kritik der reinen Vernunft*, A598–9 / B626–7.

is given with its original "creation" as an inseparable unity of spirit (=reason). This existence of this unity is felt—not perceived in Kant's narrow sense of sense perception—as the relative identity of reason and nature that is ontologically grounded in an absolute identity external to it. In immediate self-consciousness, the individual's relative identity is felt as an act of absolute positing. As such, its individuality is unique, and not reproduced by another absolute positing of any other individual. Novelty, not identity, is the result of absolute positing.

Secondly, novelty is the contribution to being that is a function of the individual's development through time. An absolute positing instantiates an individual existence that is then born by the unique history of its individual unfolding. In Kant's terminology, an entity's relative positing (*relative Setzung*) is a function of the predicates characterizing that entity's "reality" or whatness.[29] For Kant, it is relative, not absolute positing, that determines the predicates qualifying individual uniqueness. If Kant's term can be used to interpret Schleiermacher's understanding of the human individual, then, for Schleiermacher, existence that is posited as a sheer fact is unfolded through a historical-biographical sequence as the relative positing of individual novelty. Predicates of individuality are nurtured and acquired through interaction, in relative freedom and dependence, with one's environment.[30] Contours, textures, and colors to the self are fleshed out in action upon and influence from the world. To borrow a metaphor from Schleiermacher's Christology, the foundational "act of union" is unfolded through the biography of the "state of union" through influences from and actions upon the external world.[31] In the conception of the *Dialektik*, it is the organic pole, not the intellectual pole, that accounts for plurality and change.[32] Through the organic pole, the external world engages the individual in the mutual reciprocity of sense perception and categoriza-

---

29 Reality (*realitas* in Latin; *Sachheit* in German, according to Kant) or whatness is located in the table of categories under quality, indicating the entity's way of being. See KANT, *Kritik der reinen Vernunft*, A143 / B182 and A597–602 / B625–30.

30 For a detailed description of individuality in relation to relative freedom and dependence, see KRISTA DUTTENHAVER's essay, "Relative Freedoms: The Influence of Spinoza on the Systems of Whitehead and Schleiermacher," section II (on Schleiermacher and Spinoza) in this volume.

31 At this point, I am interpreting Schleiermacher's philosophical anthropology from the perspective given in his Christology in the CG; the act and state of union are christological terms Schleiermacher borrows from Protestant orthodoxy to indicate the absolute and relative positing of Jesus's person. See CG § 97 (II, 58–76).

32 DialKGA II, 460–1 (*Kolleg* 1822; 21st lecture).

tion, thereby introducing the novelty of individual sense perception into the schematization of reality. Whereas absolute positing preserves the continuity of the I through time, relative positing accounts for the changes of the individual who lives in the world of temporal flow.

Within the parameters established by absolute positing, the individual's relative positing takes place through commerce with the environment. As a work in progress, an individual's self-unfolding within the realm of relative freedom and relative dependence introduces novelty into the series of her own existence. The fundamental openness to the future of the trajectory invites novel ways of negotiating unique combinations of freedom and dependence. Given different situations, an individual will determine herself in different ways. Yet historical contingency does not preclude individual predictability by virtue of absolute positing. In the *Life of Jesus*, Schleiermacher identifies his intention to determine an original "calculus" to Jesus's individuality that would predict Jesus's individual stamp on any situation.[33] The absolute positing of Jesus's individuality is the goal of hermeneutical divination that aims to predict all cases of individual relative positing in the historical events of his life.

Novelty is introduced into the world as a function of the individual in an absolute anthropological sense with the creation of the individual per se and in a relative sense with the individual's reciprocal and temporally unfolding relations to its environment. For Schleiermacher, individual novelty is never absolute in the cosmological sense. Apart from the absolute positing of the world in the cosmological sense—which is not to be confused with a *creatio ex nihilo* that Schleiermacher rejects both in the *Dialektik* and in *The Christian Faith*[34]—all historical series, whether as distinct religions or as human individuals, posit the new in relation to the old. The ethical principle of "connecting to the preceding" (*Anknüpfung*) characterizes all world-immanent historical novelty.[35] The relation of the new to the preceding moments of a series is given for all historical processes, including the history of a religion that has an identity of its inner characteristic features through time. In the psychological terms of the human individual, immediate and sensible self-consciousness are the two dimensions of the soul that establish the identity of the individual with all human individuals; an individual instantiates the relation between immediate self-

33 LJ 9–10 (2nd lecture).
34 DialKGA II, 551–2 (*Kolleg* 1822; 46th lecture); CG § 41 (I, 198–204).
35 BO § 203 (103). (Page numbers given in parentheses.)

consciousness and her particular sensible self-consciousness in view of both the sheer existence of that individual and her development through time. In this way, an absolute positing of an individual distinguishes that individual from all others by actualizing some of the features already qualifying that series in the first place; a human individual is, by the necessity of human nature, absolutely posited as a unique mixture of immediate and sensible self-consciousness. The new is always related to the old as its modification. It is this identity amid diversity, the absolute in the relative positing, that embeds the individual in the system of the ideal and the real, the ethical and the physical. By identifying systematic structures of identity, knowledge of individual novelty is made possible. It is the more precise epistemological question of knowledge of the individual that I address in the following section.

II. Individual and Systematicity

The aim of scientific inquiry is, according to Schleiermacher, knowledge of the essence of an individual. By definition, by conceptual analysis, and by empirical-historical study, the predicates of an individual are determined in relation to each other as well as to the whole, thereby yielding the "complete concept" of the individual. It is nevertheless knowledge of an individual qua individual that presents the epistemological problem. For to know an individual qua individual, it is necessary to know the individual as it is determined by all its marks, and to know these marks as they are themselves exhaustively determined.[36] At least according to this Leibnizian understanding, knowledge of an individual is integrally related to the knowledge of the totality of reality, or in other words, to the world as an analytic judgment. It is this knowledge of the whole that conceptually situates the systematic issue raised by knowledge of the individual. In this section I show how, for Schleiermacher, the method for reaching knowledge of the individual is characterized by both a dialectical and a hermeneutical pro-

36 On Leibniz's determination of the complete concept see both HANS-PETER GROSSHANS, "Denken und Wirklichkeit: Zu den ontologischen Bedingungen von Begriff und Urteil in theologischer Perspektive," in *Schleiermachers Dialektik: Die Liebe zum Wissen in Philosophie und Theologie*, ed. Christine Helmer, Christiane Kranich, and Birgit Rehme-Iffert, Religion in Philosophy and Theology, no. 6 (Tübingen: Mohr Siebeck, 2003), 162–78; and JACQUELINE MARIÑA's article, "Schleiermacher Between Kant and Leibniz: Predication and Ontology," in this volume.

cedure. Both methods aim to tease out individual novelty as it is situated within a systematic context, respectively an epistemological and a grammatical context. Knowledge of individual novelty presupposes structures of systematicity.

The first question posed is the epistemological one. How is knowledge of the individual possible? It was Kant who problematized this question by his transcendental analytic that raised the question concerning how categories could be concepts of objects of possible experience. Kant's intention was to show how universally valid and necessary judgments could be made concerning objects of possible experience. The categories functioned to determine an object only as an object as such, not as a concrete irreplaceable and completely determined individual entity, as Manfred Frank has argued.[37] By formally determining the possible object through the categories,[38] the problem of knowledge concerning materially contingent attributes, obtained by empirical study, arises.[39] The problematic distinction between categorical knowledge and a determination of concrete individuality remained the task of Kant's posterity to solve.

As it is his intention in his theology to determine the essence of the Christian religion, Schleiermacher is confronted with the question of knowing an individual essence. It is primarily the methodological question regarding knowledge of Christianity's individual essence that is the subject matter of the Introduction (§§ 1–31) to *The Christian Faith*. In CF § 2,2, Schleiermacher asks if knowledge of the individual can be attained either categorically or empirically. Categorically, he judges that "no science can by means of mere ideas reach and elicit what is individual, but must always stop short with what is general."[40] Conversely, an empirical study of the object would yield only an average concerning all observable entities of the same species, but would not result in knowledge that requires more. Schleiermacher writes, "[a]nd the purely empirical method, on the other hand, has neither standard nor formula for distinguishing the essen-

37 FRANK has pointed this out in detail in his *Unendliche Annäherung*, 52.

38 For Kant, possibility does not mean an entity that does not exist, such as a unicorn. Rather, possibility is always the real possibility that an entity might exist. On this meaning of possibility, see FRANK, *Selbstgefühl*, 46–7.

39 Manfred Frank has shown that Kant appropriated the distinction made by the Wolffian school between the *natura formaliter* that determined an object's formal characteristics through the categories and the *natura materialiter considerata* (or *spectata*) that determined an object through its materially contingent characteristics. FRANK, *Unendliche Annäherung*, 52.

40 CF § 2,2 (3–4).

tial and permanent from the changeable and contingent."[41] Like Leibniz, however, Schleiermacher does accept individual concepts, which are more than just the materially underdetermined indexical references of pointing because each has a minimum of one predicate.[42] Hence the Introduction to *The Christian Faith* is constructed as a set of epistemological steps determining the conceptual location for the individual essence of Christianity.[43] First the general sphere of ethics, as the activities of human *Geist* providing the rules for historical agency, claims necessity for the appearances of religious communities as the manifestations of a common essence of piety (§§ 3–6). Next, the sphere of philosophy of religion classifies the various manifestations of the essence of religion as the positive religions according to kind and type (§§ 7–10). Finally, the field of apologetics minimally fixes the concept of the essence of Christianity as the "redemption fully accomplished by Jesus of Nazareth" (§ 11).[44] On the conceptual basis achieved by the Introduction, the dogmatic-theological part, or main body of *The Christian Faith*, fills in the historical predicates of the Christian religion as they reflect the situation of the contemporary Christian church.[45] If the *Brief Outline* is borrowed at this point to complement *The Christian Faith*, then exegetical theology and church history fill in the remaining historical predicates of Christianity's essence, respectively the predicates constituting the origin and the historical development of the church.[46] Schleiermacher's theological method aims to know the individual by an oscillation between conceptual and empirical determination.

It is in the *Dialektik* that Schleiermacher works out the method of scientific investigation as one of oscillation between conceptual and empirical study. The method is an interweaving between concept and judgment formation, which are the two fundamental types of thinking. A concept is

---

41 CF § 2,2 (4).

42 Schleiermacher's example of indexical reference is, "Something shines" ("Etwas glänzt," DialKGA II, 504 [*Kolleg* 1822; 34th lecture]) where "to shine" is the one predicate required for a subject concept.

43 See footnote 1.

44 In CG, apologetics (§§ 11–14) belongs to the Introduction. In the *Brief Outline*, apologetics is ordered to philosophical theology as one of two constitutive elements (the other being polemics). See BO §§ 43–53 (26–32).

45 Schleiermacher's famous definition of dogmatic theology is: "the knowledge of doctrine that now has currency in the evangelical Church" (BO § 195 [97]); see also CG § 19, proposition (I, 119).

46 See BO § 26 (16); §§ 82–90 (47–52).

formed when a judgment predicates one concept of another concept.[47] An authentic judgment (*eigentliches Urteil*) predicates a concept gleaned from experience of the subject, while an inauthentic judgment (*uneigentliches Urteil*) merely predicates a concept of the subject that is already contained in the subject concept.[48] By taking up predicates from experience through judgment formation, Schleiermacher advocates the correspondence to and basis for all concepts in reality. Although errors can be made in schematizing those predicates, these errors do not preclude the wedge held in reality by experience.[49]

The correspondence between concept and judgment formation on the one hand, and reality on the other hand, gives both the epistemological rationale and the ontological basis for seeking to know an individual within the context of the whole. Schleiermacher's theory of knowing the individual presupposes an entire system of thought, on the one hand, that corresponds to the system of reality, on the other hand. In the *Dialektik*, Schleiermacher describes the system of thought as the system of hovering concepts. Corresponding to the concepts of essences (or in other words, the system of ideas) in thought is the system of powers in reality; and corresponding to the system of causality in thought is the system of appearances in reality.[50] According to the system of causality, each predicate is linked to all others by virtue of the causal relations between them. Furthermore, these predicates are predicated of categorically derived concepts that themselves correspond to the powers serving as the principle of unity for their respective appearances. From the epistemological perspective, predication results from the modeling of predicates derived from experience into the system of categorical knowledge; judgment formation predicates a concept of another concept as either an authentic or inauthentic judgment. On the basis of judgment formation, the subject concept is formed. The epistemological procedure of oscillation between empirical judgment formation and categorical concept formation corresponds to reality because it itself presupposes the ontological interconnection between the system of pow-

---

47 A judgment is composed of joining a concept (predicate) with another concept (subject). DialKGA II, 493–4 (*Kolleg* 1822; 31st lecture). Schleiermacher can define judgment in this way because he accepts individual concepts (whereas Kant does not). See footnote 42.

48 DialKGA II, 503 (*Kolleg* 1822; 33rd lecture).

49 See DialKGA II, 593–600 (*Kolleg* 1822; 57th–59th lecture), and ibid., 637–40 (*Kolleg* 1822; 69th lecture), in the formal part that discusses the possibility of error in detail.

50 DialKGA II, 541 (*Kolleg* 1822; 43rd lecture).

ers and the system of appearances. Thus, Schleiermacher argues, that he has united idealism with realism.[51] He has connected knowledge of the system with knowledge of the individual in such a way as to account for the predicates arising from the river of life within their conceptual embeddedness. Whether as inauthentic or as authentic judgments, experience is the source of knowledge.

Schleiermacher's procedure regulating the drive to knowledge of the individual represents a unique position in the history of philosophy. With his method of oscillation, Schleiermacher moves within the bounds stipulated by Kant's critical philosophy, yet he goes beyond Kant by addressing the empirical source of knowledge in a way Kant could not, given the two-source theory of knowledge and the transcendental idealist parameters of the latter's system. For Schleiermacher, the empirical determination of the concept by judgment formation is ordered by an intensional logic.[52] With a logic of concept containment, Schleiermacher, like Leibniz, can advocate individual concepts, which have a minimum of one predicate, and thereby their place within the system. Once one predicate fixes the subject concept, the concept is open to further, and infinite, determination through judgments. An individual's conceptual determination is opened by the flow of life running through the particular individual's existence. Even if the concept is fully saturated with all possible predicates, as is the case with the inauthentic judgment, life still offers an extra predicate: movement can still be predicated of the subject.[53]

On the basis of this method of knowing the individual, Schleiermacher spells out the systematic implications of Kant's critical philosophy in an open-ended way, thereby providing an alternative to the monological and totalizing systems of his Absolute Idealist colleagues. For Schleiermacher, the maximal determination of an individual concept presupposes the maximal determination of the entire system of reality. An individual concept reaches maximal determination only when the totality of reality has also been determined in all of its aspects. For finite knowers, maximal determination remains a regulative idea; all knowledge is a process of infinite approximation to the idea of the world (*die Idee der Welt*).[54] Only for God

---

51 DialKGA II, 521 (*Kolleg* 1822; 38th lecture).

52 Manfred Frank argues for Schleiermacher's intensional logic in "Identität, Korrespondenz und Urteil," 14–15.

53 DialKGA II, 503 (*Kolleg* 1822; 33rd lecture).

54 DialKGA II, 580–1 (*Kolleg* 1822; 54th lecture).

is the world completely determined in an analytic judgment.[55] Together with the non-reducibility of life to a dead concept, infinite approximation informs Schleiermacher's critical philosophical insight into the fundamental inexhaustibility of knowledge of the individual, while at the same time, presupposing system as the condition for the possibility of knowing at all.

The systematic presupposition of Schleiermacher's dialectical procedure secures an ontological guarantee that regulates the finite search for knowledge. Other than allocating to system its function as a regulative idea, Schleiermacher orients his own decisive interest elsewhere. His focus is on a procedure for thinking that aims towards systematic knowledge, yet he keeps this method open by a fundamental stance of conceptual fallibility.[56] The search for the knowledge of system is guided by the regulative idea of a *terminus ad quem*, the idea of the world,[57] yet its location in finite thinking raises systematic openness as a function of possible errors introduced by the categorization of sense perception as well as by the linguistic articulation of those concepts. It is Schleiermacher's doctrine of schematism, located in the second or formal part of the *Dialektik*, that serves as the key to his consensus theory of truth.[58] The important aspect to the schema is

---

55 Schleiermacher borrows the idea of the *conceptus complexus* from Leibniz for whom the world is an analytic judgment in God's eyes.

56 The argument of conceptual fallibility, however, is difficult to resolve with another argument Schleiermacher puts forth in the transcendental part of the *Dialektik* on the innate concepts (DialKGA II, 524 [*Kolleg* 1822; 39th lecture]). He uses the argument of innate concepts to show that predication is not arbitrary, and that it makes advances in truth and knowledge. How the relation between conceptual fallibility and the innate ideas might be harmonized is beyond the scope of this paper.

57 DialKGA II, 580 (*Kolleg* 1522; 54th lecture). In an article, "Systematic Theology: Beautifully True," I use Kant's distinction between a regulative and a constitutive idea to conceive Schleiermacher's system as an "open system." See CHRISTINE HELMER, "Systematic Theology: Beautifully True," in *Truth: Interdisciplinary Dialogues in a Pluralist Age*, ed. Christine Helmer and Kristin De Troyer, with Katie Goetz, Studies in Philosophical Theology, no. 22 (Leuven: Peeters, 2003), 27–46.

58 The schematism is the doctrine of moveable images (*verschiebbare Bilder*) that relates predicates to their corresponding concepts. The schematism also accounts for the error in categorizing reality due to organic misinformation or to the difficulty in relating sense perception to concept, and leaves open future predications that might result in the creation of a new concept or genus. DialKGA II, 626–32 (*Kolleg* 1822; 66th–67th lecture). The schematism is developed by KANT in the *Kritik der reinen Vernunft*, A137–47 / B176–87 as the explanation for relating the content of intuition to the corresponding concept. According to B180: "Diese Vorstellung nun von einem allgemeinen Verfahren der Einbildungskraft, einem Begriff sein Bild zu verschaffen, nenne ich das Schema zu diesem Begriff." In his *System des transzendentalen Idealismus (1800)* (Philosophische Bibliothek, no. 254 [Hamburg: Felix Meiner, 1957], 176–9), FRIEDRICH WILHELM

the connection they establish between thought and language.[59] For Schleiermacher, all predication is bound to thought and hence to language as its conscious expression.[60] Given Schleiermacher's position concerning the "irrationality"[61] of language, truth is not pre-established, open for discovery and linguistically captured by instrumental means. Rather, predication occurs by virtue of consensus regarding the linguistic articulation of the concepts' terms at all. In order for predication to take place in intersubjective discussion, it must presuppose hermeneutical consensus regarding terms and their translatability into other languages.

It is with the discursive inevitability of predication that the intersubjective dimension to thinking becomes significant. Because consensus yields the concepts, hermeneutics, not the dialectic, is the science that studies the fundamental communicability and understandability of terms from one discourse to another. In a key paragraph to his *Hermeneutics and Criticism*, Schleiermacher points out that hermeneutics, which is related to rhetoric as the linguistic dimension, is dependent on the dialectic.[62] As two technical-critical disciplines, the dialectic and the hermeneutics are synchronized to each other in order to produce knowledge of the individual by virtue of the individual's thought that is expressed in language. If knowledge is the product of intersubjective consensus, hermeneutics plays a role in understanding the communication of another individual.

Like knowledge of an individual that presupposes its embeddedness in the system of the totality of reality, knowledge of the linguistic articulation of a human individual presupposes its embeddedness within a grammar, language usage, and linguistic structure shared by users of the same

---

JOSEPH VON SCHELLING also explains the schematism. I am grateful to Manfred Frank for pointing out these sources as the backdrop to Schleiermacher's thought.

59 A detailed description of the doctrine of the schematism is beyond the scope of this paper.

60 For Schleiermacher, language and thought are related to each other on a continuum; thought attains its completion in speech (*Rede*): "This leads to the unity of speech and thought; language is the manner in which thought is real." *Hermeneutics*, 8.

61 By "irrationality," Schleiermacher means the impossibility of translating one term from one language into another without translating its entire semantic field together with the term. See BO § 126 (67–8).

62 *Hermeneutics*, 7–8. On the relation between the dialectic and the hermeneutics as they inform exegetical theology, see CHRISTINE HELMER, "Schleiermachers exegetische Theologie: Urteilsbildung und Korrespondenz in der neutestamentlichen Wissenschaft," in *Schleiermachers Dialektik: Die Liebe zum Wissen in Philosophie und Theologie*, ed. Christine Helmer, Christiane Kranich, and Birgit Rehme-Iffert, Religion in Philosophy and Theology, no. 6 (Tübingen: Mohr Siebeck, 2003), 55–77.

language. In Schleiermacher's hermeneutics, grammatical interpretation is absolutely essential because it gives the possibility for understanding the meaning of an author's speech (or text) in the first place.[63] By grammatical analysis, a common linguistic bed is worked out that is shared between an author and her literary environment. On the basis of linguistic commonality, the novelty of an individual expression or work can be determined. The new is grasped in view of what is shared by other speakers of the same language.[64] Not only by psychological interpretation—which includes divination and its twin, comparison—is knowledge of the individual approximated, but together with grammatical interpretation can an authorial intention be grasped from its linguistic articulation.[65] By using hermeneutical tools designed to tease out individuality from identity, the thought of an individual is accessed.

In this section, I have argued that Schleiermacher solves, at least in one way, the problem left by Kant of knowing the individual. Knowledge of the individual presupposes system, both the system of thought that corresponds with the system of the totality of reality, and the system of language that accounts for the possibility of intersubjective communication and understanding. Without these features of system, the possibility of acknowledging the novelty introduced by the individual into the system could not even be communicated to others. By looking at the relationship between the individual and system, I have also pointed out that system can be constructed in such a way as to be truly open to novelty, while providing the stability required for novelty to be what it is.

III. Individual Novelty and System in Christian Systematic Theology

The final question addressed in this paper concerns the relation of individual novelty to the system of Christian systematic theology. In this section, I discuss the question of systematicity in view of the individual perspective focusing any systematic theology, and then show how, in spite of novelty, there are specific systematic features that remain relatively stable across var-

63 *Hermeneutics*, 8–9, 30–1.

64 *Hermeneutics*, 32.

65 The important division of Schleiermacher's *Hermeneutics and Criticism* into two parts, the grammatical and the psychological parts, gives the interpretative clue that both must be read together as contributing to the same goal.

ious systematic theologies. By describing how individuality and system fund the building of a systematic theology, I also argue for the possibility of transdiscursive features of system that make the communicability of theological insights beyond one's individuality or group identity possible.

In § 196 of his *Brief Outline*, Schleiermacher describes a key element for a dogmatic—and in his own case he means Christian, theology. The proposition reads, "A dogmatic treatment of doctrine is not possible without personal conviction."[66] The theologian who articulates a systematic theology is not to be divorced from her work. Her discursive expressions are stamped by the idiosyncrasy of her individuality, and as such, her individuality is determined by that which constitutes it at its inmost core of integrity. For Schleiermacher, that core is the absolute positing of the individual in an original relation between immediate and sensible self-consciousness, and a relative positing in the individual's biography. When this psychologically determined anthropology is viewed from a theological perspective, the positing is conceived according to the two pillars of Christian experience: sin and redemption. Because sensible self-consciousness is itself a part of the matrix of sinful corporate existence, relative positing introduces sin into individual existence.[67] A sinful structure of consciousness is counteracted by the transformative impact of Jesus's perfect God-consciousness.[68] The result of an encounter with Christ is that the intentional side of self-consciousness, sensible self-consciousness, becomes more and more saturated with the feeling of absolute dependence. The theologian's personal and immediate conviction of Christianity's truth as faith in Jesus Christ[69] is expressed in her thinking and doing. By virtue of the unity of consciousness, individual faith places its mark on engagement in and with the world. For Schleiermacher then, the personal conviction of the Christian theologian is the conviction that overlays the individual's relative positing with the redemptive potential of a transformed relation between the two sides of self-consciousness.

By reading Schleiermacher's theological anthropology from a soteriological perspective, the coherence of a system of theology can be explained. As a result of rebirth, the positing of an individual is re-oriented to the redeemed relation between immediate and sensible self-consciousness. Al-

66 BO § 196 (99).
67 CG § 68, proposition (I, 360–1).
68 CG § 100, proposition (II, 90).
69 CG § 14,1 (I, 94–6), and § 14, postscript (I, 98–105).

though this redeemed state is imperfect in its temporal extension—sin continues to struggle with redemption—, its original positing begins to be expressed in the activities of sensible self-consciousness. Although temporal expression might bear the conflict between immediate and sensible self-consciousness, the intentionality of expression, in general, exhibits the stamp of an immediate self-consciousness that has been transformed by an encounter with Christ.[70] In his *Hermeneutics and Criticism*, Schleiermacher makes this point in view of the New Testament authors. The works of these authors reflect an underlying unity that is a function of Christ's redemptive effect.[71] Coherence pervades—or should pervade in the ideal case—intentional expressions of sensible consciousness by virtue of redemptive positing in self-consciousness. The coherence of a consciousness constituted by a sensible consciousness that is permeated with immediate self-consciousness affects the whole, the expressions of thinking and doing.

As a function of self-consciousness, coherence is also related to the way in which the individual engages reality. In the *Brief Outline*, Schleiermacher refers to the individual grasp of the "dominant principle of the period" that,[72] together with "personal conviction," is used to structure the grasp of doctrine as a whole. The signs of the times are perceived through the individual's personal conviction of her redemption. As such, coherence includes the dimension of self-consciousness together with the intentional dimension of the whole of reality. A successful articulation of doctrine requires a comprehension of the contemporary world; as currently valid in the church, doctrines must adequately express the currency of their content in living religion. Furthermore, as informed by an inner self-consciousness and an intentional dimension, individual principles of coherence explain the prevalence of plural expressions of doctrine.[73] The beauty of systematic theology is its individual novelty that does not preclude other novel expressions.

If authorial individuality is the principle of coherence funding a theological system, then it must be related to structures of identity if it is to be communicated to others. Communicability presupposes identity; if a theologian selects a language to intentionally express his novel principle of coherence that is not part of shared discourse, then he has isolated himself

---

70 I explain the relation between redemption and the soul's coherence in HELMER, "Systematic Theology: Beautifully True," 37–9.

71 *Hermeneutics*, 54.

72 BO § 200 (101).

73 BO § 200 (101–2).

from the intersubjective pursuit of knowledge. This was Schleiermacher's contention in the 1833 Introduction to the *Dialektik* against the monological model of implied colleagues, Fichte and Hegel.[74] For monological novel-ists, Schleiermacher contends that scientific language and questions emerge from the common language and the reality of experience.[75] The living connection with ordinary language and with questions arising from lived reality does not entail a reduction in the complexity of scientific answers, but is established for the hermeneutical purpose of communicability. For Schleiermacher, the crucial first step in pursuing knowledge is the communicability of thought in language because it is only through communication that intersubjective progress in science can be made at all. From this starting-point then, the transdiscursive features of knowledge can be established as the condition for the possibility of knowledge. Furthermore, the rules governing the intersubjective pursuit of knowledge and truth can be derived from precisely those transcendent conditions.

Schleiermacher's dialectical-dialogical procedure of identical symbolization has implications for the transdiscursive possibilities of theological systems. The question concerns how the principle of coherence is enacted on an intentional subject matter in a way that has been relatively stable in Western Christian theology since Anselm of Canterbury first constructed a system of theological thought in the eleventh century. The relative stability has firstly to do with Christianity as a religious worldview. As such, it schematizes an understanding of self, world, and God through a christological lens as its soteriological and in some cases (i. e., Karl Barth) epistemological principle. Secondly, Western theology has shown remarkable consensus in the stability of a meta-narrative extension of doctrines from God to eschatology. The meta-narrative tends to begin with the metaphysical attributes of God and the doctrine of the immanent Trinity, and then moves to the historical, economic relation between God and humans, from creation to redemption to sanctification. Only in recent times with the advances of the historical-critical method has this relatively stable structure been re-oriented to an empirical Christology "from below," that nevertheless presupposes some speculative knowledge of a Trinity "from above."[76]

74 DialKGA I, 412–22 (*Einleitung* 1833, § 4).

75 This is especially significant in the 1833 Introduction to the DialKGA. See ibid.

76 For an example of a narratively derived Trinity that presupposes some speculative content, see ROBERT JENSON, *Systematic Theology*, vol. 1, *The Triune God* (Oxford: Oxford University Press, 1997), ch. 4–5.

Although on the surface it might appear that doctrinal concepts are fixed for all time, this is, on Schleiermacher's grounds, not the hard and fast rule. Rather, for Schleiermacher, concepts are necessary as they minimally fix the parameters of the subject matter for subsequent material determination by experience. Concepts provide the basis for their subsequent predication, yet their relative stability throughout the history of doctrine is not due to some property inherent in the concepts themselves. Rather, they are the results of consensus regarding the schematizing of the Christian worldview. As the products of a transhistorical consensus, they are open to modification by those whose Christian schematizing of reality might add new experiential dimensions and additions to the concepts. For example, the co-authors of the recently published book, *A Native American Theology*, suggest the categories of land and trickster as additions to the conceptual theological repertoire.[77] It of course remains to be seen if these categories truly add a new dimension that is not accounted for by an already established category, and if transcultural consensus will vouch for the validity of these additions.

The discussion of theological categories takes place at the intersection between the individual and system. The relation between orthodoxy and heterodoxy in Schleiermacher's sense of those terms concerns the novelty arising from a shared cultural context. For Schleiermacher, heterodoxy exhibits the novelty required for any theological proposal that truly is concerned with its contemporary location, while orthodoxy represents the ethical dimension of the way in which the new is communicable by virtue of its attachment (*Anknüpfung*) to the old.[78] Any development of human spirit is only novel to the extent that it can be communicated and can be exhibited to be in continuity with the parameters established by the preceding sequence. The question concerning the viability of the categories for future systematic theology rests with heterodox novelty on the one hand, and the degree to which novelty can be related to the whole on the other hand.

For the genre of systematic theology to have a viable future, it must be healed of its current state of fragmentation into regionally located theologies that have, until now, adopted few tools for communicability and un-

---

77 Clara Sue Kidwell, Homer Noley, and George "Tink" Tinker, *A Native American Theology* (Maryknoll: Orbis, 2001), ch. 8–9. See my review of this book: Christine Helmer, "Review of *A Native American Theology*," *Pro Ecclesia* 12, no. 2 (Spring 2003): 240–2.

78 BO § 203 (103).

derstandability across their respective boundaries. What I have proposed in this paper is a transdiscursive theology that establishes identity for the sake of the communicability and understandability of individual systems of theology. This identity must be minimally fixed, such as the theological categories, in order to maximize the plural potential of fixing by individuals. And by this maximization, novelty would provide the excitement of any theology as a living system. Theology would not be a stagnant backward looking enterprise, but a vehicle for change, inviting the participation of the heterodox and allowing its proposals to be scrutinized by the wider community that must establish valid rules for all academic pursuit of knowledge and truth. In all their perspectival manifestations, individual systems would be related to common transdiscursive features and through these relations, they could be on the one hand, deployed to promote a religious community's health and on the other hand, used to communicate theology's unique subject matter to other disciplines of the academy. Only in this way can the contributions of the parts be heard for the liberation of the whole. Transdiscursive theology does not mean an insular meta-level of discourse divorced from reality and coolly located above the fray. Rather, it aims to know the individual in the context of embeddedness within identity so that communication beyond its sphere might be possible, and by being understood, thereby introduce change into the system.

IV. Conclusion

In this paper I have looked at two features, the individual and system, in an effort to conceive of a transdiscursive theology as a possibility of overcoming the communicative and methodological impasses associated with the fragmentation of theology into various theologies. For the longevity of systematic theology into the twenty-first century, a new concept of system is required that looks at experience at the level of the individual in order to understand the privileging of the individual as the site of theological reflection and articulation. Yet the individual is metaphysically and epistemologically related to elements of identity and system that allow for the possibility of transdiscursive communication and understanding. I attempted in this essay to point out specific formal structures of system in such a way as to open up system to take into account the novelty represented by each individual participant in the discussion. Although these structures of identity are the product of consensus and are still open for their falsification, they remain an inevitable requirement for the possibility of discussing a common topic at all. System itself must be open to change, yet it is necessary at some level if knowledge of the individual is to be obtained in the first place. The degree to which this happens is the degree to which systematic theology remains a viable genre for the writing of theology, and for service to the health of religion, the academy, and the world.

Open Interpretation

Whitehead and Schleiermacher on Hermeneutics

J. R. Hustwit, Claremont Graduate University

One of the happier (and potentially dangerous) traits of a fully systematic body of work, such as those given life by Schleiermacher and Whitehead, is the ease with which one can extend that system to topics originally left unexplored. In this respect, my aim in this essay is largely experimental. Although Whitehead never explicitly formulated a hermeneutical theory, his insights into language, perception, and symbol lend themselves to an interesting project: to lay the foundations of a Whiteheadian hermeneutic. If Whitehead's system were extended to bear upon the field of hermeneutics, which I believe is a very natural application, what would the resulting hermeneutical theory look like? How would this hermeneutic vary from Schleiermacher's theory?

Hermeneutics originally materialized when scholars enlisted philology, grammar, and psychology in service of the interpretation of texts. This new discipline, at its inception, was applied solely to jurisprudence, classical literature, and biblical texts, as these were matters in which the meanings intended by the authors were most crucial and unavailable. This view of the text—as an ambiguous and sometimes opaque cipher for an author's intention—is the foundational presupposition of Romantic hermeneutics. Friedrich Schleiermacher, whose name has become synonymous with Romantic hermeneutics, was responsible for the first truly systematic account of this new science.

This paper is divided into two parts. First, I seek to weave Whitehead's remarks on language and symbol into a coherent hermeneutical position. I will first discuss the structure of an actual occasion, as it is the location of all subjective understanding. Then, I will argue that Whitehead's system makes interpretation a pervasive aspect of the universe—a metaphysical, and consequently universal relation that exists between and within all occasions of experience. To conclude the first part, I will employ Whiteheadian concepts to provide a metaphysical description of the process of interpretation.

In the second part of the paper, I will discuss points of contact, both agreement and disagreement, between a Whiteheadian and a Schleiermacherian hermeneutic. I will focus on the issue of authorial intention. Although Whitehead would agree with Schleiermacher that the authorial intention is a crucial determinant of meaning, Whiteheadian thought also demands reference to the reader's own deployment of concepts. The resulting account locates meaning in the vacillation and interplay between the authorial occasion and the reader occasion.

I. The Structure of a Subject

In order to make any headway in understanding language and meaning in a Whiteheadian context, it is necessary to explain Whitehead's own doctrine of the actual occasion and some technical terms associated with this doctrine. Whitehead rejects a substance metaphysic. He argues that the "final real things of which the world is made up" are "actual occasions or actual entities."[1] The term "actual occasions" reflects a break from orthodox metaphysical thought. According to Whitehead, the world is not made up of substances that endure through time. Instead, an individual enduring thing is not a "substance" in the formal sense of a fully actual or finally real thing, but a serially or temporally ordered society of actual occasions, understood to be experiential events, and hence called "occasions of experience." Atoms, molecules, and human souls are all, by virtue of their individuality, strings of temporally ordered and causally related actual occasions. Any account of how human thought and understanding takes place must be couched in terms of the internal workings of and relations between the serially ordered actual occasions that constitute a person's mind.

The process by which an actual occasion moves from indeterminacy and potentiality to actuality is the process of *concrescence*. This process of determination occurs by means of *prehensions*, which are an individual occasion's feelings of data. A prehension has three aspects: its datum or content, its subjective form, and the occasion that is prehending it. For example, if

1 PR 18.

a particular fact is felt, a complete description of that feeling will include its content (the datum), its subjective form (the emotive weight associated with the datum), and its prehending subject (the actual occasion feeling the datum with that subjective form). Whitehead explains that

> in a process of concrescence, there is a succession of phases in which new prehensions arise by integration of prehensions in antecedent phases. In these integrations 'feelings' contribute their 'subjective forms' and their 'data' to the formation of novel integral prehensions .... [T]he process continues till all prehensions are components in the one determinate integral satisfaction.[2]

An actual occasion, which in the case of our hermeneutical inquiry is a moment of human experience, develops by means of several stages of prehensions, or feelings (these terms will be used interchangeably).[3] The occasion begins by feeling the entirety of the past world and, based on this initial prehension, then selects and integrates aspects of those initial data in a process of continual refinement until it reaches a determinate satisfaction. Whitehead divides this process of concrescence into four stages, each of which contains a unique type of prehension. These four types of prehension will be crucial to a Whiteheadian description of human interpretation.

Every occasion of experience is dipolar. In the process of concrescence, an occasion is determined by both the external world and its own subjectivity. The initial determining of an occasion by the environment is known as the *physical pole*. The subsequent development of the occasion according to its own subjectivity is known as the *mental pole*.

The first phase in the process of concrescence makes up the physical pole, and is known as the *conformal phase*. In this phase, the subject feels all other actual entities.[4] The current actual occasion receives data from past actual occasions, which have already reached satisfaction and are positively prehended.[5] These physical prehensions, which make up the occasion's initial phase, causally determine the possibilities for the development of the subject. The physical feelings in the first phase are both the source of all potentiality for, and the limiting factor of, the concrescing subject.

---

2 PR 26.

3 Although I will use the terms interchangeably, there are two varieties of prehensions: negative and positive. Only the latter are feelings.

4 PR 23.

5 For a more detailed account of negative and positive prehension, see section II. 1 below.

After the initial data are prehended in the physical pole, the occasion begins to concresce in accordance with its *subjective aim*, or process of self-determination.[6] The remaining three phases of concrescence that are influenced by an occasion's subjective aim take place in the mental pole.

The second phase of concrescence is the conceptual phase. Whitehead explains that "[f]rom each physical feeling there is the derivation of a purely conceptual feeling, whose datum is the eternal object exemplified in the definiteness of the actual entity."[7] The concrescing occasion selects or picks out *eternal objects* (Whitehead's term for universals)[8] from the actual entities felt in the initial physical pole. This is a *conceptual feeling*. For example, suppose a book were physically prehended by a concrescing occasion of human experience.[9] That occasion, in the conceptual phase, might pick out the particular shade of blue found on the cover of the book, the rectangular shape, the ability to open and close, or its musty smell. Any of these would be a conceptual feeling.

The third phase of concrescence integrates the conceptual feelings from the previous phase with the physical feelings of the first phase. These sorts of prehensions are *propositional feelings*, the data of which are "propositions." Whitehead defines a proposition as "[a]n eternal object realized in respect to its pure potentiality as related to determinate logical subjects."[10] Propositional feelings combine an eternal object (the proposition's predicate) with a past actuality (the logical subject of the proposition). In terms of our example, the shade of blue found on the cover of the prehended book might be combined with the actuality of a stray dog that lies some-

---

6 It is important to note that all or at least most of the self-determination occurs at an unconscious level. For Whitehead, subjectivity is not to be equated with self-consciousness, but rather (at least partly) includes an unconscious *telos* at work in the occasion. For this reason, atomic and molecular occasions of experience possess self-determination, but not self-consciousness.

7 PR 248.

8 One of Whitehead's metaphysical principles states that only actualities can be causally efficacious. Universals, understood as less-than-actual by Aristotle or Plato, would be unable to have any effect upon the world. Whitehead's eternal objects are distinguished from universals in that they are all located in the mind of God. Since God is an actuality, the eternal objects are able to be causally efficacious.

9 Macroscopic examples of nexūs such as books, computers, or animal bodies are actually made up of millions or billions of actual occasions that are each individually prehended. To speak of prehending a book is actually to speak of prehending millions of actual occasions that make up a book.

10 PR 214.

where in the prehending subject's causal past. The resulting propositional feeling's datum would be "dog X as blue."

Whereas propositional feelings occur in moderately complex occasions of experience (i. e., living cells), the final or intellectual phase of concrescence takes place only in high-grade occasions of experience (i. e., human beings). Only the occasions sophisticated enough to attain the intellectual phase are capable of consciousness, which is one subjective form of an intellectual feeling.[11] The datum of an intellectual feeling is "the generic contrast between a nexus of actual entities and a proposition with its logical subjects members of the nexus."[12] This final complex level of integration of feeling combines and compares a propositional feeling with the actuality of the subject of that propositional feeling. In the present example, an occasion that had reached this level of concrescence would compare the propositional feeling "dog X as blue" with the actual dog X. The subject would then probably perceive that dog X is predominately black, and not blue at all. The proposition "dog X as blue" would be judged to be false, and consequently prehended in an intellectual feeling with the subjective forms of "disbelief" and "consciousness," among others.

The prehensions in the four phases of concrescence describe the human experience as it develops from occasion to occasion. Prehensions oscillate between efficient causation, which occurs between occasions, and final causation, which occurs within an occasion during concrescence. Upon reaching satisfaction, or full determination, an occasion of experience achieves *objective immortality* insofar as it is now an object for all future occasions to prehend, and consequently continually influences the future.

These phases of concrescence have striking ramifications for a Whiteheadian account of interpretation. According to Whitehead's model, the vast majority of human experience is unconscious. Even when we are most aware, "there is a small focal region of clear illumination, and a large penumbral region of experience which tells of intense experience in dim apprehension."[13] Whitehead's account of human experience stresses unconscious experiences, rather than consciousness, as primary. Unlike Schleiermacher, who conceives of hermeneutics as a task originating in discursive thought, a Whiteheadian account of interpretation will attribute the bulk

---

11 See section III. 3. for a discussion of subjective form.
12 PR 266.
13 PR 267.

of interpretation to unconscious determination.[14] Whiteheadian hermeneutics will be biased towards the descriptive rather than prescriptive, as the majority of information that informs the conscious entertainment of a text by a reader will have been determined preconsciously. This is not to deny the practical side of hermeneutics; conscious interpretation does occur. The need for the prescriptive aspect of hermeneutics never vanishes, but is secondary.

II. Experience as Pervasively Interpretative

Heidegger and Gadamer redefined the hermeneutical task as a discipline that is properly ontological in nature. Heidegger's characterization of human existence as *Geworfenheit*, or being-thrown-into-the-world describes being as necessarily being in the midst of a world one did not create.[15] Humans find themselves immersed and penetrated by the world, unable to untangle themselves from it. Likewise, Gadamer's category of *wirkungsgeschichtliches Bewußtsein*, or consciousness exposed to the effects of history, establishes the interpreting subject as a thoroughly historical being.[16] Any act of understanding is contingent upon the prejudices[17] and traditions of the reader's cultural moment. This ontological turn in hermeneutics transformed the discipline into a universal enterprise since pre-understanding, tradition, and prejudice were inescapable and the notion of experience unmediated by interpretation was considered a myth of modernity. Though he proposes a radically different ontology, Whitehead will side with Gadamer and Heidegger on this point: the interpreting subject is both thrown into the world and historically conditioned. For Whitehead, however, it is the thoroughly relational character of the world that necessitates hermeneutics as a universal enterprise.

---

14 Although this interpretative integration is unconscious, it is still teleological insofar as it is driven by the subjectivity of the concrescing occasion.

15 Martin Heidegger, *Being and Time*, trans. John Macquarrie and Edward Robinson (New York: Harper & Row, 1962), 220–4.

16 Hans-Georg Gadamer, *Truth and Method*, trans. Joel Weisenheimer and Donald G. Marshall, 2nd rev. ed. (New York: Continuum, 2000), 300–2.

17 Gadamer's concept of prejudice is distinct from its pejorative popular usage. Gadamerian prejudice is much broader and includes a wide variety of presuppositions and attitudes, both good and bad—not just those that might be considered inappropriate or bigoted.

Whitehead points out that if a poet is composing a verse about trees, he may take a walk in a forest for inspiration. For her, the trees are symbols for words.[18] Concrete actuality is primary and only later gives rise to language. As soon as the poet's verse is read, the symbolism is reversed, and her lyrics, in the process of interpretation, become symbols for trees. Whitehead describes the use of language as "double symbolic reference:—from things to words on the part of the speaker, and from words back to things on the part of the listener."[19] In Whitehead's example, a tree is prehended by a poet. The poet interprets the tree as pointing to some formulation of language, and she then externalizes verse in writing. This verse is then read, and the reader realizes the poetic language as a symbol for an actuality. The words evoke a prehensive feeling in the reader, which may or may not be similar to the poet's original prehension of the actual tree. Hermeneutics is primarily concerned with the latter half of this process—the reader's flow of understanding.

If one considers this process while focusing on the role of the communicated message, the message, whether a single fact, or complex collection of propositions, seems to flow from the world to the author, back into the world, and to the reader. In this case, the beauty of a tree was prehended by an author, transmitted to a reader via language, and back into the world, as the experience of the tree's beauty is one of the prehensions that constitutes the reader as an object for future occasions. The causal connections found in this particular example characterize Whitehead's entire cosmology. The world is made up of a tangled web of relational bonds, which connect every actuality to every other past actuality. This pervasive interconnectedness is known as the *principle of relativity*.[20]

The principle of relativity gives rise to a complication for an interpreting subject. Because every actuality is also a subject, which refines and selects the data of its own experience, there is always an unavoidable distortion of the message at two moments in its flow through the world: during the process of concrescence and at the moment of expression. In terms of human subjects that communicate with each other, this distortion most commonly occurs first during thought, and then when the thought is externalized in language.

---

18 S 12.
19 Ibid.
20 PR 22.

This double distortion rules out the possibility of an exact correspondence between an author's expression and a reader's understanding at the ontological level.[21] Interpretation is doubly necessary to make intersubjective relationality possible. This insight is reinforced by Lyman Lundeen's work. In his *Risk and Rhetoric in Religion*,[22] Lundeen demonstrates that analogy is the basis of perception; interpretation is the only means by which information is transmitted between and within occasions.

1. Perspective

In the mental pole, the process of concrescence conceals and selectively emphasizes much of the initial data that enter into the physical pole of an actual occasion. Whitehead argues that "the selective character of the individual obscures the external totality from which it originates" and that it "has attained its individual depth of being by a selective emphasis limited to its own purposes."[23] The aggregate of all past actual entities is known as the *initial data*. The vast amount of information in these initial data undoubtedly contains contradictory facts and propositions. Those bits of data that are contradictory or incompatible with the concrescing occasion are not felt, but are negatively prehended.[24] Although the data still have some (vanishingly small) impact upon the concrescing occasion, their content is ignored. After the strictly contradictory elements of the world have been negatively prehended, the many positively prehended initial data are converted into one complex objective datum in the conformal phase. Because the process of concrescence, in terms of information, is essentially a process of discrimination, selection, and integration, the occasion refines a huge amount of data into a relatively small amount of data. Even then, many of the data that *do* get through are ignored or at least obscured in the subsequent phases of the mental pole.

The selection in the mental pole is not without purpose or benefit. In reference to the concrescence of an actual occasion, Whitehead remarks, "intellectual coördination is more readily achieved when the primary facts are selected so as to dismiss the baffling aspects of things into intellectual sub-

---

21 Despite the impossibility of *exact* correspondence, frequently this distortion is vanishingly small, and near-perfect understanding occurs, which can be considered perfect for all practical purposes.

22 Lyman T. Lundeen, *Risk and Rhetoric in Religion: Whitehead's Theory of Language and the Discourse of Faith* (Philadelphia: Fortress, 1972).

23 PR 15.

24 PR 83, 231–2.

ordination."[25] The drive towards coherence in the coordination of data in the phases of the mental pole prompts the concrescing occasion to obscure and ignore the problematic facts, and emphasizes those that are amenable to the subjective aim. These problematic facts are not strictly contradictory; all contradictory data were negatively prehended in the conformal phase. These problematic facts participate in contrasts that, if brought into consciousness, would be too intense for the concrescing occasion to integrate.

This process of selection allows for individuality and coherence in the world. Selectivity, however, comes at a cost, one dimension of which is the loss of exactitude. The world cannot be prehended identically by any two subjects, and no actualities will prehend the world absolutely, without some selection of data.[26] No two occasions will feel the world in the same way.

Every actual entity (except for the divine actuality) feels only a portion of the data in its universe because of negative prehension and selection. Consequently, no finite actuality is capable of an absolute perspective. The selection of data both in the transition from initial data to objective datum in the physical pole and in the phases of the mental pole establish perspectivalism and epistemic relativism as pervasive features of the universe.

Roughly put, epistemic relativism is the doctrine that human thought is contingent upon a variety of subjective factors, such as culture, location, and perspective. The degree to which human thought is conditioned is a matter of a much larger debate. This controversy is best conveyed by the two oversimplified characterizations that represent the opposite poles of the debate. The first position argues that human thought is independent of any external determination and capable of perceiving the world objectively. The opposite position argues that both thought and perception are wholly determined by history, culture, and other factors. According to the latter position, claims to either objectivity or unmediated reality are untenable.

Whitehead clearly rejects both of these caricatured positions. Mild relativism, or acknowledgment of finite perspective, is a necessary feature of Whiteheadian metaphysics. Finitude frequently leads to distortion and error if a subject, by means of Whitehead's *fallacy of misplaced concreteness*, mistakes the feelings in its partial perspective to have absolute signifi-

---

25 AI 47.
26 PR 210.

cance.[27] Nevertheless, it still makes sense to speak of objectivity. Before the occasion's subjectivity manipulates the data, the world is objectively given to every occasion as the objective datum in the physical pole. As long as the fallacy of misplaced concreteness is avoided, epistemic relativism of the extreme variety is not a worry in the Whiteheadian system.

2. The Imprecision of Language

The second point of distortion and imprecision comes when an author expresses a prehensive feeling by putting it into language. As Anna Case-Winters has explained, language, for Whitehead, is persistently inadequate to the task of describing reality. Even when "honed to a greater precision, it is not the case that there will be some kind of direct correspondence between reality and the linguistic representations of it."[28] Whitehead calls this *the fallacy of a complete dictionary*: language is always capable of greater specificity, and will consistently fall short of a complete description of reality.[29] For this reason, cases in which a person seeks to understand the intention behind an expression, such as in most everyday conversation, are precluded from an interpretation-free understanding. Because the language a person uses to express herself will never coincide with her actual sentiments, even the simplest bits of communication require some degree of interpretation.

Language not only suffers from chronic imprecision, but also suppresses the relational character of facts and propositions. In Whitehead's account of the harmful presuppositions of much philosophy, he criticizes those who "presuppose that language does enunciate well-defined propositions."[30] This is not the case at all because "[l]anguage is thoroughly indeterminate, by reason of the fact that every occurrence presupposes some systematic type of environment."[31] The purpose of language is to express and communicate prehensive feelings, usually propositional feelings, to other subjects. Prehensive feelings arise in an environment in which they are related to every other feeling in the author, as well as the entire past actual world.

---

27 The fallacy of misplaced concreteness occurs when one takes an abstraction for a fully concrete reality. Whitehead frequently attributed this fallacy to those who argue for a substance metaphysic. See PR 7–8.

28 See Anna Case-Winters, "System and Dynamism in Whitehead's Thought: The Category of the Ultimate and the Concept of God," 142 in this volume.

29 MT 173.

30 PR 12.

31 Ibid.

The prehensions that lie behind expression have their meaning in a complex tangle of relations to other prehensions and actualities. The act of expression, however, cuts a feeling out of its context of relatedness to concrete fact. It abstracts the feeling from its original relational environment and presents it in sterile isolation for the reader to consider. Whitehead warns that "single words, each with its dictionary meaning, and single sentences, each bounded by full stops, suggest the possibility of complete abstraction from any environment."[32] To assume that any unit of language (or expression) accurately represents a prehensive feeling is to commit another Whiteheadian fallacy—that of *simple location*. Simple location is the mistake of assuming that an entity "does not require for its explanation any reference to other regions of space-time."[33] The actual significance of any feeling involves its connection to everything else; the proper locus of meaning is in the life of an experiencing subject.

David Ray Griffin has highlighted this problem with language by emphasizing the distinction between verbal statements and propositions.[34] He uses the sentence "Caesar crossed the Rubicon" to illustrate the fact that one verbal expression could represent multiple propositions. This sentence could potentially refer to a would-be emperor of Rome crossing a river. It might also, I would add, describe a criminal named Caesar betraying his organization, which is nicknamed "the Rubicon." It could also be used metaphorically to describe a person making an irreversible decision. This statement, divorced from the actuality of its referents, lacks the relational connections to its surrounding world that enable one to interpret its meaning correctly.

The isolating character of language, when considered in light of the principle of relativity, raises a perplexing question. If every actual occasion is related to every other actual occasion, then how can the act of expression excise meaning from its relational character? Are there not causal chains connecting the text to both author and reader? One might suggest that the actual occasions that make up the medium of expression could provide the concrete relatedness required for meaning, and save us from simple location. The problem with this suggestion lies with the sophistication of the medium's occasions. The actual occasions that make up paper and ink or

32 MT 66.

33 SMW 49.

34 David Ray Griffin, *Reenchantment Without Supernaturalism: A Process Philosophy of Religion* (Ithaca: Cornell University Press, 2001), 324–6.

vibrating air molecules are not as sophisticated as human occasions of experience. The actual text is not capable of achieving the complex contrasts and syntheses of feeling that are constitutive of human thought.[35] This discrepancy in sophistication means that human experience loses much of its richness when contained in a non-human occasion. For this reason, language relies on symbols for the passing along of complex propositions and concepts.[36]

As an author uses language, the prehensive feelings expressed (propositional or otherwise) are necessarily ambiguated and distorted. Likewise, the very process of concrescence tends to obscure and distort the totality of the world as a subject coordinates data into patterns of coherence. These two moments of distortion entail that communication is doubly reliant upon interpretation. Readers must take into account both the indeterminacy of the means of expression, and the particularity of their own perspectives.

3. Analogy and Experience

In his study of Whitehead's philosophy of language, Lundeen supports the notion that interpretation is a pervasive characteristic of the universe. He argues that the causal relationships between actualities are by nature interpretative. Specifically, Lundeen argues that analogy is "fundamental in the coordination of societies and enduring entities, so it plays a crucial role."[37] He continues to define analogy as a "similarity which is qualified by differences which are real, but which can be excluded as of negligible relevance in respect to a specific interest or context of significance."[38] In an analogy, two elements that are in fact not identical are compared with special attention paid to their similarities, so that the significance of their differences is marginalized. The expression "John was as hungry as a bear" is a striking example. John, who is human, has many differences from a bear—body hair, tooth size, cognitive capacity, metabolism, and so on. Upon awakening from hibernation, bears are notoriously ravenous, and so is John. For the purposes of the analogy, the two elements are judged to be similar with

35 Although this observation—that texts cannot think—seems uninteresting, the notion of a significant gap dividing occasions that can use language, and those that cannot raises interesting questions, i. e., "Exactly how homogenous is Whitehead's panexperientialism?" See section III. 3. below.

36 See section III. 2.

37 LUNDEEN, *Risk and Rhetoric in Religion*, 171.

38 Ibid., 50n.

regard to their appetites. The differences between John and a bear are not significant in relation to the meaning of the analogy. Such is the function of analogy: to intensify similarity while dampening difference.

The Whiteheadian metaphysical category that Lundeen points out as most analogical is that of *transmutation*. This species of feeling is "a transmutation of simple physical feelings of many actualities into one physical feeling of a nexus as one."[39] Suppose, for example, a human prehends a puddle of water. Every actual occasion of every molecule of water is slightly different from the next; each prehends differing initial data, and each reaches a unique satisfaction according to its unique subjective aim. Yet, all these molecules are presented to an individual as a unity. We do not prehend the billions of molecules as billions of individuals. Instead, we perceive them as a relatively undifferentiated unity. The same is true for most of our experience. We experience the world as groupings, or nexūs, of actual occasions: a table, an apple, air, a person, a tree. Without transmutation, we would experience a vast multitude of unique actual occasions. This is the power of analogy; it explains the tendency of a subject to experience the world by emphasizing similarities over differences relative to a specific criterion. Lundeen argues that this is why

> Analogy is not a temporary or secondary linguistic expedient to be used only until it can be replaced with straightforward statements requiring no qualification. Experience is qualified by pervasive analogies, and its most adequate expression requires the type of qualification associated with interpretive models and illuminating metaphors.[40]

The universe is riddled with persistent analogies at many levels of experience. Existence at its most fundamental and simple level involves prehension and experience. Every level of existence participates in interpretation on some level by virtue of the selection of concrescence and ambiguity of language. Lundeen's example of analogy and Whitehead's category of transmutation are two examples of the many ways that the relations between occasions of experience are essentially interpretative. For this reason, hermeneutics, as an account of how interpretation occurs, takes on radical significance in a Whiteheadian metaphysic. Hermeneutics is properly defined as a metaphysical investigation into the relations between and

39 PR 251.
40 LUNDEEN, *Risk and Rhetoric in Religion*, 184.

within occasions. Whitehead states that "[t]here is no going behind actual entities to find anything more real."[41] Whitehead's ontological principle requires that only actual entities can be given as a reason for things.[42] They are the only finally real things. Because the universe is made up exclusively of actual entities, and each entity interprets, a Whiteheadian hermeneutic is universal in scope.

III. A Metaphysical Account of Interpretation

How does a reader perform interpretation? How does one move from expressions in the world to a determinate prehensive feeling? This is the primary question that hermeneutics seeks to answer. Whitehead's theory of symbolism provides the means by which subjects make the translation between language and prehensive feeling.

1. Perception as Symbol

An understanding of Whitehead's doctrine of three modes of perception—the pure modes of *presentational immediacy*, *causal efficacy*, and the impure mode of *symbolic reference*—is crucial to a Whiteheadian account of interpretation. At the phenomenological level, these modes establish a subject's constant reliance on symbol and interpretation.

Perception in the mode of presentational immediacy is "our perception of the contemporary world by means of the senses."[43] It is the immediate experience of, for example, visually perceiving spatial fields of color, without any reflection or interpretation of the sense data. Although presentational immediacy is prominent in conscious experience, the most fundamental mode of perception is perception in the mode of causal efficacy. Perception in this mode is an actual occasion's physical prehension of the world in its physical pole. The most common example is the power of memory. As a subject, we have access to memories, yet these memories are not presented to us through sensory phenomena.

Imagine the example of a dog running towards you while barking. In terms of presentational immediacy, a black shape in your field of vision is getting larger, and a noise is getting louder. In terms of causal efficacy, you

41 PR 18.
42 PR 19.
43 PR 311.

are prehending other causally efficacious actualities. Symbolic reference is the transition between one mode and the other. It is the recognition of increasing loudness and growing black shapes as representing a causally efficacious actuality: an approaching dog.

This movement from sign (presentational immediacy) to referent (causal efficacy) is "so habitual in human experience that great care is needed to distinguish the two modes."[44] Phenomenologically, human conscious experience does not begin with causal efficacy—which occurs in the conformal phase before consciousness arises. We do not feel the raw efficient causation of other occasions. Nor does it begin with blotchy fields of color and uninterpreted sounds. Rather, we see and hear *things*. The interpretative act by which we infer the existence of causally efficacious actuality from sensory experience is accomplished by perception in the mode of symbolic reference. Perception in this mode is largely automatic. It constitutes the given character of experience prior to conscious reflection.

2. Language as Symbol

Language, both written and spoken, operates as symbolic reference. As Whitehead points out, "[s]poken language is merely a series of squeaks."[45] When prehended (either visually or audibly), words and sentences do not actually convey their meaning in the prehension. There is nothing inherent in the actual occasions making up the ink formations in the written word "dog" that convey the idea of an animal. A word is a "symbol, and its meaning is constituted by the ideas, images, and emotions, which it raises in the mind of the hearer."[46] Sensory phenomena point to a more concrete actuality in symbolic perception. In the case of language, bits of sound or ink blots point to the world.

For Whitehead, as for Schleiermacher, the process of understanding is an act of interpreting symbols. Lundeen argues that some common ground must exist as the basis of a symbol-meaning relationship.[47] He points out that most symbols in sense perception contain a "high degree of causal conformity" and that, as a result, interpretation occurs almost automatically. As symbols become more abstract and complex, as they do in language, however, "this causal relationship is reduced," and consequently, "[t]he

---

44 PR 121.
45 PR 264.
46 S 2.
47 LUNDEEN, *Risk and Rhetoric in Religion*, 157.

ground for symbolic reference becomes somewhat artificial and conventional."[48]

Whereas perception in the mode of symbolic reference relies heavily on natural similarities between symbol and referent, words only resemble their meanings insofar as they have a history of association.[49] Readers are not isolated and autonomous agents, who freely assign meaning to language in accordance with their will. Rather, they are immersed and penetrated by pre-established norms of interpretation. The reader occasion is constituted by what has come before; it is radically historical in Whitehead's ontology. The past makes up the content of a contemporary occasion. History is the common ground that acts as the basis of the language-meaning relationship. When a reader interprets language, she prehends the symbol to be interpreted as well as past occasions of experience in which that same symbol was associated with particular meanings. Such associations have occurred both in past occasions of her own experience and in the subject's larger environment.

In terms of prehensions, written or spoken symbols are felt as past actualities: ink-stained paper or vibrating air molecules. Conceptual feelings, which are feelings of eternal objects, are combined with those symbols to form propositional feelings. Although Whitehead makes no special designation, I will call the species of propositional feelings that combine an actual human symbolic expression and a conceptual meaning for that symbol a *semantic feeling*. A simple example of a semantic feeling is that of a lexical meaning assigned to a particular word, prehended from past successful interpretations of that symbol. Semantic feelings, however, would include any prehension of a past use of language. These semantic feelings are compared to the actuality of the symbol being interpreted. Meaning is then applied to a symbol in an intellectual feeling. For example, the proposition "the word 'dog' as meaning furry" is a semantic feeling, as is "'dog' as meaning companion" and "'dog' as meaning bloodthirsty." The reader processes these semantic feelings in light of other semantic feelings and selects those which are appropriate for application to the symbol being interpreted. This application occurs by means of an intellectual feeling. Almost

48 Ibid., 176.

49 One notable exception being the literary device of onomatopoeia (i. e., "splash"), where the pronunciation of a word closely resembles its meaning: the character of the auditory presentational immediacy it represents. In this case, there is a natural similarity.

all cases of interpretation are much more complex than the case of assigning meaning to one word. Real interpretation involves more than single words in isolation. The basic process, however, is the same. Interpretation relies on past applications of meaning to symbols and applies them to contemporary symbols. The power of the past comes to bear on an individual through these semantic feelings.

Intersubjective communication relies on consensus within a community of language users of what a word or phrase means. If a reader applied a definition to a symbol in too much of an unexpected way, the communicability of that symbol would be undermined. Semantic feelings are always felt with the subjective form of appropriateness or suitability. It is important to note that the prescriptive forces of semantic feelings are always subordinated to the occasion's subjective aim; interpretation is not *wholly* determined by the past. The potential for novelty is present in every occasion of experience's application of meaning to symbol. The power of past usage on the interpreting subject is largely persuasive, but is never deterministic. Due to the idiosyncratic subjective aim of a reader, slight and subtle shifts in semantic application frequently occur. This flexibility accounts for the gradual evolution of language over time. Also, innovation in language is presupposed in poetry, and other lyrical genres in which unexpected and surprising applications of meaning contribute to the aesthetic quality of the work.

The reader is constituted by and manipulates the sedimentation of past language use, to borrow Ricoeur's metaphor.[50] The past not only provides the tools of grammar and usage, but a tradition of other past interpretations. History serves as a rich accumulation of potential meaning. The sedimentary character of linguistic tradition is not rigid—a brittle shale, but rather like thick clay—retaining its own shape and form, but ultimately moldable, especially under gradual pressure over time.

3. Form and Content

As previously mentioned, every prehensive feeling consists of three aspects: its objective content, its subjective form, and the concrescing subject that feels the feeling. In terms of these three aspects, hermeneutics can be described as an account of how the first two aspects, objective content

50 PAUL RICOEUR, *Hermeneutics and the Human Sciences*, ed. John B. Thompson (Cambridge: Cambridge University Press, 1981), 52.

and subjective form, are transmitted between subjects (the third aspect) by means of symbolic expression.

The previous section described how the first aspect of feeling, objective content, is determined. Semantic feelings are prehended and applied to a symbol in an intellectual feeling. There is more to meaning than austere dictionary definition. Emotive force and emphasis in expression are conveyed in the second aspect of a prehensive feeling: subjective form. Subjective form is determined alongside objective content in the interpretative process. Whitehead argues that the subjective form of a prehensive feeling has both a qualitative aspect and a quantitative aspect.[51] In addition to the "flavor" of a subjective form, it is also felt in some quantity. A conceptual feeling can be felt with a great deal or small amount of appetition. Likewise, it can be felt with varying amounts of aversion.

When a reader interprets an expression, she is constructing both the definition of an expression and its emotive force. The combination of objective content and subjective form determines the meaning of an expression. In some forms of discourse, however, the objective content is more important than the subjective form. For example, scientific and legal texts usually place minimal importance on the subjective form of the expressed feelings.

The opposite can also be said of poetry, or religious expressions. In these cases, the subjective form arguably plays a much more important role than the content of the prehensive feelings.[52] The emotive force of what is said or written frequently takes precedent in the mind of the author. Whitehead points out that music, ceremonial clothing, and ceremonial smells are all examples of expressions in which the importance of the objective content of the interpreted feeling is minimal and the subjective form is maximal.[53] In addition to assigning content and form to symbols, the interpretative process includes a judging of how much importance is to be placed on content, and how much is to be placed on form.

There are also cases of interpretation in which no content is entertained. All that is evoked is a subjective form. Lyrical poetry, music, and abstract

---

51 PR 233.

52 It is important to note that Whitehead makes no clean break between discourse in the humanities and scientific discourse. Although the evaluation of form and content will be different in a novel than in a scientific treatise, only one methodology is required to evaluate both types of expressions. Whitehead's panexperientialism makes no distinction-in-kind between conscious actualities and non-conscious actualities. Because all actual occasions are experiential, the familiar hermeneutical distinction between explanation in the natural sciences and understanding in the humanities is weakened, if not collapsed.

53 AI 249.

paintings, for example, do not seem to evoke any propositional content that contains a subject and predicate. Instead, all that is induced in the reader is a vague emotion. For example, consider a poem that employs non-propositional language. The phrase "burning buzzing dizziness," as a stanza in this poem, is nonsensical if one is trying to extract a proposition from these three words. This is predication with no logical subject. Likewise, listening to music does not evoke discursive thought, only emotion. Authors frequently use these sorts of expression to consciously evoke various emotive responses in readers. A subjective form divorced from its objective content would simply be the feeling of an eternal object: a conceptual feeling.[54] If certain forms of expression evoke form without content, then a consciously held conceptual feeling would arise.

Although not strictly ruled out, the notion of consciously entertaining conceptual feelings would not be a regular occurrence in Whitehead's system. Consciousness arises as a feature of the later phases of concrescence. The subjective form of consciousness appears in conjunction with the "affirmation-negation contrast" of comparing a possibility, as present in a propositional feeling, with an actuality, as present in the physical feeling.[55] Consciousness is usually associated with intellectual feelings in the fourth phase of concrescence.

It is important to note, however, that boundaries of consciousness in the phases of concrescence are not sharp. Whitehead qualifies his description of consciousness as limited to the intellectual phase by pointing out that upon satisfaction, all subjective form in an occasion is shared. He writes, "all feelings acquire their quota of irradiation in consciousness."[56] Conscious entertainment is not strictly limited to intellectual feelings; it is only centered there. The unity of the subject diffuses consciousness throughout all other feelings in an occasion upon satisfaction. Abstract forms of expression stretch and push consciousness towards the regions of experience that are normally left in the dark. By partially illuminating raw objectless emotion, abstract and lyrical language[57] highlights the more primitive aspects of human experience that are cognized by abstracting them from more sophisticated comparative feelings.

---

54 PR 232.
55 PR 267.
56 Ibid.
57 Perhaps music can be included here as an extreme example of an underdetermined and plurivocal language.

To summarize the way symbolic interpretation functions according to Whitehead's doctrines: language and other external expressions are prehended by a reader. Possible meanings for a symbol are prehended from the past by the reader, who then selects one, or a combination of those meanings, and applies them to that symbol in an intellectual feeling. In every application of meaning, there are two components of every definition that must be evaluated: objective content and subjective form, each of which may have varying significance in the expression.

In a Whiteheadian system, interpretation takes place on all levels of experience. To speak of a Whiteheadian hermeneutic is to discuss a broad theory that includes any relationality between any two occasions of experience. The discussion so far has focused on humans interpreting concrete symbolic expressions of other humans. The scope of a truly Whiteheadian hermeneutic would be much broader. It must account for the interpretative experience of every species of actual entity. Molecules, squirrels, and humans all interpret the data of their experience, albeit not necessarily consciously. Because it only applies to the most sophisticated actual occasions and to their most conspicuous symbolic externalizations, hermeneutics, as traditionally conceived, would be a "macrohermeneutic" in Whitehead's system. Whitehead argues that there is no cognitive difference-in-kind between human and non-human actualities, only a difference of degree.[58] This uniformity establishes the universality of interpretation. Despite this universality, there seems to be a gap as one moves from non-human animals to humans. Only humans seem to be saddled with the questions of language, tradition, history, and culture. This tension raises an issue for future exploration. To what degree is hermeneutics uniformly applicable to both molecules and human beings?

IV. Points of Contact

A hermeneutic derived from Whitehead's metaphysical principles engages many issues with which Schleiermacher also grappled. In this section, I compare Whitehead and Schleiermacher on three key controversies: the relation of thought to language, nominalism, and the importance of authorial intention. Schleiermacher and Whitehead disagree on the first two issues. Despite this disagreement, I claim that they stand together against

58 PR 18, 107–9.

the majority of contemporary hermeneutical theories in their endorsement of authorial intention as a viable locus of meaning. Specifically, I argue that Whitehead can revive the importance of authorial intention, which has been largely discredited by contemporary hermeneutical theories. Nevertheless, unlike Schleiermacher, a Whiteheadian hermeneutic compels us to consider meaning as also located in the interaction between reader and author, rather than solely as a function of the author's intention.

1. The Relation of Thought and Language

Whitehead's distinction of thought from language is clear.[59] Thought and feeling, as experienced, are clearly primary. Language is secondary, and is an abstraction from the contextual and relational characteristics of experience.

According to Whitehead, linguistic concepts are derived from the eternal objects.[60] These objects are mediated to human subjects externally from God. Although the eternal objects are the basis of linguistic concepts, they are not linguistic in origin. Because eternal objects enjoy permanent existence in the mind of God, they are metaphysically real apart from any human capacity for communication or thought.[61] The eternal objects are the basis of all determinateness—ontological, epistemological, and linguistic. As the source of all differentiation in the world, they precede both thought and language, and place the two on even footing.[62] Thought and language are two distinct phenomena, neither of which depends on the other. Instead, both rely on the eternal objects, which are causally prior and independent of an agent's subjectivity or capacity for language.

For Whitehead, the application of semantic feelings to symbols, which is an intellectual feeling, only takes place in the final phase of concrescence.[63] The more primary modes of experience that occur in the earlier phases of concrescence are free from language. Because the vast majority of experience is non-linguistic, language is predominantly relegated to the external aspects of life—to intersubjective communication.

---

59 PR 4, 12.

60 PR 24, 158.

61 PR 33–4.

62 This "even footing" is only even in terms of causal priority relative to the eternal objects. Whitehead clearly privileges thought and feeling over language as more concrete, specific, and influential. See PR 4, 12.

63 PR 161–2.

Schleiermacher's view on the relation of thought to language is not simple identity or disjunction, but much more nuanced. Language "determines the progress of the individual in thought."[64] Language is something that is particular and differently formulated for every individual. Yet for Schleiermacher, thought must be structurally homogenous from individual to individual if understanding is to ever take place.[65] Christine Helmer has carefully described this interaction between the individual particularity of an agent (the organic pole) and the uniformity of reason (the intellectual pole).[66] Linguistic concepts enter the individual through the organic pole, which is responsible for all intercourse with the empirical. Because these concepts are

> originally gleaned from empirical experience, they are not arbitrary cultural constructs. And because they originate in the organic pole's wedge in reality, they are fundamentally revisable.[67]

It is the dichotomy between the universality of the intellectual pole and the particularity of individual language at the organic pole that necessitates hermeneutics. Specifically, Schleiermacher's scheme of the thought-language relation would look more like a continuum. At the intellectual end is indeterminate or nebulous thought, which employs only the broadest and most general concepts. Schleiermacher's doctrine is that all thought employs concepts, and so must be linguistic through and through. Yet, as thought develops in the individual, it is fixed and increasingly made determinate by linguistic concepts, which refine and differentiate as they are applied to thought. As a thought develops, it increases in linguisticity, understood here as conceptual specificity. Whitehead views language and thought as two non-intersecting activities; the former is a wan substitute for the latter. Schleiermacher's view of the relation is one of interaction between the universality of reason and the particularity of human experience.

2. Nominalism and Conceptual Anchoring

Whitehead's system agrees with most contemporary hermeneutical theories insofar as meaning is largely determined by culture, history, and environ-

64 *Hermeneutics*, 9

65 *Hermeneutics*, XXI–XXII (Introduction by ANDREW BOWIE).

66 See CHRISTINE HELMER, "Novelty and System in Schleiermacher's Thought," 174–5 in this volume.

67 In the unpublished version of ibid.

ment. Despite this, Whitehead's position breaks from them in a meaningful way. Most hermeneutical philosophers, Schleiermacher included, presuppose some version of nominalism: the doctrine that universals, or the recognizable qualities in objects, do not exist independently of human minds. A nominalist holds that universals are only real insofar as we have created names for them. Any identification of "redness" or "squareness" in objects is a matter of social convention. To speak of universals apart from human perception and language makes no sense.

Whitehead, however, rejects nominalism. His doctrine of the eternal objects asserts that universals are real. The eternal objects are defined as "[a]ny entity whose conceptual recognition does not involve a necessary reference to any definite actual entities of the temporal world."[68] Anything abstracted from a particular actuality is an eternal object. So, the "roughness" and "rectangularity" of a book, when abstracted from the actual book, are eternal objects, as are the "wittiness" and "economy" of its prose.

An occasion prehends the eternal objects as mediated by past actual entities, whether finite or divine.[69] Eternal objects make up the content of conceptual feelings, which are the basis for all higher-level prehensions. Consequently, the eternal objects are the sole means by which definiteness occurs in the universe. Any difference between actual occasions ultimately arises from the eternal objects felt in those occasions.

In a Whiteheadian hermeneutic, the application of semantic feelings to symbols may be arbitrary and conventional, but the actual content of the semantic feelings are not. Lexical definitions and other patterns of language use are reducible to eternal objects, which are permanent metaphysical features of the world. The eternal objects act as fixtures or anchors around which language can be oriented. When interpreting language, a reader appeals to her own experience. She prehends semantic feelings in order to potentially apply them to a symbol. Likewise, the meaning any author wishes to express consists of eternal objects available in her experience. Both the author and the reader are drawing from the same, albeit very large, pool of eternal objects in their attempts to create and interpret symbols.[70] Eternal objects make translation possible, regardless of cultural or historical discrepancies between reader and author.

68 PR 44.
69 PR 31.
70 PR 23, 40.

This anchoring of definitions in eternal objects does not go very far at all towards closing the gap between the meaning intended by an author and the interpretation of a reader. It is a necessary, but not a sufficient condition for the translatability of language. It cannot ensure correct interpretation. There is no guarantee that a reader positively prehends the same eternal objects as the author, nor can it guarantee that the reader would apply the same eternal objects to a symbol in the same way as the author. The rejection of nominalism is only significant in that it rules out the possibility of incommensurability between a reader and author. It precludes the possibility that the concepts that make up an author's definition of a symbol are *qualitatively* different than the concepts available to the reader, thus making reasonably accurate interpretation and understanding futile.

Imagine the world as an undifferentiated whole. This whole is experienced according to the way an individual, by means of concepts, differentiates and organizes world-experience. The historical gap between author and reader has left the two with worlds that are aggregated differently. The conceptual scheme that the reader uses to pick out aspects of the world would be qualitatively different in the author. For the reader, the task of assigning meaning to the author's linguistic symbols amounts to fitting square pegs into round holes. Despite the best efforts, the concepts do not match up. Whitehead's doctrine of eternal objects acts as a unifying principle; it guarantees that all parties involved share a common conceptual ground. Although the application of meaning to symbols is a matter of convention in Whitehead's system, the actual conceptual content of potential symbolic meanings are metaphysically determined.

3. Authorial Intention Revived

Schleiermacher defines the discipline of hermeneutics as the "art of understanding particularly the written discourse of another *correctly*."[71] The notion of correctness implies that there is some sort of relationship between a reader's interpretation and another factor that determines whether correctness obtains. The factor that guarantees correct interpretation is the author. Schleiermacher views the individual as the locus for understanding any expression: "every utterance is to be understood via the whole life to which it belongs."[72] The goal of Schleiermacher's hermeneutics is not so

71 *Hermeneutics*, 3 (my italics).
72 *Hermeneutics*, 9.

much to understand a text itself, but its author. Specifically, Schleiermacher seeks to understand a text insofar as it represents a moment in the life of an individual.

This concern with the totality of the author comes through in Schleiermacher's discussion of the *Keimentschluß*, or seed-decision.[73] This seed-decision is defined as the authorial intention that unites the expressive life of an individual. All texts written by an author exist as moments in the life of a genius. The *Keimentschluß* is the defining essence that binds all of the disparate intentions, texts, and moments of an author's life into a coherent unity. Schleiermacher's vision of the hermeneutical task is concerned first and foremost with extracting the intentionality behind the text—to truly understand the author in light of the *Keimentschluß*.

Schleiermacher's concern with the relation of a particular expression to the totality of an author's life is very congenial to a Whiteheadian understanding of expression. Whitehead complains that language distorts feeling by cutting away the relational web that surrounds it in the world.[74] Consideration of a fact isolated from its context in the life of an experiencing subject is bound to lead to misunderstanding. Whitehead also views language as an activity in which prehensive feelings are externally symbolized for the purposes of communicability.[75] The expression is a symbol, and as a symbol, it points to something else. It points to prehensive feeling in a concrescing occasion of experience: an author. Whitehead and Schleiermacher share the view that a proper description of a text is a symbol for something that is constitutive of the life of an author. Whitehead's principle of relativity, however, will eventually drive us to look elsewhere for meaning. But first, let us examine Whitehead's endorsement of authorial intention as a locus of meaning.

For a foil, we can again look to Gadamer, who explicitly rejects authorial intention as a normative guide for hermeneutical inquiry. He argues for this by emphasizing the futility of attempting to reconstruct authorial intention. The radical historicity of the individual precludes the possibility of an accurate reconstruction of the author's mind.[76] One can never overcome the temporal and cultural gap between the reader and author. No matter how hard a reader tries to understand the cultural context of the au-

73 *Hermeneutics*, 109–11.
74 MT 66.
75 PR 182–3; S 11–13.
76 GADAMER, *Truth and Method*, 166–7.

thor, the reader's reconstructed authorial intention will still be conditioned and affected by the intervening years.

Whitehead, to some degree, would agree with Gadamer on this point. No two occasions of experience share the same actual world.[77] Consequently, a reader's experience will always be conditioned differently than the authorial experience. Whitehead, however, would also assert that this interpretative gap does not preclude the possibility of correspondence between the author's intention and the reader's interpretation.

Whitehead's rejection of nominalism enables him to assert both the radical historicity of the subject and the viability of authorial intention as a locus of meaning. Whereas Gadamer's interpreting subject would be interpreting a text with a set of concepts that are different-in-kind than the author, the ontological nature of the eternal objects, and their externality to human minds, provides a common source of concepts to both author and reader. An interpretative gap will remain. The disparate actual worlds of reader and author require that the eternal objects constitutive of the author's prehensive intention will be different than those invoked in the reader prior to conscious reflection. Despite this actual difference, the eternal objects entertained by the authorial occasions are still available to the reader occasions.

Also, because the author is a part of the past actual world of the reader, the concepts employed by the author would be prehended in the reader's initial data. Though the mental pole of the authorial occasion does not play much of a role in the concrescing reader occasion (it might even be negatively prehended), it would still be an ingredient in the reader. Whitehead's principle of relativity revives the possibility of recovering authorial intention. Contrary to Gadamer, the author is not lost to us forever, but lives on in objective immortality as a *superject*: a causally efficacious element in the lives of all future occasions.[78]

Whitehead and Schleiermacher both argue for a realist account of hermeneutics, one in which it is possible to have real conformation between interpretation and the meaning intended. Whitehead's rejection of the notion that authorial intention is beyond epistemic reach or unimportant deserves closer inspection, for a text's meaning is not only related to its author, but also to each reader.

---

77 PR 22–3.
78 PR 45.

4. Beyond the Author

Ronald Farmer, in his work on process biblical hermeneutics, has argued that meaning is not found solely in the author, but in the interactions between author, text, and reader.[79] Farmer points out that the indeterminacy of language creates the possibility for novel applications of semantic feelings to language in every reader, and that the life of the reader is as viable a ground for meaning as that of the author.

Farmer is right on this point. Whitehead's prescriptive emphasis upon novelty and creative advance would preclude a theory that locates meaning solely in a moment of authorial expression, as Schleiermacher's does. To describe the meaning of a text with reference to its causal past (an author), but not its causal present (the reader), is to commit the fallacy of simple location. Texts, as actualities, must be described in terms of their causal relations, which extend to contemporary audiences[80] as well as backward in time. Potential meaning is only fully concrete in the life of individuals.

Interpretation is tricky business. On the one hand, language is symbolic for prehensive feelings, which are properly located in the life of an individual. Meaning must be located within the relational context of an occasion of experience. To avoid misplaced concreteness one must, in the act of interpreting a text, refer back to the life of an author. The authorial occasions are the first source of an expression's meaning. The danger of simple location, however, prompts us to consider a text in light of its relation to the reader as well as the author. This reference to contemporary readers introduces novel and surprising applications of semantic feelings.

In the attempt to navigate twin Whiteheadian fallacies, a reader collects meaning in both her own life-context and that of the author. While misplaced concreteness drives one away from the sign (the text) and towards the referent (the author's life), simple location then urges the reader back toward her own construction of meaning. Interpretation has a back-and-forth motion—it relies on mutuality, dependence, and the interplay between the nodes of author and reader, which are meaningful in virtue of their living relation to the rest of the world.

Up until this point, my discussion of interpretation has largely been descriptive. This is appropriate considering that prescriptive statements are

---

79 Ronald L. Farmer, *Beyond the Impasse: The Promise of a Process Hermeneutic* (Macon, Ga.: Mercer University Press, 1997), 121–2.

80 Contemporary is not meant in the technical sense here. Whitehead's system prohibits causal efficacy between two strictly contemporary occasions. See PR 318.

only relevant to conscious agents; Whitehead's system requires the majority of interpretation to occur at the unconscious level. It is still necessary to ask what role conscious discrimination plays in interpretation. The dialectical process of collecting meaning between the poles of reader and author can be illuminated by the method Whitehead endorses for all philosophical endeavors.[81] Anna Case-Winters pointed out that Whitehead saw the philosophical task as one of imagination and construction, controlled and limited by the criteria of logic, applicability, and adequacy to experience.[82] When interpreting an expression, a reader begins with her own construction: unique applications of semantic feelings to symbols. Yet the reader's own interpretative judgments are subject to checks against reality. They must do justice to the objective data that inform her own experience. These data include the intentions of the author. Whitehead's method calls for a dialectic between imagination and verification. Such a dialectic offers the unique and valuable combination of novelty and enrichment with a firm grounding in reality.

V. Summary and Conclusion

I have first aimed to imaginatively construct a hermeneutic according to Whitehead's metaphysical principles. This theory posits interpretation as immanent in every relation between and within actual entities. In terms of Whitehead's metaphysics, human interpretation consists in the application of past prehensions of linguistic usage, or semantic feelings, to contemporary expressions. I have also endeavored to compare my constructed Whiteheadian hermeneutic to Schleiermacher's own. Although Whitehead and Schleiermacher disagree on the issues of the thought-language relation and nominalism, both privilege the author as an authoritative source of an expression's meaning. This agreement, however, must be qualified by noting that Whitehead also privileges the reader's idiosyncratic interpretation as well. The reader must adjudicate between her own construal of an expression, which conforms to the facts of her own experience, and the original feelings expressed by the author, of which the expression is a symbol.

---

81 PR 5.

82 Case-Winters, "System and Dynamism," 141–1 in this volume.

The dialectical relationship found between the poles of author and reader is a reflection of a much more general characteristic that runs through Whitehead's system. In *Adventures of Ideas*, Whitehead states that

> the Enduring Societies with their rise, culmination, and decay are devices to combine the necessities of Harmony and Freshness. There is the deep underlying Harmony of Nature, as it were a fluid, flexible support; and on its surface, the ripples of social efforts, harmonizing and clashing in their aims at ways of satisfaction.[83]

Here we find two opposing virtues: harmony and freshness. The former stands for coherence, agreement, and conformity. The latter indicates novelty, dynamism, and surprise. Whitehead maintains that these two are not antithetical, but complementary—and the dialectic between the poles of author and reader in the construction of meaning is an example of this complementarity. Authorial intention serves as a guide that grounds and limits the meaning of an expression, whereas the reader's construal introduces the principle of freshness and relevance into an interpretation.

Assertions of meaning must be considered hypothetical, tentative, and "open" toward their own finality. This allows expressions to take on the character of a "fluid, flexible support" for the complexities of life. Whereas theories that rigidly endorse univocal interpretation frequently shatter when exposed to the pounding tumult of lived experience, systems that exhibit openness—such as those of Schleiermacher and Whitehead—flex and stretch, while remaining adequate to the facts of experience. This plasticity allows the hope that hermeneutical theories, and on a broader level, systematic philosophies and theologies, need not sacrifice originality and innovation while searching for adequacy and consistency.

83 AI 286.

Systematizing Agency

Toward a Panentheistic-Participatory Theory of Agency

PHILIP CLAYTON, Claremont School of Theology

I. COMPARING SCHLEIERMACHER AND WHITEHEAD

Schleiermacher and Whitehead are both systematic thinkers, even if their two notions of system sometimes diverge dramatically. At the same time, the writings of both reveal a keen awareness of the tensions between the drive toward system on the one hand and the spontaneity of agents in the world on the other. Agency represents one of philosophy's most fascinating problems; not surprisingly, the problems increase sevenfold for those of us who find ourselves compelled to introduce the concept of a divine agent as well. Moreover, "agency" implies a number of features—spontaneity, intentionality, freedom, creativity, novelty—which stand in prima facie tension with philosophical systematization. One does not have to construe "system" in the strict sense of, say, the Hegelian system in order for these tensions to become manifest. As we will see, the conundrums of "agency and system" arise in a generic manner *whenever* one attempts to give an account of what it is to be an agent or to exercise agency.

At first glance the juxtaposition of Schleiermacher and Whitehead on this topic seems disastrous. Schleiermacher wrote in response to the thought-world of German Idealism, albeit with a stronger glance toward Spinoza and a stronger fixation on theological topics than is evident in (say) Kant or Fichte. He, like they, was fascinated by the problem of the enduring "I," the subjective unity of apperception. Along with the Idealist philosophers, he contributed to a thoroughly dialectical philosophy that conceived God, world, and other subjects in light of the self-constituting activity of the subject and her experience. Whitehead, as is well known, was a panexperientialist who used the resources of atomism and the doctrine of internal relations to construct his own account of agents. The result was a comprehensive metaphysical system based on agents or "actual occasions" as the fundamental units of reality. With breath-taking vision, Whitehead's metaphysics applies a single, (more or less) consistent concept of the becoming and perishing of agents across the entire spectrum of existence, from the smallest particles up to God as the Comprehensive Experiencer and Universal Influencer.

Schleiermacher shared Kant's view—as Whitehead most certainly did not—that metaphysics in the classical sense was no longer possible. After Kant the only credible metaphysics is regulative; its statements represent an *ideal* for reason, not an achievement. This contrast is worthy of a paper on its own, especially in a dialogue between German-trained and America-trained philosophers. I repeat: Schleiermacher held that the Kantian critique spelled the demise for all traditional metaphysical systems—including, had he known of it, Whitehead's. With the irritating insistence of Nietzsche's madman, come before his time, I must stress the significance of this great divide, which threatens to turn apparent agreements on "Schleiermacher and Whitehead" into actual equivocations.

(Of course, the end of metaphysics does not mean the end of systematic thought. As Eilert Herms has shown in an important monograph, Schleiermacher still pursued the idea of a *System der Wissenschaften*.[1] Yet, Schleiermacher recognized, no system can ever successfully capture what it is that makes the individual unique. Herein lies Schleiermacher's fundamental break with the early Fichte, with Hegel—and with Whitehead. In the end, as Günter Meckenstock notes, giving a systematic account of human action has to imply sameness, which means the elimination of particularity.[2] These are not conclusions that Whitehead would endorse.)

How, in light of all this, can one speak of a "wedding" of Schleiermacher and Whitehead, as some have done? The answer lies in the fact, well known to married persons, that weddings do not dissolve differences. A successful marriage establishes common ground between two people without requiring them to be identical. In the best case, their diversity strengthens the

---

1 See Eilert Herms, *Herkunft, Entfaltung und erste Gestalt des Systems der Wissenschaften bei Schleiermacher* (Gütersloh: Gütersloher Verlagshaus Gerd Mohn, 1974).

2 Günter Meckenstock, "Schleiermachers Auseinandersetzung mit Fichte," in *Schleiermacher's Philosophy and the Philosophical Tradition*, ed. Sergio Sorrentino (Lewiston: The Edwin Mellen Press, 1992), 44–5: "Fichtes Berufsgedanke ... sei aber nicht hinausreichend, weil die Vernünftigkeit des Menschen ja gerade die Gleichartigkeit, die Aufhebung der Besonderheit bedeute."

value of their union. One would much rather hear, "Oh, they work so well together!" than "Oh my god, he seems to imitate her in everything she does!"

Despite their differences, there are *some* grounds for optimism. Schleiermacher spoke out vociferously against any system that would constrain individual spontaneity and feeling. His *Dialektik* did not exclude but included the contrasts between conceptual systems, the contrasts between thinkers' opinions, even the ultimate contrast between thought and being. The *Hermeneutik* fought to preserve a place for the particularity of each reader of a text. And the *Glaubenslehre* brooked no opposition between Christian doctrine and the lived experience of the church, insofar as it made the experience of dependence on God into the starting point and cornerstone of theology.[3] (One might also explore the parallels between the alleged panpsychism of the first-edition *Speeches* and what used to be called Whitehead's panpsychism, though I do not do so here.)

If Schleiermacher is the nineteenth-century advocate of a systematic theology that remains in constant process and change, Whitehead is the champion of the same ideas among twentieth-century philosophers. Whitehead's philosophy is equally experience-based, since it is "the endeavor to frame a coherent, logical, necessary system of general ideas in terms of which every element of our experience can be interpreted."[4] His is not a metaphysics of process that itself stands above process. Rather, he realizes, the world of continual process demands a metaphysical system that is never closed, never reaches its end, never attains certainty. Anna Case-Winters in her paper quotes *Process and Reality*: "Metaphysical categories are not dogmatic statements of the obvious; they are tentative formulations of the ultimate generalities."[5] She comments, "He is very clear that this is an 'endeavor,' one that is humble, tentative, imaginative, and readily and necessarily correctable."[6] To be dogmatically Whiteheadian makes as much sense as being a "supply side" Marxist (or being insanely rational).

---

3 Schleiermacher may actually be more helpful to theologians as a philosopher than as a theologian. Thus Michael Theunissen writes, "Gerade auch die theologische Relevanz Schleiermachers liegt in seiner Philosophie, nicht in seiner Theologie." See MICHAEL THEUNISSEN, "Zehn Thesen über Schleiermacher heute," in *Schleiermacher's Philosophy and the Philosophical Tradition*, ed. Sergio Sorrentino (Lewiston: The Edwin Mellen Press, 1992), IV.

4 PR 3.

5 PR 8.

6 ANNA CASE-WINTERS, "System and Dynamism in Whitehead's Thought: The Category of the Ultimate and the Concept of God," 138 in this volume.

So there is some reason to think that the stars will smile on this union. The partnership of Schleiermacher and Whitehead, *if* we can draw on their complementarities as well as their similarities, may point toward a more adequate theory of agency than any of the other options available in contemporary philosophy. I propose to evaluate the potential partnership less via an exegesis of their texts than by putting it to work on a specific task. That is, in these pages I wish to give the hoped-for marriage of their two theories of agency a sort of trial by fire, subjecting it to a rather difficult test (no honeymoon period here!). I wish to ask whether the panentheistic-participatory theory of agency, which (I claim) results from their collaboration, is able to do justice not only to human agency but also to the interaction of human and divine agency. Using the resources provided by these two thinkers, can one conceive a divine agent who created the cosmos and who in some sense remains active within it? Can one conceive divine action such that one preserves the autonomy of the natural world *and* the freedom of the finite agents who arise within it?

II. Agency in Whitehead and Schleiermacher

In order to employ the theories of agency from these two thinkers, we must first understand the distinctive features and emphases of each thinker. Whitehead defines agency in terms of the creative synthesis of the "many" of input data in the "one" of the agent's experience at some moment. Agency is exclusively present-tense; once the synthesis is accomplished, the agent passes out of existence *qua* agent, although the results of her synthetic activity now become available to other agents. Of the three elements of Whitehead's Category of the Ultimate (the one, the many, and creativity) Anna Case-Winters writes,

> A closer look at the Category of the Ultimate as Whitehead describes it reveals the fundamental dynamism among these three elements. The many and the one continually flow into one another in the creative process. Wherever a unity is achieved, it is in fact an incorporation of multiplicity, a complex unity embracing contrasts. It does not remain a closed unity, but it becomes, in its turn, a datum of experience for the multiplicity of other actual entities. Each droplet of experience entails a new integration of multiplicity into a new unity.[7]

[7] Ibid., 146.

Schleiermacher, for his part, stands much closer to the idealist theory of the synthesizing subject and, through it, to Augustine's notion of the enduring subject or soul. Kant had taught that each subject is presented with a manifold of percepts and synthesizes them into the relative unity of experience. His argument that we must posit a "transcendental unity of apperception" to unify the diversity of perceptions and to make sense of the synthesizing activity remains among the most compelling passages of the *Critique of Pure Reason*.[8] Indeed, there are unmistakable parallels between Kant's account of this synthetic process and the theory of concrescence which underlies Whitehead's metaphysics of actual occasions.

Beyond Kant, Fichte also contributed to the answer that Schleiermacher developed. As is well known, Fichte is the great champion of the self-constituting subject, a center of pure activity that synthesizes both itself and its world. Schleiermacher may have opposed Fichte on a variety of points in the 1790s; as Christine Helmer reminds us, he was still attacking Fichte as late as 1833![9] Nonetheless, during those early years, in his own proto-Romantic way, Schleiermacher appropriated Fichte's notion of the ego as the basis of the experienced world, and it remained a framework notion for the rest of his work.[10] In his system, the enduring subject serves as a necessary condition for any adequate account of experience, since the subject and the world are together responsible for the individuality of each person's experience. As Herms writes, "In its mediated reciprocal relationship with the world the self finds itself pre-given, without mediation, as reality."[11]

Like Whitehead, Schleiermacher realized—far better than most of his contemporaries and certainly than Hegel—that *the particularities of individual experience* are crucial to an understanding of agency. General theories of the subject stand in permanent tension with the uniqueness of each

8 Immanuel Kant, *Kritik der reinen Vernunft*, newly ed. Raymund Schmidt, Philosophische Bibliothek, no. 37a (Hamburg: Felix Meiner, 1956), A 106–114.

9 Christine Helmer, "Novelty and System in Schleiermacher's Thought," 181 in this volume.

10 A study of Johann Gottlieb Fichte's *Anweisung zum seligen Leben*, Philosophische Bibliothek, no. 234 (Hamburg: Felix Meiner, 1954) and of the later editions of the *Wissenschaftslehre*, in which Fichte sought to place the self-constituting ego within the context of the whole (*das Ganze*), reveals further affinities to Schleiermacher. (The 1804 edition of the *Wissenschaftslehre* is contained in Immanuel Hermann Fichte, ed., *Johann Gottlieb Fichtes nachgelassene Werke* [Bonn, 1834–35], vol. 2.)

11 Herms, *Herkunft, Entfaltung und erste Gestalt des Systems der Wissenschaften*, 156. (Translation mine.)

individual's experience of the world. In this sense, Schleiermacher was indeed the spiritual father of Wilhelm Dilthey's method of the empathetic understanding (*Verstehen*) of "the Other"[12] and of Windelband's "idiographic" approach to social science.[13] It is no exaggeration to say that the focus on individuality and experience serves as one of two *Grundmotive* for his entire opus.[14]

These similarities notwithstanding, there are at least three contrasts that separate Schleiermacher and Whitehead on our topic, which is the relationship between the moment of creativity or autonomy and the general or structural features of agency. First, as noted earlier, Schleiermacher did not claim knowledge of a metaphysical subject, as a metaphysician might. The subject remained for him, as for Kant, a transcendental postulation, a condition of the possibility of morality and freedom. Now Whitehead certainly recognized the contingency of all metaphysical reflection; we owe to him, as I have argued elsewhere, the idea of "metaphysical hypotheses." *But metaphysical hypotheses are not the same as transcendental postulates.* I do not share the Kantian conclusion that language about "God, freedom, and immortality" remains regulative at best, but here is not the place to make that case. (Let's just say that the possibility of metaphysics will itself be a regulative assumption in what follows!)

Second, Schleiermacher postulated an enduring agent, a subject whose creative activity extends over many moments of time. (For him, as for Kant, the whole of time precedes its division into moments, whereas for Whitehead temporal moments are primary.) For Whitehead, by contrast,

---

12 See WILHELM DILTHEY, *Hermeneutics and the Study of History*, ed. Rudolf Makkreel and Frithjof Rodi (Princeton: Princeton University Press, 1996); IDEM, *Introduction to the Human Sciences*, ed. Rudolf Makkreel and Frithjof Rodi (Princeton: Princeton University Press, 1989).

13 ARNOLD RUGE, ed., *Encyclopädie der philosophischen Wissenschaften*, together with Wilhelm Windelband, vol. 1 (Tübingen: J. C. B. Mohr [Paul Siebeck], 1912).

14 In taking this position, I break with those who would divide Schleiermacher's later work more sharply from his early, *frühromantische* phase. Although there is no doubt that Schleiermacher broke from certain tenets of the Schlegels and the Athenäum period, that his style became more academic, and that theological concerns came to weigh more heavily, this focus on individuality and experience continued to characterize all of his major writings. The growing influence of Schelling, the so-called philosopher of Romanticism, on his work after 1803 supports this interpretation; see HERMANN SÜSKIND, *Der Einfluss Schellings auf die Entwicklung von Schleiermachers System* (Tübingen: J. C. B. Mohr [Paul Siebeck], 1909; reprint, Aalen: Scientia Verlag, 1982). See also MANFRED FRANK, *"Unendliche Annäherung": Die Anfänge der philosophischen Frühromantik*, 2nd ed., Suhrkamp Taschenbuch Wissenschaft, no. 1328 (Frankfurt a. M.: Suhrkamp, 1998).

what we call an agent or person is actually a society of actual occasions, a construct built up out of moments of experience that are metaphysically basic. Finally, Schleiermacher presupposed a sharp contrast between human agents and the natural world; agency is what makes us unique. Whitehead's metaphysic, by contrast, makes this contrast merely a matter of degree: nothing in the world exists that is not either a prehension or the result of a prehension.

III. Panentheism: Human Agents and the Divine Agent

Our discussion would be difficult enough if the problem were limited to finite agents. But, in fact, both Whitehead and Schleiermacher held that human agents exist in interaction with the divine; indeed, much of what is most fundamental to agents, they owe to God. We turn now to the theological problem of agency.

For both thinkers, the individual's relation to God is constitutive of her individuality. For Whitehead, as Case-Winters notes, "God has particular relevance to each creative act and is therefore, for each, the 'initial object of desire' establishing the initial phase of each subjective aim."[15] Schleiermacher does not similarly imagine an input from God at each moment of concrescence. For him, human subjectivity is a reality extended over time, so that God's influence does not need to be constant. But God's role for the authentic individual is no less dramatic. As Giovanni Moretto notes, "for Schleiermacher authentic individuality is born that moment in which a determined consciousness of Transcendence begins and the finite is united with the Infinite."[16] Nor is the individual self ever matted out in the process: "Only one who has found oneself again," the *Speeches* assert, "can discover the living God more clearly than others can."[17] Here indeed is a break from the early Fichte: the individual's power, though crucial, is never the power of absolute spontaneity or self-sufficiency; it is always "power appropriated (*angeeignet*) from the universe."[18]

---

15 Case-Winters, "System and Dynamism," 150.

16 Giovanni Moretto, "The Problem of the Religious in Fichte and Schleiermacher," in *Schleiermacher's Philosophy and the Philosophical Tradition*, ed. Sergio Sorrentino (Lewiston: The Edwin Mellen Press, 1992), 56. Moretto interprets *den Augenblick* as "the generating moment of religious individuality" in idem, *Etica e storia in Schleiermacher* (Napoli: Bibliopolis, 1979), 205.

17 Quoted in Moretto, "The Problem of the Religious," 56.

18 So Albert Blackwell, *Schleiermacher's Early Philosophy of Life: Determinism, Freedom, and Phantasy* (Chico, Calif.: Scholars Press, 1982), 138.

Both thinkers reject the model of God and world as external to one another and hence as externally related. Whitehead's doctrine of the internal relations of God to each actual occasion is well known. Previous actual occasions are immediately present to each new occasion as "data," and each occasion values those data in a unique way. As Marjorie Suchocki writes, "Internally, every actual entity is a confluence of creativity as it unifies the discrete manyness of its past into the unique unity of itself."[19] Among the many possibilities available at any moment are the possibilities that God values and makes available, as it were, to the entity. These are not merely abstract possibilities, however. Because God is an agent who responds specifically to events in the world, God holds out to each occasion a specific valuing of its possibilities, a specific recommendation for its becoming.

Likewise, Schleiermacher's understanding of the immediate presence of the divine to each subject remains an equally fundamental commitment in his thought. One thinks of the sense of the presence of the infinite in the *Speeches*, as well as of the sense of absolute dependence in the *Glaubenslehre*. We return to both positions below.

I construe both positions as panentheistic in structure. We might define *panentheism*, following Steven Knapp, as the view that "there is a divine reality that is the ultimate explanation of the existence of our universe. The observable or 'natural' universe is inseparable from that divine reality, although the divine reality itself is not simply identical to that universe."[20] Beyond these core tenets (or something like them), panentheism is best understood as a family-resemblance term.[21]

On the question of panentheism, one finds some rather major differences between our two thinkers. To put the point sharply: Whitehead conceives the divine agent according to the same metaphysical principles with which he conceives all other agents—except that God never perishes. If the world is "within" God on this view, there is a related sense in which God is "within" every actual occasion at the moment of its becoming. By contrast, the Schleiermacher of the *Speeches* sometimes does not seem to conceive the ultimate as an agent at all! Instead, the Infinite (*das Unendliche*) is

19 Marjorie Hewitt Suchocki, "System without Certainty," 122 in this volume.

20 Steven Knapp, personal correspondence.

21 A good sample of the range of positions, and their common features, is provided in Philip Clayton and Arthur Peacocke, eds., *In Whom We Live and Move and Have Our Being: Panentheistic Reflections on God's Presence in a Scientific World* (Grand Rapids: Eerdmans, 2004).

the "whole" or the "Universe," much in the sense of Spinoza's philosophy, rather than a person or actor. At the times when Schleiermacher does not posit God as a distinct agent and does not attribute agency to the universe, he stands closer to pantheism than to panentheism.

These two views of God cannot be bridged; they are simply different. I am forced therefore to detour from the discussion for a moment and to specify how I will use the term "God" in what follows. (A quick roadmap for specialists: I have moved from Schleiermacher's *Speeches* to *The Christian Faith* and have extrapolated in particular from his comments about the Trinity, although I do not develop that argument here. Whitehead I have modified under Hartshorne's guidance, and Hartshorne in turn has been pushed in some of the directions that Joseph Bracken has suggested in his various publications.) For present purposes it will suffice to list the conclusions as postulates.[22] One might think of the project as a sort of thought experiment: if the God I'm about to describe existed, could that God exercise agency in the world without abolishing all finite agency?

I begin with a larger quotient of difference between God and world than Whitehead is ready to admit. I posit God as a not-less-than-personal reality whose existence and essential nature is not contingent on the existence of a world, even though God's actual experience is. On this view, God freely created the world; the world and the creatures who inhabit it are essentially contingent; and not only each individual, but also the cosmos as a whole, might not have existed. As long as God can be conceived as a divine community of persons, prior to and apart from creation, the metaphysical requirement that persons exist in a society of interrelation can still be met.[23]

What is implied by a freely self-limiting God? Although finite agents are semi-autonomous by nature, the divine agent is semi-autonomous only because of a prior decision to self-limit its creative activity in order to allow other autonomous centers of activity. This thesis contrasts with Whitehead's view that God as highest occasion is necessarily limited by other agents: God would not exist as a center of activity if God could not respond to the concrescences of other occasions. Hence for Whitehead the

---

22 See PHILIP CLAYTON, *The Problem of God in Modern Thought* (Grand Rapids: Eerdmans, 2000), and IDEM, *God and Contemporary Science* (Edinburgh: Edinburgh University Press, 1997) for more detail.

23 I and others have spelled out some of the details of the argument for a process trinitarianism in JOSEPH BRACKEN and MARJORIE HEWITT SUCHOCKI, eds., *Trinity in Process: A Relational Theology of God* (New York: Continuum, 1997).

existence of a world of occasions is a metaphysical necessity for the existence of God; for, as Case-Winters writes, quoting Whitehead, "apart from the experiences of subjects there is nothing, nothing, nothing, bare nothingness."[24]

This view of God that I have just outlined, which remains widespread in Christian philosophical theology today, constitutes the challenge for the remainder of this paper. How can one construct a theory of agency adequate both to the God "in whom we live and move and have our being" (Paul in Acts 17:29) *and* to our more mundane, everyday experiences as human agents? Can Schleiermacher and Whitehead help us here?

IV. Toward a Theology of Participatory Agency

Process thought conceives divine participation in the becoming of each entity in a robust and differentiated way. The divine input is emphatically not limited to human agents: God contributes to each moment of becoming for each occasion within the universe, without, however, compelling the outcome. Whitehead's universal solution is open, however, only to those who accept an atomized theory of time and who conceive every physical object as its own center of experience. What options remain for those who do not endorse these two premises? Is there any other way to conceive an action both as the act of a finite agent and as an act of God?

I suggest that the introduction of panentheism (as defined above) radically changes the framework for discussion from that with which traditional theism struggled. For panentheists, *every* action, being contained within the divine, represents an act of God in some sense. The challenge is to integrate this belief with the various types of finite agency. Traditionally, theists have been committed to three different types of occurrences in the world:

(a) natural regularities in the world of the sort that physicists study, which are not the products of direct (intentional) action either by humans or by God;

(b) free actions initiated by human agents, which are not determined either by natural regularities or by God;

---

24 Case-Winters, "System and Dynamism," 143, citing PR 167.

(c) direct influence by God on the world—on the minds of humans, on the actions of other living things, and perhaps also on the outcome of strictly physical processes.[25]

Sorting out these various modes of influence, I suggest, requires a partnership between two different kinds of intellectual specialization. We turn to systematic philosophers for help in finding a conceptual framework adequate to the challenge, and to scientists in order to understand what kinds of causal agency are exercised at different levels of the natural world. My argument for a panentheistic-participatory theory of agency thus proceeds in three stages.

1. Stage One: Schleiermacher's Contribution to the Theory of Participation

Participation is an inherently dialectical notion. Applied to God and creatures, it expresses the creature's dependence on God for existence (through creation), for ongoing being (through ontological participation or *creatio continua*), and for agency (as in the double-agency doctrine in Thomas Aquinas). I have elsewhere traced the Neo-Platonic sources for the notion of participation and their influence on early modern metaphysics and will not repeat that discussion here.[26] A complete discussion would also have to include a theory of language adequate to the dialectical dimensions of participation. Coleridge's theory of the symbol as something that "always partakes of the Reality which it renders intelligible"[27] supplies a necessary component for such a theory. A symbol "is always an essential part of that, of the whole of which it is the representative."[28] An adequate theory of participation requires symmetries between philosophy of language, theory of agency, and metaphysics, in which each mirrors the other two.

Schleiermacher's theory of participation in the first edition of *Speeches*, while often implicit and clearly not complete, provides some crucial components for the theory I seek. It is worthy of examination in some detail.

---

25 Only the last of these types of direct influence represents a miracle in the traditional sense. See THOMAS AQUINAS, *Summa Contra Gentiles*, qq. 100–3; C. S. LEWIS, *Miracles: A Preliminary Study* (New York: Macmillan, 1947).

26 See CLAYTON, *The Problem of God in Modern Thought.*

27 SAMUEL TAYLOR COLERIDGE, *Lay Sermons*, 30, quoted in STEVEN KNAPP, *Personification and the Sublime: Milton to Coleridge* (Cambridge: Harvard University Press, 1985), 15.

28 COLERIDGE's marginalia on the seventeenth century divine Jeremy Taylor, quoted in ibid., 17.

On Schleiermacher's view, finite agents would not exist without continual participation in the Infinite of which they are parts. A core claim underlies the argument in the crucial second speech:

> Everything finite exists only through the determination of its limits, which must be "cut out" from the Infinite. Only in this fashion can a finite thing itself be infinite within these limits and formed on its own [or: as an individual].[29]

Schleiermacher's premise is the *sine qua non* for relating parts to the infinite whole. He writes of a chaos, "out of which nothing individual can be separated, except insofar as it is 'cut out' in time and space."[30] One can understand any individual agent only from the perspective of "infinite, undivided humanity," which religion alone is able to intuit as revealed through each individual.[31] The reason for religion's special status is that it understands the universe as active and as revealing itself to us; religion, in thematizing this action, apprehends "everything individual as a part of the whole, everything limited as a representation of the Infinite."[32] One will recognize that there is only one world, one nature, only when one is ready to consider the possibility that all is the manifestation of God, "a work of this Spirit and an expression and carrying out of these laws"; only for such a person "is everything that is visible also genuinely 'world'—formed, permeated by divinity, and one."[33]

True, in 1799 Schleiermacher did not yet possess an adequate theory of divine agency. But he knew that the divine had to be conceived both as the all-encompassing whole (*das Ganze*) and as capable of action. He writes

---

29 Friedrich Schleiermacher, *Über die Religion: Reden an die Gebildeten unter ihren Verächtern*, 1st ed., Philosophische Bibliothek, no. 255 (Hamburg: Felix Meiner, 1958), 53 (speech two). I cite the pagination from the first edition (my translations), given as marginal numbers in the Philosophische Bibliothek reprint: "Alles Endliche besteht nur durch die Bestimmung seiner Grenzen, die aus dem Unendlichen gleichsam herausgeschnitten werden müssen. Nur so kann es innerhalb dieser Grenzen selbst unendlich sein und eigen gebildet werden."

30 Ibid., 126–7 (speech two) : a "Chaos ..., woraus nichts Einzelnes gesondert werden kann, als indem es willkürlich abgeschnitten wird in Zeit und Raum."

31 Ibid., 90 (speech two).

32 Ibid., 56 (speech two): "alles Einzelne als einen Teil des Ganzen, alles Beschränkte als eine Darstellung des Unendlichen."

33 Ibid., 87 (speech two): all is "ein Werk dieses Geistes und eine Darstellung und Ausführung dieser Gesetze, nur [für ihn] ist alles Sichtbare auch wirklich Welt, die durchs Universum unmittelbar gewirkt sind."

often of the "actions of the universe"[34] and notes that religious feelings are religious "only insofar as they are immediately caused by the universe."[35] He also realized that theology requires a dialectic that retains the agency of the part as well as the agency of the whole, even while the parts remain *parts of* the whole. Thus Schleiermacher speaks of the human as "a free part of the whole, who is active through his own power."[36] Our awareness of ourselves as separate individuals is not an illusion. It's only that in the final analysis the priority must go to the One, the All.[37] The claim is not, *You have no individuality*, but rather, *Seek to realize your connectedness with the infinite agency, for it is the ground of your own existence as finite agent.* Does not Schleiermacher's stress on the feeling of absolute dependence (*das schlechthinnige Abhängigkeitsgefühl*) in the *Glaubenslehre* imply the same dialectical position? I recognize my utter dependence on God; paradoxically, in recognizing this dependence, I also recognize that I am an individual who must act, and who is capable of acting, either consonant with the whole or in contradiction to it.

Just as Schleiermacher does not allow the "individual agency" pole of the dialectic to collapse, so he also fights to retain the "divine agency" pole intact—despite the strong gravitational pull of Spinoza's monism. He may insist that a religion without God *can be* better than one with God. Still, the highest notion of God is that "of a highest being, a Spirit of the universe, who rules it with freedom and understanding."[38] The goal of the process is *here and now*, "in the midst of the finite, to become one with the Infinite."[39]

The author of the *Speeches* may have been hesitant to require a theistic interpretation of religious intuition—probably, I suggest, because the only theism available to him at the time was one that separated God from world. How might the conclusion have come out differently if a panenthe-

---

34 Ibid., 67 (speech two): "die Handlungen des Universums."

35 Ibid., 119 (speech two): "Alle religiösen Gefühle sind übernatürlich, denn sie sind nur insofern religiös, als sie durchs Universum unmittelbar gewirkt sind."

36 Ibid., 71 (speech two): "ein freier, durch eigene Kraft tätiger Teil des Ganzen." Note that Schleiermacher does not place human concerns at the pinnacle of his theology. At several points he explicitly relativizes the place of humanity to a "Mittelglied" between the lowest and the whole (104–5, speech two).

37 Ibid., 132 (speech two): "Strebt danach, schon hier Eure Individualität zu vernichten und im Einen und Allen zu leben."

38 Ibid., 125 (speech two): the concept "von einem höchsten Wesen, von einem Geist des Universums, der es mit Freiheit und Verstand regiert."

39 Ibid., 133 (speech two): the striving is, "mitten in der Endlichkeit Eins werden mit dem Unendlichen."

istic theology, with its strong stress on immanence, had been available to Schleiermacher? Indeed, I suggest that by the time of the *Glaubenslehre* Schleiermacher had come to hold a *de facto* panentheism.[40] Panentheism is present in the *Speeches* in rudimentary form, namely in the guise of Schleiermacher's (controversial and often derided) notion of a world-soul or *Weltseele*. Adopting a sort of eschatological panpsychism, he suggests that the ultimate goal of the universe is that all should be pervaded by living spirit: "Nothing should be 'dead mass;' everything should be ... life."[41] It is hard not to be struck by the obvious parallels with Whitehead's thought at this point.

Moreover, the following passages about the "world-spirit"[42] could also be reinterpreted and incorporated within a dual-agency theory of divine and human agents. In the moment of apprehension, Schleiermacher writes, my spirit *becomes* the world-spirit, and the world in this moment *is* my body, "for I permeate (*durchdringe*) its muscles and its limbs as if they were my own."[43] The Romantic exuberance of Schleiermacher's formulation makes for the more poetic position, and I in no way wish to undercut it.[44] Still, it takes only a mild reformulation of his text to make it consistent with the panentheistic-participatory theory of agency that I am advancing

40 See CF § 8, postscript (39): "Let us think of the highest stage of religion, and let us accordingly hold Pantheism fast to the usual formula of One and All: then God and world will remain distinct at least as regards function, and thus [the one who holds this view], since he reckons himself as belonging to the world, can feel himself, along with this All, to be dependent on that which is the corresponding One. Such states of mind can scarcely be distinguished from the religious emotions of many a Monotheist .... [N]othing can strictly be said about God in terms of the antithesis between internal and external without imperilling in some way the divine omnipotence and omnipresence."

41 SCHLEIERMACHER, *Über die Religion*, 103–4 (speech two): "Nichts soll tote Masse sein ... alles soll eigenes, zusammengesetztes, vielfach verschlungenes und erhöhtes Leben sein. Blinder Instinkt, gedankenlose Gewöhnung, toter Gehorsam, alles Träge und Passive, alle diese traurigen Symptome der Asphyxie der Freiheit und Menschheit sollen vernichtet werden."

42 Ibid., e. g. 103, 107 (speech two): "den Weltgeist."

43 Ibid., 74 (speech two). Cf. the famous description by Wordsworth in "Tinturn Abbey" of "a sense sublime / Of something far more deeply interfused, / Whose dwelling is the light of setting suns, / And the round ocean and the living air, / And the blue sky, and in the mind of man; / A motion and a spirit, that impels / All thinking things, all objects of all thought, / And rolls through all things" (in WILLIAM WORDSWORTH, *Poetical Works: With Introductions and Notes*, ed. Thomas Hutchinson [Oxford: Oxford University Press, 1969]).

44 Quite the contrary: I am working on a defense of Schleiermacherian Romanticism, tentatively entitled *The Romantic Worldview: How Science, Spirituality and Metaphysics Contribute to the New Romanticism*.

in this paper. Could we not say that, in the moment of apprehension, my spirit acts in harmony with the world-spirit; the world at such moments is Spirit's body, for Spirit permeates its muscles and its limbs as if they were its own?

In the *Speeches*, Schleiermacher would have us concentrate on the points "where everything that is in conflict unites itself again, where the universe manifests itself as totality, as unity in multiplicity, as system, and thus first earns its name" as uni-verse.[45] Even if in 1799 he did not yet find himself compelled to use the notion of God to express this intuition of the One and All, by the time of the 1811 *Dialectic*, he was clear that the concept of God is required to comprehend the intuition:

> Since we posit the absolute as the fundament of all thinking, we have to assume that the idea of deity is present in everything; but in this way we are in dispute with those who separate God from the world.[46]

Reading the *Speeches* back through the eyes of Schleiermacher's later work, we realize that what he there attributed to human agents—

> Your personhood encompasses in a certain sense all of human nature, and this, in all its appearances, is nothing other than your own I—many times multiplied, more clearly identified, yet in all of its changes still eternal[47]

—can finally be attributed to human agents only as they participate within the divine Spirit. Religion, Schleiermacher notes, perceives this process "as one of the greatest actions (*Handlungen*) of the universe."[48] For the highest calling of religion is to tie together the various appearances and actions of Spirit, "and as a consequence to recognize the Spirit by whom the whole is led."[49] Herein lies the core of a panentheistic-participatory theory of agency.

---

45 SCHLEIERMACHER, *Über die Religion*, 128 (speech two): "wo alles streitende sich wieder veeinigt, wo das Universum sich als Totalität, als Einheit in der Vielheit, als System darstellt und so erst seinen Namen verdient."

46 *Dialectic(1811)* 39.

47 SCHLEIERMACHER, *Über die Religion*, 99 (speech two): "Eure Persönlichkeit umfaßt in einem gewissen Sinn die ganze menschliche Natur und diese ist in allen ihren Darstellungen nichts also Euer eigenes vervielfältigtes, deutlicher ausgezeichnetes, und in allen seinen Veränderungen verewigtes Ich."

48 Ibid., 99 (speech two).

49 Ibid., 100 (speech two).

2. Stage Two: A Crucial Whiteheadian Correction

The theory is not yet complete, however. It still struggles under certain limitations that are inherent to the part-whole concept. When one *only* emphasizes that the ultimate is the whole, the universe, the *hen kai pan*—as Schleiermacher tended to do in the *Speeches*—then one's understanding of the agential dimension of the ultimate is truncated. The whole can have certain characteristics, including values, and perhaps it can even "broadcast" these general values, in some sense or another, to the parts. But the whole as such cannot offer specific ideals to specific agents at specific times.[50]

A Whiteheadian doctrine of participation helps to overcome this limitation in Schleiermacher. Each entity participates in the divine experience, insofar as that experience is made available to it as a datum in its moment of becoming. But whereas the early Schleiermacher focused on agential aspects of the *universe* or *das Ganze*, Whitehead's consequent nature of God remains agent through and through. For this reason, Whitehead can consistently maintain that God holds out to each occasion a specific valuing of its possibilities, a specific recommendation for its becoming. Schleiermacher's "universe" could only broadcast a message to all agents, say: "Experience deeply! Live authentically! Rest on the bosom of the Infinite!" By contrast, Whitehead's God, who is an agent with intentions, can intend specific goods for specific individuals at specific moments. God never causes another agent's response. But the divine "lure," the vision that God offers, can be as differentiated and as specific as is needed for any given occasion.

Much more could be said about the details of the process vision. For now, perhaps it is sufficient to note that it has received strong presentation and defense in the literature, and that I am dependent on those arguments here.[51] More urgent in the present context is the question of how to synthesize the diverse sources. In this and the last section I have argued that thinking together these two dimensions—God-as-whole and God-as-agent—remains one of philosophical theology's most perplexing challenges. The early Schleiermacher offers us assistance with the former concept and Whitehead contributes to our understanding of the latter. Holding the two notions in dialectical tension, I suggest, offers a richer concept of

---

50 Schleiermacher did make major advances over the *Speeches'* position in his later theology, but that's a story for another day.

51 See Lewis Ford, *The Lure of God* (Philadelphia: Fortress, 1978); David Ray Griffin, *Reenchantment without Supernaturalism: A Process Philosophy of Religion* (Ithaca: Cornell University Press, 2001).

God than either one alone. It is, however, a tension rather than a contradiction, since there is no reason that the whole cannot be conceived simultaneously as having some of the characteristics of agents. Indeed, the central insight of panentheism lies in the realization that the whole is not just the sum of all agents but, beyond that, also agent itself.

3. Step Three: The Scientific Study of Agents in the Natural World

Philosophical theories should not trump scientific studies in scientific domains. (Incidentally, the converse also holds.) If you want to know precisely what type of influence is exercised by various causal agents in the world, you must turn to the disciplined study of entities of that type. Only empirical study can determine what it is empirically meaningful to assert of a given entity. Thus with a panentheistic-participatory theory of agency in place as one's overarching conceptual framework, one must then join forces with physicists, biologists, primatologists, neuroscientists, social scientists—and, where necessary, with historians, literary theorists, and artists—to complete the account of what kinds of entities do what kinds of things in the natural world.

This is not the place to enter into a detailed listing of the scientific data that bears on this question.[52] Instead, more important in the present context is to specify the implications that follow once one makes the methodological commitment to include careful empirical study *as a constitutive component of metaphysics*. I list the eight most significant implications in outline form[53]:

(1) At the lowest level, physical particles and forces exercise causal powers in a completely law-like manner. For panentheists, such particles and forces are not external to God; hence their causal powers are still ultimately modes of divine activity. Still, since science reveals this domain to obey strict mathematical laws (e. g., every electron is strictly identical), metaphysics is more justified in treating physical interactions as efficient causes than as the results of intentional actions by agents, finite or divine.

---

52 See PHILIP CLAYTON, *Mind and Emergence* (Oxford: Oxford University Press, forthcoming 2004), ch. 3.

53 I gratefully acknowledge Steven Knapp's contributions in personal correspondence to the formulation of points (1) and (5), as well as his important influence on the position developed in the latter half of this paper.

(2) At the biological level, variations between individual organisms influence outcomes to a greater degree. I return to the question of spontaneity in a moment.

(3) At the level of the higher primates, including humans, actions result from internal motivations and choices developed in a quasi-autonomous manner.

(4) Physical determinism would exclude free agency (*pace* compatibilism). Hence *some* degree of spontaneity in the behavior of natural entities is necessary for agency. Fortunately, quantum indeterminacy appears to leave room for the required spontaneity in the physical world.

(5) Indeterminism alone is however not sufficient for agency. Since the other conditions required for agency are lacking in the realms studied by physics and chemistry, one should not apply the category of agency to causal interactions at this level. Atoms and electrons are not themselves agents, nor are their behaviors the result of direct intentional interventions by God. One should thus be skeptical about direct divine influence on the outcome of such processes. Theologically, we infer from the strict regularity of these forces and processes that God has decided to behave non-intentionally at this level, that the divine creative intent was to create a universe that is self-regulating at the physical level. God would in a sense be defeating or contradicting, or perhaps just unnecessarily complicating, God's own creative method by performing intentional actions at this level. (It's a different matter, as we will see, when God is responding to the spontaneous agents that eventually emerge from this regularity.)

(6) Human thought (but presumably not *only* human thought) is not reducible to algorithms (cf. Donald Davidson's "anomalous monism"). What regularities are manifested in conscious agency are the result (*inter alia*) of character, language, culture, intellectual training, the attempt to be rational, the quest to be moral, etc. Hence no natural laws are abrogated if and when the divine agent influences the outcome of conscious processes—just as no laws would be broken if you actually come to hold some new beliefs based on reading the argument in this paper.

(7) The anatomy and behaviors of at least some animals are sufficiently analogous to humans that one should postulate that they, too, exercise spontaneous agency. To the extent that this is true, there is no obstacle to a divine influence on their actions as well.

(8) More generally, wherever study of the biological world reveals sufficient spontaneity, room remains for a divine influence on behavior. (Exactly what level of spontaneity is sufficient is a matter for further reflection and study.)

V. Conclusion

In this paper I have outlined a panentheistic-participatory theory of agency in dialogue with Schleiermacher and Whitehead. On the view defended here, the actions of all finite agents participate in the divine act, yet in such a fashion that their (partial) autonomy of action is preserved. One expects the degree of autonomy to vary widely across the levels of creation. Establishing the degree of autonomy is a matter not of philosophical stipulation but of empirical study, on which we remain permanently indebted to our scientific colleagues.

I have not been able to press the ethical implications of this view. Perhaps a word on this topic will be allowed in closing. I have spoken of *quasi*-autonomous action on the part of finite agents because only in the ideal case—a case that may or may not ever be actualized—is full or perfect autonomy exercised. Paradoxically, on this view the fully autonomous action would be the one in which the agent's act is perfectly in accord with the creative intents of the divine agent in whom she participates. Given the dual framework of participation and panentheism, to act in accord with the divine intent is to participate in the ongoing divine act. To act *perfectly* in accord with divine intent would be to *fully* share in the divine act, the result being an action that is fully divine and fully human. This perfect doubling of agency exposes the moral, or even christological, dimension to the theory of agency I have defended. Perfect or "pure" agency represents the ideal of one who could say, "Not my will, but thine be done." Obviously, such a state represents an ideal rather than a given for most, if not all, agents. To understand agency as ideal rather than as given helps to explain both why system and agency remain in permanent tension, and why that tension need not be ultimate or unresolvable.

III

System and Contemporary Themes

The Force of Dialogue and a Dialogue of Forces

Resources for Open Theological Systems

JULIA A. LAMM, Georgetown University

The theme chosen for this constructive paper is that of systematic theology and whether it is still possible. There is no little irony that Whitehead criticized modernity for, in the words of David Ray Griffin, its "loss of commitment to systematic thought."[1] For now the deadening refrain of postmodern critiques of modernity is that it was over-committed to system. The perceived evils of system are that its intentions and pretensions are totalizing; it forgets, ignores, controls, and elides the particular, the historical, the other; it refuses to acknowledge gaps and cleavages, exceptions and threats.

In light of such criticisms and suspicions, any attempt to engage in systematic theology naturally comes under suspicion. Theologians have so internalized the postmodern critique that the field of theology (at least in the United States) has become fragmentary and fragmented. The few who do dare to entertain the possibility of a systematic theology usually focus on one doctrine, rather than trying to systematically relate several doctrines. Typically, they also begin with a new brand of apologetics by self-imposing limits on any claim. For example, Roger Haight writes in his Introduction to *Jesus: Symbol of God*:

> This is an essay in systematic theology and christology. As a work of systematic theology, however, it does not intend to contain the subject matter in a totalizing manner and does not

1 DAVID RAY GRIFFIN, "Feeling and Morality in Whitehead's System," 265 in this volume.

> result in 'a system.' Generally speaking, postmodernity's historical and pluralistic consciousness have taken the teeth out of the totalizing pretentions of systems of thought. Rather the systematic character of this work is defined, first, by a *consistent perspective and method.* I attempt to present a continuous argument from the first to the last chapter .... Second, the work is systematic also because it deals with a certain range of topics that are deemed relatively adequate to constitute a *broad treatment* of the subject matter.
>
> But these same two indicators underline the severe limitations of this and any other systematic work. There are other legitimate perspectives and methods in Christian theology and christology that will yield other genuine insights.[2]

However important his contribution to Christology, Haight here puts forward an unnecessarily impoverished view of the task of systematic theology.[3] The first mistake of such a view is its uncritical acceptance of postmodernity's critique of system and its claim that it (and seemingly it alone) honors the "historical." The second, and much more serious, mistake is the admission of only two criteria—*consistency* and *breadth*. These two criteria in themselves neither make a system nor guard against the very dangers any system might yield. Totalitarian systems, after all, can be both consistent and broad.

I maintain that a much more robust understanding and practice of systematic theology is both possible and required. Certainly, any systematic theology would have to be both consistent and broad, but those are necessary rather than sufficient criteria. Christian theology necessarily works out of the convictions that there are universals, however difficult to discern; that values are grounded in something besides preference, despite the temptation to prefer one's own values; that there are genuine relations and interconnections, or at the very least that there are possibilities for new relations and interconnections. A robust systematic theology would dare to ask the question of truth and of ontological grounding—otherwise it is no more than ideology.

---

2 Roger Haight, *Jesus: Symbol of God* (Maryknoll, N.Y.: Orbis Books, 1999), xiii (emphases added).

3 My criticism of Haight is here directed at the level of form rather than content.

The perceived dangers of systems are not inherent to the notion of *system* but to a kind of system—*closed systems.* Closed systems are produced by and produce ideologies. There is another type of system—*open systems*—that not only try to avoid, but by nature try continually to expose, the dangers of being "totalizing" and "absolutizing." "Open systems" are *systems* insofar as they (1) seek connections and interconnections, (2) strive toward something universal, (3) look for the universal in and through the particulars, (4) can coherently and adequately account for the relations of particulars to particulars and of particulars to universals, (5) allow the question of the real, thereby resisting the temptation to see everything as *only* construction, and (6) develop a consistent method by which we can come to understand these relations. "Open systems" are *open* insofar as they (1) are open to novelty, the unexpected, and difference, (2) attune themselves to the historical, being always aware of how history envelops us in ways we cannot begin to comprehend, (3) welcome critique, being willing to give an account of as well as to relinquish cherished principles, and (4) are committed to self-critique. According to this understanding, both F. D. E. Schleiermacher and A. N. Whitehead can be said to have produced open systems.

I suggest that the possibility of developing Christian systematic theologies in the twenty-first century is closely tied to this notion of *open systems.* Friedrich Schleiermacher—the author of the first modern theological system—can provide us with significant resources. His *Glaubenslehre* (*The Christian Faith, Systematically Presented according to the Basic Tenets of the Evangelical Church*) is itself one obvious resource, but I want to look behind, within, beneath, and beyond its particular content and structure to something more elemental, to what makes it a system. Before elaborating on what those elemental resources are, I need to make clear what I shall *not* do in this essay.

I shall not examine, defend, or revise Schleiermacher's *The Christian Faith* as a pre-eminent example of an open theological system. That is a task worth doing, but such a task would not be constructive, and my assignment is to write a constructive paper. Nor shall I enter into recent and current debates about Schleiermacher's dogmatic theology (whether it is a "system"), about the fundamental character of his theology (whether it is "cultural-linguistic" or "expressivist," foundational or non-foundational), or about consistencies and inconsistencies between his theology and his philosophy.

What I propose doing instead is to look in, behind, and beyond Schleiermacher's theology and philosophy for elements that make possible and can sustain an open theological system. Two of these elements, in particular, are promising; it just so happens that they have certain affinities with Whitehead. The first is Schleiermacher's *dialogical-dialectical method*, which is confident of the lure of truth at the same time that it remains empirical and provisionary. The second is his dynamic ontology, according to which reality—particular entities, more universal entities, and the many degrees and kinds of relations between and among them—is understood primarily in terms of *force* rather than substance. These two principles together—dialogue and force—are promising resources for engaging in a robust systematic theology—robust not in the sense of closed and totalizing, but in the sense of art. A systematic theology should be a composition that dares to ask the question of truth, dares even to answer it, without claiming to control it. For writing open systematic theologies in the twenty-first century, what Schleiermacher understood about the force of dialogue and a dialogue of forces just may prove to be instructive.

First, however, I need to examine my own irritation with what I have labeled the standard post-modern criticisms of modernity and its systems. Rather than engage in a full-scale critique of such criticisms, I shall simply rely on the carefully made arguments of two philosophers, citing one and building on the argument of the other. James Schmidt writes in his essay, "What Enlightenment Project?,"

> The Enlightenment project is largely a projection of the Enlightenment's critics, a projection that fastens onto a few thinkers or tendencies within a broader period and, having offered an account of what it sees as the failings of these thinkers or these tendencies, prides itself on having demonstrated the failures of the entire age.[4]

Stephen Toulmin, who shares some of the legitimate concerns of the post-modern critique of modern system building, is more interested in unraveling the mystique of modernity than in taking to task certain of its post-modern critics; nevertheless, his re-contextualizing of modernity exposes the sloppiness and ignorance of many other critics of modernity. More im-

4 James Schmidt, "What Enlightenment Project?," *Political Theory* 28, no. 6 (Dec. 2000): 737–8.

portantly, his *Cosmopolis* provides an illuminating set of criteria for what *open systems* might look like.

In section one, I shall therefore look more closely at the issue of modernity, using Stephen Toulmin's analysis as a suggestive guide. In section two, I shall discuss Schleiermacher's own suspicions and critique of 'system.' These critiques of modernity and of system will supply us with four *limiting and determining criteria* for open systems. In section three, I shall explore Schleiermacher's dialogical dialectics, testing to see whether it meets two of the limiting and determining criteria. And in section four, I shall in turn explore Schleiermacher's notion of *force*, testing whether it can meet the remaining criteria.

I. Toulmin's Analysis of "Modernity"

Published in anticipation of the new millennium, Stephen Toulmin's *Cosmopolis: The Hidden Agenda of Modernity*[5] seeks to understand the present transition from the modern to the post-modern (or, from the second to the third phase of modernity[6]). Toulmin sees the postmodern age as both inevitable and necessary: inevitable historically, in that we are already there[7]; necessary intellectually, in that we need (scientifically, politically, philosophically) to acknowledge and, in his words, "to protect *diversity and adaptability*."[8] Hoping to "salvage what is still humanly important in its projects," Toulmin undertakes an intensive analysis and re-contextualization of modernity. Thus, in pursuing a constructive agenda, he employs an historical argument and, in the process, uncovers the depth, complexity, and tensions of modernity, which has all too often been assumed to be a great monolith. Toulmin concludes, "we can neither cling to Modernity in its historic form, nor reject it totally—least of all despise it. The task is, rather, to reform, and even reclaim, our inherited modernity, by *humanizing* it."[9]

---

5 Stephen Toulmin, *Cosmopolis: The Hidden Agenda of Modernity* (New York: Free Press, 1990).

6 Ibid., 203.

7 See for example ibid., 3.

8 Ibid., 183.

9 Ibid., 180.

Toulmin begins his historical investigation by challenging the conventional dating of the beginning of modernity. Modernity, he argues, began not with the rationalism of Descartes, which tried to extricate itself from history and its contingencies, but with Renaissance humanism a century before. With this new starting point, Toulmin goes on to describe several historical stages of modernity, stretching from the sixteenth century to the third quarter of the twentieth century. Throughout these stages, he traces what he variably calls "two distinct origins,"[10] "two traditions,"[11] "two aspects,"[12] a "dual trajectory,"[13] "twin legacies,"[14] and the "two rival agendas"[15] of modernity. The dominant trajectory he refers to as "high" or "unreconstructed Modernity"; the other trajectory, eclipsed but not irretrievable, he calls a "humanized Modernity."[16] The former Toulmin traces back to the seventeenth century; the latter, to the sixteenth century. Always, Toulmin carefully offers sympathetic explanations as to how each arose and why one would become dominant. A quick glance at descriptive categories for the two trajectories captures the differences between the two modernities (see Table One). Note that Toulmin associates system with "high Modernity" (the right-hand column).

Toulmin laments that "these two traditions were not seen from the beginning as complementary, rather than in competition."[17] Yet his own prescription for the future does not always appear to be consistent with this diagnosis. Sometimes his project appears to be one of integrating the two traditions[18]; at other times, he suggests that rationalism's commitments are just no longer tenable and what is needed is a recovery or re-appropriation of humanism's commitments.[19]

10 Ibid., 23.
11 Ibid., 43.
12 Ibid., 167.
13 Ibid., 172.
14 Ibid., 175.
15 Ibid., 192.
16 Ibid., 182, 183.
17 Ibid., 43.
18 "The 'modern' focus on the written, the universal, the general, and the timeless … is being broadened to include once again the oral, the particular, the local, and the timely" (ibid., 186).
19 See ibid., 200–1.

Table One: Toulmin's Two Trajectories of Modernity[20]

"HUMANIZED MODERNITY"	"HIGH MODERNITY"
"experiential" "scientific"	"doctrinal" "metaphysical" (167; cf. 172)
"commitment to intellectual modesty, uncertainty, and toleration" (174)	"dream of *foundationalism* ... the search for a permanent and unique set of authoritative principles for human knowledge" (174)
"Aristotelian attention to times and places"	"Platonist drive toward universal theory" (181; cf. 192)
"humane wisdom of the Renaissance"	"Cartesian philosophy and the exact sciences" (174)
"reintegrates nature and humanity"	"saw nature and humanity as distinct and separate" (183)
"*diversity and adaptability*"	"*stability and uniformity* of Science or State" (183)
"respect for complexity and diversity" (28)	"rigor, exactitude, and *system* (184; emphasis added)
"*context bound* issues"	"*context free* issues" (184–5)
"concrete, practical issues" (24)	"three pillars" of "certainty, *systematicity*, and the clean slate" (179; emphasis added)
the oral, particular, local, and timely	the written, universal, general, and timeless (see 30–6, 186–92)

The last set of contrasts in Table One becomes the hinge on which Toulmin's argument (both historical and constructive) turns. Because these four polar coordinates (written / oral, universal / particular, general / local, and timeless / timely) together comprehend all other contrasts in them, they become for Toulmin the criteria by which he both measures the movement from humanism to rationalism and anticipates the retrieval of "humanized Modernity."[21] Or, to shift images, they form an axis of polar coordinates. This axis of polar coordinates (see Table Two), I suggest, is especially illuminating for understanding not only modernity but also open

20 Schematization of TOULMIN's *Cosmopolis*.
21 See ibid., 30–6, 186–92.

systems. I want to use it to argue two points, one historical, the other constructive. *First*, the two strands were not always and everywhere in competition in modernity. Schleiermacher quite successfully held the two together in theory as well as in practice. Thus the task Toulmin enjoins upon us—to broaden so as to retrieve and include, thereby to integrate—has already been done. There is at least one precursor, one example from which to learn. *Second*, this axis of polar coordinates provides a promising way for understanding "open system." The one set of coordinates is determining for "system": a system should pertain to the general, universal, and timeless. The other set serves to limit the system, ensuring that it remains "open": it attends to the particular, the local, and the timely.

Table Two: Axis of Polar Coordinates[22]

the oral	the written
the particular	the universal
the local	the general
the timely	the timeless

II. Schleiermacher's System Beyond "System"

Schleiermacher stood on this side of what Stephen Toulmin has identified as the rationalistic era of modernity. A product of the Enlightenment, Schleiermacher was nevertheless fundamentally and irreversibly a Renaissance humanist. The Renaissance in which his intellect was first formed was the German Renaissance of the late eighteenth century. In the last quarter of that century, a confluence of three movements resulted in an amazing rebirth of the ancient as well as birth of the new German literature. German philosophy had come into its own, in part because philosophers dared to break with Latin as the authoritative language, crafting instead philosophical distinctions and terms in German; philologists had returned to the classics—modern as well as ancient—brimming with the confidence that the German language alone could translate and communicate their riches; German literature itself reflected the genius and newness of the age. There was an effusive confidence that the German language had entered a new

22 My term, not Toulmin's.

stage of development, which in turn allowed the German genius to grasp new things. In this context, Schleiermacher studied the ancient Greeks at the University of Halle. He was intent, for a time, on translating Aristotle (see the left-hand column of Table One). That project never came to fruition, since someone else beat him to it; as it turned out, he translated Plato instead, changing the course of Plato studies as a result. Studying the ancients—and studying them with the aim of translating—informed his fundamental approach: to start with the particular, to understand the complexity, to accept ambiguity, to seek clarity nonetheless, and always to appreciate beauty.

Schleiermacher thus provides one example of how the dual trajectories of modernity were not always as cleanly separate as Toulmin suggests. Although a humanist, Schleiermacher loved his rationalists. Or, perhaps more accurately, he preferred both / and over either / or. He saw Spinoza as both rationalist and humanist; he tried to de-rationalize Kant's moral anthropology. He did mistrusted systems and dismantled them (often unconsciously) whenever he could.

Schleiermacher's first break with "system" occurred in the early 1790's in a series of essays he wrote while still a student. The young Schleiermacher, influenced by his reading of the ancients, boldly resisted the confines of the new dominance of Kant's philosophical anthropology. He rejected Kant's view of nature as a system of mechanical, efficient causes and, consequently, Kant's extraction of the human person from such a system in order to protect and guarantee morality. Schleiermacher had to reconceptualize the system of nature, a system in which human beings were integral. He insisted on the "universal causal connection" and "total interconnection."[23] Thus fully engaged in the task of practical philosophy, Schleiermacher understood the human person and her or his actions precisely in terms of the particular, the timely, and the local. Because his Renaissance humanism did not precede but followed (or was contemporaneous with) what Toulmin calls "high Modernity," and because therefore he was critical of some of the delusions of rationalism at the same time that he shared some of its aspirations and applied its tools of analysis, Schleiermacher could glimpse a possibility of something new. This possibility became realized with his participation in early German Romanticism, which broke the old molds.

---

23 See FRIEDRICH SCHLEIERMACHER, *On Freedom (1790–92)*, trans. Albert Blackwell, Schleiermacher: Studies and Translations, vol. 9 (Lewiston: The Edwin Mellen Press, 1992), 120, 53, respectively.

The first wave of Romanticism railed against system and theory. The early Romantics—Schleiermacher among them—saw themselves as revolutionaries who consciously and decisively made a break from the *Aufklärung*. The very term *romantic* (*romantisch*) was coined by Friedrich Schlegel, the leading theorist of early German Romanticism, to distinguish the new movement from what was "modern." Their "philosophy of poetry" dissolved the typical modern categorization of literary forms. It likewise challenged modern philosophy's entrenched habit of isolating reason from the rest of human experience. The early German Romantics were thus the first self-conscious post-moderns. They were willing to abandon the modern slavish preoccupation with form, system, and clarity; they encouraged the disruption and even dissolution of particular forms so that the infinite could break through; they relished the excess and chaos that often resulted; they abandoned the notion that ideas and art were imitations or representations. They embraced the tenuousness, the paradox, and the necessary incompleteness that the Enlightenment had tried so vigorously to eradicate. That necessary incompleteness of finite existence, which is continually forming and re-forming and which can never calmly contain the teeming infinite, is best captured by a literary form that is un-formed—the fragment, which, because of its fragmentation, is universal. And so the pages of the *Athenaeum*, the main literary vehicle of the Romantics, were filled with fragments, letters and dialogues. "A dialogue," Schlegel explained, is really "a chain or garland of fragments."[24]

Schlegel, at least in his writings, remained with the fragment, although he would eventually despair of the project and abandon it, eventually opting for closed systems—political and, some would say, religious. His one-time friend, Schleiermacher, continued to seek that "higher" system, a system beyond "system." For most of the early Romantics, *system* meant something closed, finite, contained, constrained, forced, and in the end dead. Such systems were what Schleiermacher variously referred to as "transcendental philosophy," "rounded idealism," or "speculation." Yet Schleiermacher was not as willing as his Romantic friends to let go of the notion of system entirely. He proposed to them an entirely new way of understanding system: "You are not accustomed to connect the concept of something infinite with the term *system*, but rather the concept of something that is

[24] FRIEDRICH SCHLEGEL, "Fragment 116," *Athenaeum* 1 (1798); trans. PETER FIRCHOW, *Friedrich Schlegel: Philosophical Fragments* (Minneapolis: University of Minnesota Press, 1991), 27.

limited and completed in its limitation."[25] Here we have the beginning of a distinction between "closed" and "open" systems.

Schleiermacher's critique of "system" (the prime example of which at the time of the *Speeches* was Fichte's) can be read as a critique of a "closed" system—that is to say, systems that are constructed and then imposed, drained of life and genuine possibility. Such a system "classifies the universe and divides it into this being and that, seeks out the reasons for what exists, and deduces the necessity of what is real while spinning the reality of the world and its laws out of itself"[26]; it "will destroy the universe by appearing to fashion it; it will degrade it to a mere allegory, to an empty silhouette of our own limitedness"[27]; it has "the tendency to posit essences and to determine natures, to lose itself in an infinity of reasons and deductions, to seek out final causes, and to proclaim eternal truths."[28]

He proposed instead another kind of system, a *higher* system or unity. In this system beyond system, genuine knowledge is possible, but it can be neither naive nor certain; rather, it must yield to the recognition of the possibility of an even "higher" knowing or succumb to mockery on the part of the universe. Schleiermacher offered an example from astronomy: "the perturbations in the course of the stars indicate a *higher unity*, a bolder combination than that which we have already proved true from the regularity of their paths, and the anomalies, the superfluous touches of malleable nature, compel us to see that it treats its most definite forms with an arbitrariness, with an inventiveness, as it were, whose principle we can discover only from a higher standpoint."[29] With no little irony, he called this system beyond system "infinite chaos":

> This infinite chaos, where of course every point represents a world, is as such actually the most suitable and highest symbol of religion. In religion, as in this chaos, *only the particular is true and necessary*; nothing can or may be proved any thing else. Everything universal under which the particular is supposed to be treated, each collection and combination of this sort, either exists in a different territory, if it is to be referred to the inner and essential realm, or is only the work of play-

25 *On Religion*, 26 (speech two).
26 *On Religion*, 20 (speech two).
27 *On Religion*, 24 (speech two).
28 *On Religion*, 20 (speech two).
29 *On Religion*, 36 (speech two; emphases added).

> ful imagination and freest caprice .... Individual persons may have their own arrangement and their own rubrics; the particular can thereby neither win nor lose.[30]

This *higher unity* or *infinite chaos* is a "system" precisely because it is characterized by the fluid interconnectedness of all of its constituent parts, which is all of reality; it accounts for the one and the many, it does not lose sight of the particular, and it does all this because it is steeped in the infinite. It is "open" because it allows for genuine novelty and change as it resists attempts to essentialize or over-determine it; the causal relations are mutual and pluriform; it is knowable, but infinitely knowable and resistant to attempts to be absolutized. It is both "open" and a "system" (or, better, a system of countless systems) because it is *living* and because our knowledge of it, too, must be *living*.

Two elements, in particular, stand out as sustaining this as both "open" and "system." The first is its dialectical method, described in the *Speeches* as oscillation:

> While intuiting a universal relationship your glance is so often led back and forth directly from the smallest to the greatest and from the latter back again to the former and moves between the two in living vibrations until it becomes dizzy and can distinguish neither great nor small, neither cause nor effect, neither preservation nor destruction any longer.[31]

The second is its view of the universe as an organic system of forces. In typical Romantic fashion, Schleiermacher appealed to the metaphor of the organic body to describe this universe and our place in it:

> ... chemical powers, the eternal laws according to which bodies themselves are formed and destroyed, these are the phenomena in which we intuit the universe most clearly and in a most holy manner. See how attraction and repulsion determine everything and are uninterruptedly active everywhere, how all diversity and all opposition are only apparent and relative, and all individuality is merely an empty name. See how all likeness strives to conceal itself and to divide into a thousand diverse

30 *On Religion*, 27 (speech two; emphases added).
31 *On Religion*, 43 (speech two).

> forms, and how nowhere do you find anything simple, but everything is ornately connected and intertwined.[32]

These two—the oscillating movement and the world viewed as an organic system of forces—are correlated. The method arises from and reflects the organic universe and is attuned therefore to its secrets. Elsewhere I have argued that these are two aspects (*higher realism* and *organic monism*) of Schleiermacher's post-Kantian Spinozism that can be traced from his earliest essays through to his mature theology.[33] Nevertheless, recognizing continuities is not to deny discontinuities. Here I want to argue that the method of oscillation and the organic world view which Schleiermacher presented in his *Speeches* are still too parochial to serve as resources for twenty-first century open systems. His later development of them, however, can provide us with vital resources. For our present purposes, the primary (but not sole) importance of Schleiermacher's *Speeches* is its critique of "modern" systems—a critique which does not dismiss the possibility of system altogether. Schleiermacher would have appreciated Toulmin's analysis of modernity, but, unlike Toulmin, he would not have relegated *system* to the right-hand column.

III. The Force of Dialogue: Schleiermacher's Dialogical Dialectics

1. Schleiermacher's Original Insights

Philip Clayton has argued convincingly that Schleiermacher's lectures on *Dialectics* prove to be an invaluable resource for re-thinking the category of the infinite in terms of limit.[34] I, too, want to retrieve an aspect of Schleiermacher's dialectics. While, for philosophical theology, the notion of limit is clearly important, for systematic theology it needs to be balanced by, or perhaps infused with, a notion of limit as a lure that guides method. A dialectical method in itself does not guarantee openness and may, in fact, serve to tighten the closure. There are dialectics, and there are dialectics.

---

32 *On Religion*, 36 (speech two).

33 See Julia A. Lamm, *The Living God: Schleiermacher's Theological Appropriation of Spinoza* (University Park, Penn.: University of Pennsylvania Press, 1996).

34 See Philip Clayton, *The Problem of God in Modern Thought* (Grand Rapids: Eerdmans, 2000). See esp. ch. 6: "On Using Limit Notions: First Steps after Kant," 347–83.

I therefore propose to experiment and take a step back behind Schleiermacher's first lectures on *Dialectics* to a decisive preparatory stage when he described the dialectical method as impelled by *eros*. That is to say, I want to go back to the decade or so between the first edition of the *Speeches* (1799) and the first lecture on *Dialectics* (1811), to the time between his Romantic critique of "system" and the construction of his own systematic thought, to the time when he was chiefly preoccupied with Plato.[35] Schleiermacher's monumental translation and interpretation of Plato pertains to our consideration of "system" for the twofold reason that in it Schleiermacher almost singlehandedly dismantled the basic structure of one of the most ancient and powerful forms of systems (neo-Platonism), and he did so with subtlely but decisively through his reading of Plato's dialectical method. In his reading of Plato's dialectics, we find the rudimentary oscillating dialectical movement of the *Speeches* developed into a sophisticated dialectical method—a decidedly dialogical form of dialectics.

Schleiermacher began his "General Introduction" with a critique of previous interpretations of Plato. They failed in large part because they assumed Plato had a system, and they wound up imposing a system (their own) onto Plato. Schleiermacher changed our assumptions about Plato. He agreed that there was a *unity* to Plato's thought, but unity, as we have already seen in the *Speeches*, is not the same as *system*. He located the source of the unity of Plato's works in Plato himself—in his moral integrity, artistic genius, and pedagogical wisdom. Much more than ridding Plato's thought of any explicit system, Schleiermacher dismantled the very scaffolding that supported various Platonic systems. He unceremoniously rejected Aristotle's claim that the Platonic forms had independent existence.

---

35 See PLATO, *Platons Werke von F. Schleiermacher*, trans. into German by FRIEDRICH SCHLEIERMACHER (Berlin: In der Realschulbuchhandlung, 1804–1828), vol. I / 1 (1804); vols. I / 2 and II / 1 (1805); vol. II / 2 (1807); vol. II / 3 (1809); vol. III / 1 (1828); 2nd ed. (1817). Schleiermacher's *Introductions to the Platonic Dialogues* have recently been made available in SCHLEIERMACHER, *Über die Philosophie Platons*, ed. Peter M. Steiner, Philosophische Bibliothek, no. 486 (Hamburg: Felix Meiner, 1996). The introductions and translations appear together again in *Platon im Kontext*, vol. 4 of *Literatur im Kontext auf CD-ROM* (Berlin: Karsten Worm InfoSoftWare, 1998). An English translation of Schleiermacher's *Introductions* can be found in FRIEDRICH SCHLEIERMACHER, *Introductions to the Dialogues of Plato*, trans. William Dobson (Cambridge and London, 1836; reprint, New York: Arno Press, 1973). Unless noted otherwise, translations are my own and page references for the Introductions are to the Meiner edition. For more on the significance of Schleiermacher's Introductions, see JULIA A. LAMM, "Schleiermacher as Plato Scholar," *The Journal of Religion* 80, no. 2 (April 2000): 206–39.

This had many consequences for his interpretation of Plato—the most important of which, for our present purposes, was his resistance to more speculative aspects of Plato's philosophy (those associated with closed Platonic systems) and his clear preference for dialogical dialectics.

Schleiermacher viewed Plato's dialectics primarily, indeed almost exclusively, in terms of a dialogical form of dialectics. "Necessarily and by its very nature," he wrote, "Plato's method was a Socratic one."[36] The importance of this requires a distinction between two kinds of dialectics. In the words of Julia Annas, one kind of Platonic dialectics is "articulate" and "looks like a development of the Socratic conversations."[37] This is what I am calling *dialogical dialectics*. Another kind of Platonic dialectics involves "constant use of the imagery of vision and grasping"; it is "something non-verbal and non-articulate that each person has to do for himself or herself."[38] This latter form of dialectics I refer to as *speculative dialectics*—and it is this method, I suggest, that is more closely associated with closed systems. Schleiermacher indirectly acknowledges this kind of dialectics when he refers to a "separate investigation which properly precedes communication," an investigation in which one gazes purely at the truth[39]; a glimpse into the "the higher sphere(s) of speculation,"[40] where the "higher spheres" are those more informed by the identity of thought and being than by the oppositions of the empirical sphere(s). Yet he never admitted to or identified two methodological strands in Plato, since that would have challenged the unity of Plato's thought. Instead, he either suppressed those passages and texts which contain an undeniable strain of speculative dialectics or integrated them with more dialogical-dialectical passages and texts.[41]

This determined preference for the dialogical over the speculative is an element of Schleiermacher's method that makes his thought such a valuable resource for thinking about open systems. I shall now provide three brief examples of how this preference actually functioned in his interpretation of

36 Schleiermacher, "General Introduction to the Platonic Dialogues," 40.

37 Julia Annas, *An Introduction to Plato's Republic* (Oxford: Oxford University Press, 1981), 282.

38 Ibid., 283.

39 Schleiermacher, "Introduction to the *Parmenides*," 129.

40 Schleiermacher, "Introduction to the *Sophist*," 250, and "Introduction to the *Parmenides*," 139.

41 I argue this point in detail in Julia Lamm, "Reading Plato's Dialectics: Schleiermacher's Insistence on Dialectics as Dialogical," *Zeitschrift für neuere Theologiegeschichte / Journal for the History of Modern Theology* 10, no. 1 (2003): 1–25.

Plato's dialogues and how each in turn can be a resource for thinking about systematic theology.

My first example of Schleiermacher's determination to read Plato's dialectics as dialogical is found in his "Introduction to the *Phaedrus*." In addressing the notoriously difficult problem of the two parts of the *Phaedrus*, Schleiermacher began in the second part, with Socrates' definition of dialectics as a "systematic art" and his redefinition of rhetoric in terms of it. The dialectical method of dissecting and then collecting together is the Socratic-Platonic method. Schleiermacher, stressing the collecting aspect of dialectics, argued that the real significance and power of the dialectical method cannot be fully understood unless and until we re-collect or reconnect the two parts of the *Phaedrus*. When we do that, when our attention is turned to the three speeches and especially Socrates' second speech of part one, we learn that method does not stand alone apart from *eros*. In other words, in insisting on the necessary and natural connection of the two parts of the dialogue, Schleiermacher discovers the necessary and natural relation of *eros* and dialectical method. The *eros* of which Socrates speaks is interpreted as an original "*impulse*" in the human soul and as the divine eternal for which we yearn. The very purpose, therefore, of the art of dialectics, according to Schleiermacher, is to draw forth and stimulate the impulse in another, being guided by and moving towards eternal love.

This is a valuable resource for understanding open theological systems, I maintain, because it underscores how the dialectical method —the basic method of proceeding for any genuine inquiry, for any knowledge—is inherently *dialogical*, that is to say, *communicative*. The impulse, lure, and goal of all knowing and all processes of knowing is love—love that draws us to another, that characterizes the dialogue, that draws us beyond ourselves. Following this line of thought, the tone and purpose of an open theological system would be not a lone attempt to claim certain truths but an active engagement in communicating through dialogue. It is a dialogue that is with direction; it proceeds with a confidence that there are truths to be discovered and that we may discover something about those truths, even if asymptotically. Schleiermacher's dialogical dialectics is essentially a relational view of method and knowing. *Eros* is the impetus and the lure.

A second example is from the "Introduction to the *Parmenides*." The *Parmenides* presented a particular challenge to Schleiermacher's determination to read Plato's dialectics as dialogical, for here he encountered an undeniably speculative form of dialectics. Yet he stuck to his main line of interpretation by viewing this dialogue not as representing the Socratic-

Platonic method but as a polite acknowledgment of Socrates' predecessors. The *Parmenides* is an example of an earlier dialectical method, one which had influenced Socrates as a young man but which he went beyond. Plato's purpose in writing the *Parmenides* was to acknowledge his and Socrates' debt to both Parmenides and Zeno. In addition, Schleiermacher argued that Plato uses the *Parmenides* as a way of formulating the "highest philosophical problem."[42] That problem—finding "the original unity of thinking and being"[43]—would not be solved until the later dialogues. In short, the *Parmenides*, with its cold dialectics, does not contain a Platonic form of dialectics; it rather serves to highlight the difference between a speculative dialectics and Plato's dialogical dialectics. It also serves to show Plato's honesty and graciousness: although Schleiermacher's Plato is very much opposed to Parmenides' philosophy, he nonetheless acknowledges himself to be in conversation with Parmenides (and Zeno) and therefore to be indebted to him. This serves to expand the notion of dialogue. Dialogue takes place not only between Socrates and his various interlocutors, but between Plato and his predecessors. Schleiermacher may have been wrong in his interpretation of the *Parmenides*, and at the very least it is fair to say that he avoided the major interpretative and philosophical problems of that dialogue, but his interpretation of it is telling of how extensively he views dialogue: that we are always in conversation with those who precede us, and that the tone taken should be one of respect and acknowledgment of debt where that is due, despite profound differences.

A third example is taken from his "Introduction to the *Sophist*." In the *Sophist* we encounter what clearly seems to be a speculative form of dialectics. What is more, Schleiermacher himself seemed completely caught up in it: "For practically the first time in Plato's writings, the innermost sanctuary of philosophy is unlocked purely philosophically."[44] And yet Schleiermacher still managed to rein in the speculative dialectics by insisting on its integration with dialogical dialectics. He did this by reminding us of the *Sophist's* place in the corpus—that is to say, its relation with other dialogues. The *Sophist* is the first part of a small trilogy (*Sophist-Statesman-Symposium*) that is at the center of another trilogy which is itself the center of the entire Platonic corpus. Literary structure and internal relations are everything to Schleiermacher. According to his method of interpretation,

42 SCHLEIERMACHER, "Introduction to the *Parmenides*," 133.
43 Ibid.
44 SCHLEIERMACHER, "Introduction to the *Sophist*," 248.

to be a part of a trilogy means that a dialogue cannot be understood alone or apart from the other component parts. Indeed, Schleiermacher went so far as to say that the *Sophist* and *Symposium* form "One Dialogue" together, the *Symposium* being the culmination both of that "One Dialogue" and of the small trilogy. The point here is that Schleiermacher thereby fully integrated speculative dialectics with practical philosophy. Schleiermacher here reminds us that the knowledge extolled in the *Sophist* and the wisdom described in the *Statesman* are in themselves incomplete and inadequate if they are not joined to the "joyous communication"[45] that Socrates embodied.

This, I suggest, can be a resource for an open theological system insofar as it confirms the communicative nature of dialectics. In his interpretation of the *Symposium*—its place, role, and content—Schleiermacher emphasized the ideal of friends gathered and cherished, as well as the conviction that wisdom and knowledge can only be gained and lived in communication and exchange with others. Socrates' "tireless enthusiasm for contemplation"[46] (which hints at a speculative dialectics done in isolation, a private grasping of the truth) cannot be done without engagement with others and is not in itself the final goal.

These three examples illustrate just how far-reaching Schleiermacher's dialogical dialectics is. The dialogue form, according to Schleiermacher, was not just a covering or embellishment that could be replaced by another form or removed. It operates at several different levels, he thought, in Plato's philosophy itself and thus is indispensable as a interpretative key to Plato's writings.

Reviewing Schleiermacher's reading of Plato's dialectics in light of our discussion of open systems, it occurs to me that the two forms of dialectics—dialogical dialectics and speculative dialectics—conform neatly to Toulmin's categorization of humanism (dialogical dialectics) and rationalism (speculative dialectics). Schleiermacher's elision of speculative dialectics might suggest that he belonged strictly to the humanist strand of modernity and thus would not be a good resource for understanding system. Elision, however, is not elimination. In consistently privileging the dialogical, Schleiermacher wound up transforming the speculative kind of dialectics by reining it in—by integrating it with the dialogical.

---

45 SCHLEIEMACHER, "Introduction to the *Symposium*," 274.
46 Ibid.

Two corollaries follow. For Schleiermacher's Plato, (1) the soul can never be separated from the body, and (2) the ideas always and only manifest themselves in and with the empirical. Schleiermacher asserted that our knowledge must necessarily be empirical. Yet the strictly empirical approach of dialogue is not sufficient. There is the truth of how things are related—and pursuing that and desiring that helps to ensure that we discover and reconstruct a "living being,"[47] rather than arbitrarily weaving unrelated things together. Thus speculative dialectics is necessary as "the intuition of the life of that which exists and of the necessary unity and being-in-one-another of being and knowing."[48] Never, however, does Schleiermacher allow it to be separated from dialogical dialectics. The dialectical method, we learn from Socrates in the *Phaedrus* is an "art," a "systematic art"[49]; *eros* is its impetus and lure. As an art, it is an on-going activity that approaches the truth asymptotically.

2. The Test of the Polar Coordinates

How, then, does Schleiermacher's integrated Platonic dialectics stand up to the determining and limiting criteria we derived from Toulmin's analysis? For brevity's sake, I shall here focus on two sets of polar coordinates (and address the other two in the next section). I take as illustrations Schleiermacher's interpretation of two notoriously difficult Platonic texts, both of which would seem to force Schleiermacher to one side or the other of our axis of polar coordinates; yet, in both cases, he holds the poles together.

The written and the oral. Schleiermacher recognized that his claim that Plato's philosophy is presented fully and completely in the written dialogues (as opposed to a separate or supplementary unwritten, oral tradition) appeared to contradict Plato's own suspicion of writing.[50] Plato, he noted, warns about "how uncertainty always accompanies written communication of thoughts."[51] Even so, Schleiermacher insisted that, despite these dangers, Plato still thought "writing must be risked."[52] Why? Be-

47 Schleiermacher, "Introduction to the *Phaedrus*," 71.
48 Schleiermacher, "Introduction to the *Sophist*," 249.
49 See Plato, *Phaedrus*, 265 d–e, 276 e.
50 See Ibid., 275 a, e.
51 Schleiermacher, "Introduction to the *Phaedrus*," 39.

cause for Schleiermacher the fundamental issue has to do not with writing or orality *per se*, but with the sole criterion that determines the authenticity of both—namely, the "form of the dialogue."[53] Plato was not against all writing, only against that which is not dialogical; for that reason, he renounced any method that undermines genuine discourse. Instruction, whether oral or written, must imitate "that original and reciprocal communication"[54] which took place between Socrates and his students. Plato, because he could "express his thoughts purely and perfectly,"[55] succeeded in making "written instruction as similar as possible to that better kind."[56] Without question, Schleiermacher thus holds together the oral and the written.

The local and the general. Taken out of context, Schleiermacher's interpretation of Plato's *Sophist* might suggest that he promotes a philosophical system that claims direct access to that "higher sphere of speculation," where the philosopher is freed from the weight of what is local and circumscribed. The *Sophist* celebrates the "purest dialectics."[57] Yet, as we have seen, read in the context of his other Introductions, it is clear that he does not allow us to separate the *Sophist* from the other dialogues, especially from the *Statesman* and the *Symposium.* It is in the *Symposium*, Schleiermacher wrote, that we are first presented "with an image of the philosopher in the person of Socrates."[58] Socrates is the true philosopher because of his "tireless enthusiasm for contemplation" *and* the "joyous communication" of what he has contemplated.[59] What we learn in the *Symposium* is that Socrates' object is

> not some idea of knowledge and wisdom, but is rather the philosopher, a man like any other who, although godlike in comparison with the lowly life of most, nevertheless walks as a human among humans; thus *what should be presented is not the absolute being and nature of wisdom, but its life and its appearance in the mortal life of the visible person*, in whom wisdom

52 Ibid.
53 Ibid.
54 Ibid., 40.
55 Ibid., 42.
56 Ibid., 40.
57 SCHLEIERMACHER, "Introduction to the *Sophist*," 250.
58 SCHLEIERMACHER, "Introduction to the *Symposium*," 274.
59 Ibid.

> itself—for this is clearly Plato's main point in all his explanations of philosophy—has put on mortality.[60]

This view of knowledge and wisdom as necessarily embodied and communicative, rather than as generalized as an "idea," is nothing less than an insistence that wisdom is of its very nature *local.*

In summary, Schleiermacher's dialogical dialectics recommends itself to us because it is in fact an integration of Toulmin's two trajectories of modernity. As fundamentally dialogical, it is, it reflects, it accomplishes, it respects, and it protects all those characteristics of "humanized Modernity" (listed in the left-hand column of Table One); as modestly "speculative," for lack of a better term, it does not relinquish the infinite, absolute, or the general, but experiences them as lure, as presupposition, and as embodied. Although he never abandoned the One, the infinite, the eternal, etc., Schleiermacher refused the temptation to over-define it or any pretense of grasping it. Its presence is there, both as impetus—that which creates the desire in us—and goal; but it is never separate, never realizable apart from the particular, the empirical, and the community. His elision of it and emphasis on the dialogical was not an attempt to be rid of it, but to tame and to restrain the speculative impulse to claim itself and impose itself.

Schleiermacher's dialogical dialectics relates to *system* because it works out of the assumption that things are interrelated and can be combined and recombined, but not in any arbitrary way. Schleiermacher saw the Platonic universe, not just the Platonic *corpus*, as Socratic. It is a living creature, with real, vital connections. There is a higher system of relations grounded in the higher relation of being and thought. This same dialogical dialectics, however, remains *open* in that it is asymptotic. It approaches the universal through a combining and relating of the particular, but it never claims it has perfectly captured any identity of thought and being. It is an art, a composition. It is tentative, provisional, yet not cynical. The impulses Schleiermacher developed in his interpretation of Plato would be developed even more explicitly and systematically in his lectures on *Dialectics*: there cannot be either a strict empiricism or a strict idealism; philosophy is an art; the philosopher must have a love for what is real; the concept "is always an identity of general and particular."[61]

---

60 Ibid., 275–6 (emphases added).

61 *Dialectic(1811)* 63.

IV. A Dialogue of Forces: Individuality as a Center of Powers

1. Schleiermacher's Original Insight

There is no system without an accounting of how individuals or particulars are determined, how they are related to one another, and how they are related to larger wholes. In a system, the particular and the universal, the many and the one, need to be coherently explained. One problem with modern systems was that the particular tended to get lost in the absolute universal; a problem with older, medieval systems was that the particular and the universal were understood in different terms—their distinction resting on the differentiation of particular, species, genus—and different rules applied; a problem with postmodern rejections of system is that the universal is altogether denied. Schleiermacher avoided all such problems by rooting his understanding of individuality, relation, and causality in the notion of force, which he univocally applied to all of reality.

Schleiermacher's employment of this notion of force—from his early essays on freedom and on the Pantheist Controversy through his *Speeches* to his mature systematic theology—allowed him to drop the substance-based understanding of the individual without abandoning a strong notion of individuality: the individual, the particular, is "nothing other than the cohesion, the identical combination of forces of a certain measure at a single point."[62] Schleiermacher's earlier "organic monism," however, is not in itself sufficient to protect against the dangers of closed systems insofar as the somatic metaphor it employed is applicable to only a portion of what is real. The aspect of Schleiermacher's organic monism that does guarantee the possibility of an open system is the notion of causality as "force" or "power," which in turn allows for internal as well as immediate and mediated relations and which apply universally. This allows for real universals without their having a separate existence. Just such an "open system" is what in *The Christian Faith* Schleiermacher called the "nature-system" (*Naturzusammenhang*), which is grounded in and determined by, but never usurped by, the divine causality.

As Schleiermacher himself was careful to point out, such a description of the nature-system is not merely a world view—it is a specifically Christian view of nature as dependent on God. There is a complete coincidence of natural causality and divine causality. From the viewpoint of divine causal-

62 Friedrich Schleiermacher, "Spinozistisches System (1793–94)," in KGA I/1, 574. (My translation.)

ity, God sustains the whole and "sustains individual things in their existence and their powers as long as He wills."[63] Because everything is absolutely dependent on God, "everything is equally direct and equally indirect, some in one relation and some in another."[64] To a significant degree, Schleiermacher applied this univocal language of forces to God as well. God, too, is understood in terms of activity rather than being. Schleiermacher, however, retained the divine transcendence through another dialectic of the divine attributes.

In *The Christian Faith*, Schleiermacher described the universe as a totality of living forces as opposed to a "dead mechanism."[65] In this system of forces, "forces" or "powers" (*Kräfte*) apply universally and univocally—they account for and constitute individuality and particularity at every level, and they account for and constitute the universal. Every thing that is, *is* insofar as it is a "center of power."[66] In the nature-system, particulars and universals are not fixed. Every particular is understood in terms of a whole; and wholes are always parts of larger, more universal wholes. Thus what is particular or universal depends on how forces cohering in a given place at a particular time are viewed—that is to say, how far one wants to expand or restrict the sphere of influence being considered. The spheres of influence can always be reconfigured, so that what was particular can perhaps be expanded into a universal, and what is universal can be broken down into particulars. In other words, the rejection of a substance-based view of reality and individuality, and the replacement of that with a dynamic view of reality and individuality in terms of relations and cohering forces, allows for a much more fluid and relational view of particulars and universals—a dynamic, co-determining, and reciprocal relation between particulars and universals.

This understanding of reality, of the nature-system grounded in the absolute livingness of the divine causality, can be a resource for understanding open theological systems—not only because of its rejection of substance-based understanding of individuality (which has been closely allied with closed systems), but also for understanding systematic theology as a cohering, a coming together at a particular time of different religious, theological, philosophical, political, doctrinal, moral "forces" and relations. Thus

63 CF § 46, postscript (175). (Page numbers given in parentheses.)
64 Ibid.
65 CF § 34,2 (139).
66 See CF § 46, postscript (178).

systematic theology is not just possible albeit always incomplete; it is also necessary. It is incumbent upon a community that claims to embody a living tradition, that wants to understand where it stands and where it has been thrown, that is committed to the marginalized, that wants to discern the times, that wants to respond to the divine offer and divine command—it is incumbent upon such a community to attempt to bring together in a self-conscious way the many forces that press upon it, form it, and energize it.

Schleiermacher's understanding of reality and individuality and universality in terms of force—and along with it, his insistence on the complete coincidence of divine and natural causality—meet the other two of Toulmin's determining and limiting criteria.

2. The Test of the Polar Coordinates

The particular and the universal. In Schleiermacher's account of the nature-system, the particular or individual is not lost in the system of nature, nor is the universal a mere abstraction from the particular: "Everything is referred to the ordered exercise of power which establishes the dependence of each individual on the totality of existence eternally, and for the maintenance of the general interaction makes use of the forces of individual things."[67] Human existence, especially, is "an integral part of the world."[68] This makes a system in that it can account for particulars as well as universals, it is both intensive and extensive, and it allows for novelty.

The timely and the timeless. Schleiermacher, in resisting the two-headed theological sin of anthropomorphism and speculation, essentially employed a univocal way of discussing divine causality: there is no distinction between being and activity, or power and essence, in God; the divine causality is equal in scope to natural causality; they are coextensive. In proposition 54 of *The Christian Faith*, Schleiermacher treated the divine attribute of omnipotence, which he says contains two ideas:

> first, that the entire system of Nature ... is founded upon divine causality, which as eternal and omnipresent is in contrast to all finite causality; and second, that the divine causality ... is

67 CF § 54,4 (215).
68 CF § 58,1 (236).

> completely presented in the totality of finite being, and consequently everything for which there is a causality in God happens and becomes real.[69]

The co-extensiveness of the divine causality and natural causality is not a collapsing of one into the other because the transcendence (absolute distinction) of the former is emphasized by the dialectical relation of the doctrine of divine omnipotence with the doctrine of divine eternity. In proposition 52, Schleiermacher wrote, "By the Eternity of God we understand the absolutely timeless causality of God, which conditions not only all that is temporal, but time itself as well."[70] Thus, both structurally in his text and substantively, Schleiermacher maintained both the timeliness of divine causality (the divine immanence) and the timelessness of divine causality (the divine transcendence) without interfering with or abrogating natural causality—or the relation of universal and particular.

V. Concluding Remarks

Schleiermacher's dialogical-dialectical method and his ontology of "forces" not only serve as elements of how he accounts for reality and our ability to know it; they are the connective tissues, so to speak, of his theological system. Both elements serve as connectors among the various doctrines of his theological system—from the christological doctrines, to the formation of community and its co-existence in the world, to the exposition of the experience of sin and grace, to his interpretation of creation, and even back to the Introduction. How might these basic elements of Schleiermacher's thought help us understand open theological systems in a postmodern age?

The Force of Dialogue. Christian systematic theology is not to be understood as a claim to completeness or finality, or as a closed system written in isolation. Its very force and power—its authority—comes through its dialogical nature. An open theological system is one whose aim and motive is communication; the impetus and lure of that communicative activity is love. This love is *eros*: a desire that leads us out of ourselves and towards others, a desire that craves the eternal. This love is also *agape*: the generous, respectful, and un-self-interested love that is given to us as gift and that recognizes our common source and goal.

---

69 CF § 54, proposition (211).
70 CF § 52, proposition (203).

The dialogue is sometimes between teacher and pupil, but even then, following Socrates' example, the teacher herself or himself is not all-knowing but is maybe a few steps ahead of the student, learning as much from the conversation as the student. (Here it is necessary to reject Schleiermacher's idealized view of Plato as having everything perfectly in mind before he sat down to write his dialogues.) We must retrieve the Socrates who leads the interlocutor in thought but who is also comfortable with inconclusiveness. In this kind of dialogue, progress is sometimes made and genuine knowledge sometimes gained, but such knowledge is neither complete nor owned. A dialogue characterized by *eros* is a move towards wisdom that is asymptotic in character. The systematic theologian is thus teacher. Yet dialogues are also between and among the teachers. Thus the force or power of an open systematic theology is the degree to which the theologian engages other contemporary thinkers, schools of thought, disciplines—and, as Schleiermacher's interpretation of the *Parmenides* reminds us, of earlier thinkers and systems, etc. Dialogues can also be a way to give voice to the voiceless. This, too, should be the task of constructing an open theological system: to include the powerless, the marginalized; to speak on their behalf, to bring them into the conversation.

All of this suggests that using Schleiermacher's dialogical dialectics as a resource for open theological systems means that systematic theology is not a dialogue in a narrow and literal sense of a conversation between two interlocutors. It is a process that brings together many, often very different and very much opposed, voices in a coherent and meaningful way. The force or authority or "openness" of a systematic theology resides in its dialogical character.

A Dialogue of Forces. Any systematic theology is particular. Yet while it needs to be ever mindful of its locality, it by definition strives to be more than local, to reach beyond to something more universal. It may be helpful, I suggest, to think of systematic theology as a dialogue of forces. The forces that are taken up, combined, rejected, and re-combined are unique to a present time: concerns, issues, anxieties, insights; operative symbols, and emerging ideas; the voices of individuals, of peoples, or schools of thought; the practices and experiences of a community; a community's sense of loss regarding some practice or experience, etc. Those "present" forces are taken up with forces from the past which exert influence on us, whether we are conscious of them or not: the symbols, doctrines, sacred texts of the tradition; past theologies, whether retrieved or rejected; the impulses of love, faith, and hope on and within the community. The systematic theologian draws these together, or explicates how they have on their own coalesced, with the hope of illuminating the present with the aid of the wisdom of the past in the service of creating a better future. A systematic theology may fail utterly. It may be heralded for a time as being successful in describing and illuminating our times. And it may become a classic. Time will tell whether the particular has some universal force and whether therefore it will be taken up as part of the conversation in future theologies.

Feeling and Morality in Whitehead's System

David Ray Griffin, Claremont School of Theology

The purpose of this essay is to explain how Whitehead sought to account for the presuppositions of our moral life, and to point to an ideal resolution of our basic moral tension, within a system that embodies a thoroughgoing naturalism, albeit a *theistic* naturalism. In the first section, I discuss Whitehead's ideal of a philosophical system in relation to his concern with the inevitable presuppositions of daily life, including moral presuppositions. The second section deals with his conviction that an adequate system must be both naturalistic and theistic. The third section introduces the central feature of his worldview, panexperientialism with organizational duality, according to which all individuals have feelings. The fourth and fifth sections show how Whitehead's "critique of pure feeling," based on this panexperientialism, allows him to defend three presuppositions of morality: human freedom, normative values, and ultimate importance. In the final section, I explicate Whitehead's treatment of the way in which our feelings of God, including our intellectual feelings *about* God, can provide a basis for a humanitarian ethic and an ideal resolution of our basic moral tension.

I. Systematic Thought, the Presuppositions of Life, and Morality

At the heart of Whitehead's critique of modernity is modernity's anti-rationalism, which involves the loss of commitment to systematic thought. "[T]he men of the Middle Ages were in pursuit of an excellency of which we have nearly forgotten the existence," namely, "the attainment of a harmony of the understanding."[1] Illustrating his own commitment to this

---

1 SMW 76.

ideal, Whitehead declares that "an attack upon systematic thought is treason to civilization."[2]

Whitehead did recognize, however, that there were valid reasons for dissatisfaction with most examples of systematic philosophy and theology. One of these is the common claim to dogmatic finality, which fails to understand that metaphysical categories are tentative formulations and that our approach to truth can only be asymptotic.[3] Another is "the dismissal of experience in the interest of system."[4] "Failure to include some obvious elements of experience in the scope of the system is met by boldly denying the facts."[5] This tendency of systematic thought is closely related to "the chief error in philosophy," namely, "overstatement"—the formulation of truths one has seen in an "over-assertive" way that implies "an exclusion of complementary truths."[6] Still another valid cause for complaint involves "narrowness in the selection of evidence."[7] All of these failings involve inadequacy, the worst sin of systematic thought. A cosmology "should above all things be adequate."[8]

Distinctively *modern* thought's chief inadequacy is that it "fails to justify the practice of daily life."[9] And this is to fail with respect to that to which philosophy should first of all be adequate—the presuppositions of ordinary human practice. "[Philosophy's] ultimate appeal is to the general consciousness of what in practice we experience."[10] In Whitehead's most important statement on this issue, he said:

> the metaphysical rule of evidence [is] that we must bow to those presumptions which, in despite of criticism, we still employ for the regulation of our lives. Such presumptions are imperative in experience. Rationalism is the search for the coherence of such presumptions.[11]

2 AI 162.
3 Ibid.; FR 53.
4 MT 3.
5 PR 6.
6 PR 7; RM 145, 149.
7 PR 337.
8 FR 86.
9 AI 224–5.
10 PR 17.
11 PR 151. As this statement shows, Whitehead rejected the assumption that there is a necessary tension between rationalism and empiricism. Being fully rational, he argued, involves reconciling all the ideas inevitably presupposed in experience.

But modern philosophy, far from assigning such authority to these presumptions, has accepted a dualism of theory and practice, as in Hume's appeal to "practice" to justify various "natural beliefs"—such as our beliefs in an external world and causation as real influence—that his philosophical theory could not justify. Hume's contentment with a system of thought that admittedly failed to do justice to ideas he necessarily presupposed in his life constituted "the high watermark of anti-rationalism in philosophy."[12] Whitehead, by contrast, said:

> Whatever is found in "practice" must lie within the scope of the metaphysical description. When the description fails to include the "practice," the metaphysics is inadequate and requires revision. There can be no appeal to practice to supplement metaphysics .... Metaphysics is nothing but the description of the generalities which apply to all the details of practice.[13]

Although modern philosophers have treated many of these generalities, or general notions, they have failed to "weave[ ] these notions into one coherent system of cosmology."[14]

Whitehead, in continuity with the commonsense tradition associated with Hume's opponent Thomas Reid, sometimes used the term "common sense" for these inevitable presuppositions of practice. This term, however, is now often used to refer to ideas that, while widely held at a certain time and place, have been shown by the advance of science to be false. Indeed, Whitehead himself often used the term in this sense.[15] It is necessary, accordingly, to distinguish between these different kinds of common sense terminologically. Calling the latter type "soft-core common sense," I use "hard-core common sense" to refer to the inevitable presuppositions of practice.[16]

We can say, accordingly, that our hard-core commonsense notions provide the primary basis for testing the adequacy of any system of thought. These notions do not, it should be noted, provide a *foundation* from which

---

12 PR 153.

13 PR 13.

14 PR 128. With regard to this volume's concern with system, it is interesting to note that Whitehead, in this passage, included "system" among the notions that need to be included within a cosmology.

15 SMW 114; PR 9, 17; MT 129.

16 See David Ray Griffin, *Unsnarling the World-Knot: Consciousness, Freedom, and the Mind-Body Problem* (Berkeley: University of California Press, 1998), ch. 3.

other ideas can be deduced.[17] They instead provide a *compass*, letting us know when our adventures of ideas have gotten off course.

The reason that modern thought has been especially inadequate to the presuppositions of our practical life is that whereas earlier cosmologies were based on aesthetics, ethics, and religion as well as science, modern cosmology has been based almost exclusively on science. It has, accordingly, not had categories that could do justice to our religious, moral, and aesthetic presuppositions. Philosophy, however, must insist on "the retention of the whole of the evidence in shaping our cosmological scheme," and the whole of the evidence includes the fact that "the impact of aesthetic, religious and moral notions is inescapable."[18] In this essay, I focus on moral notions.

II. Theistic Naturalism

Another criterion for a rational system of thought is that it be fully naturalistic. Although Whitehead did not use this term, he did say that the "full scientific mentality" presupposes that "all things great and small are conceivable as exemplifications of general principles which reign throughout the natural order," so that "every detailed occurrence can be correlated with its antecedents in a perfectly definite manner, exemplifying general principles."[19] There can be, therefore, no appeal to a supernatural cause to explain any of the world's features. Commenting on the appeal to God by Descartes and Leibniz to justify our knowledge of an external world, Whitehead said:

> It is a device very repugnant to a consistent rationality. The very possibility of knowledge should not be an accident of God's goodness; it should depend on the interwoven natures of things. After all, God's knowledge has equally to be explained.[20]

The same conflict between supernaturalism and rationalism is reflected in Whitehead's rejection of the "appeal to a *deus ex machina* who was capable

17 PR 8.
18 SMW vii; MT 19.
19 SMW 5, 12.
20 PR 190.

of rising superior to the difficulties of metaphysics."[21] "[T]he easy assumption that there is an ultimate reality which, in some unexplained way, is to be appealed to for the removal of perplexity," he declared, "constitutes the great refusal of rationality to assert its rights."[22]

Whitehead's rejection of supernaturalism, however, did not prevent him from endorsing a version of theism, once he became convinced that it was a necessary part of a truly rational, adequate system. Whitehead affirmed, in other words, a position that can be called either "theistic naturalism" or "naturalistic theism." The theism is naturalistic in that God is not a being outside the world that can occasionally interrupt its normal causal relationships, because what exists naturally, and hence necessarily, is not simply God, but God-with-a-world. Although *our* world—what Whitehead calls "our cosmic epoch"—has been created in time, *the* world—in the sense of a realm of finite actualities—has always existed. Whitehead thus rejected the doctrine that our world was created out of absolute nothingness, reaffirming Plato's view that the creation of our world "is not the beginning of [finite] matter of fact but the incoming of a certain type of social order."[23] Unbeknownst to Whitehead, he was thereby returning to the biblical view.[24]

In any case, one implication of this affirmation of creation out of chaos is the idea that beneath our cosmic epoch's contingent "natural laws" are some principles that are *truly* natural, lying in the very nature of things. These metaphysical principles include the basic principles of causation, which apply to causal relations not only between finite things but also between finite things and God. Seeing himself as here going beyond Plato,

---

21 SMW 156.

22 SMW 92. Whitehead did not reject all appeals to an ultimate reality, only those that do so "in some unexplained way," meaning a way that does not exemplify the general categories of the system.

23 PR 96.

24 Whitehead said, in line with the biblical scholarship of the time, that in affirming the "theory of a wholly transcendent God creating out of nothing an accidental universe," "Newton held the Semitic theory" (PR 95). But more recent scholarship has shown that the doctrine of creation out of absolute nothingness, as distinct from chaos or relative nothingness, is post-biblical. See Jon D. Levenson, *Creation and the Persistence of Evil: The Jewish Drama of Divine Omnipotence* (San Francisco: Harper & Row, 1988), and Gerhard May, *Creatio Ex Nihilo: The Doctrine of "Creation out of Nothing" in Early Christian Thought*, trans. A. Stanley Worrall (Edinburgh: T & T Clark, 1994). I have employed their work in David Ray Griffin, "Creation out of Nothing, Creation Out of Chaos, and the Problem of Evil," in *Encountering Evil: Live Options in Theodicy*, ed. Stephen T. Davis, 2nd ed. (Philadelphia: Westminster / John Knox, 2001), 108–25.

Whitehead said that for Plato the relationship of God to the world was based on the divine will, but that "metaphysics requires that the relationships of God to the World should lie beyond the accidents of will, and that they be founded upon the necessities of the nature of God and the nature of the World."[25] Whatever be the correct interpretation of Plato, the substantive point is that for Whitehead the basic causal principles of the universe are metaphysical in character, so they cannot be interrupted. Calling these principles *metaphysical* also means that they apply to all actual entities, including God: "God is not to be treated as an exception to all metaphysical principles, invoked to save their collapse. He is their chief exemplification."[26] This statement, which rejects the kind of *deus ex machina* affirmed by Descartes and Leibniz, shows that Whitehead did not intend his theism to compromise his naturalism.[27]

III. Panexperientialism, Organizational Duality, and the Critique of Pure Feeling

With regard to his affirmation of a version of theism, Whitehead compared his system to that of Aristotle, who "found it necessary to complete his metaphysics by the introduction of" a divine reality because "the general character of things requires that there be such an entity."[28] Whitehead's metaphysical account of this general character of things can be summarized as "panexperientialism with organizational duality."[29] This doctrine, Whitehead suggested, allows us to do justice to our moral as well as our scientific beliefs without violating naturalism. Whereas Kant, after writing his critique of pure theoretical reason to deal with our scientific presuppositions, wrote a critique of practical reason to treat our moral presupposi-

25 AI 168.

26 PR 343.

27 It is another matter, of course, whether Whitehead's doctrine of God actually avoided violating any of his metaphysical principles. I have argued that it violates several, but that this problem can be rather easily remedied; see David Ray Griffin, *Reenchantment without Supernaturalism: A Process Philosophy of Religion* (Ithaca: Cornell University Press, 2001), ch. 4. The present essay presupposes this remedy. For more on this, see notes 71 and 119 below.

28 SMW 173–4.

29 See Griffin, *Reenchantment*, ch. 6.

tions, Whitehead suggested that "a critique of pure feeling" can provide for both types of presuppositions.[30]

Whitehead's panexperientialism is the doctrine that the actual world is composed of *creative, experiential, physical-mental events*. To begin with the second of these four concepts, *experiential*: At least some iota of experience is to be attributed to all individuals, so that there are no "vacuous actualities," meaning things that are fully actual and yet wholly devoid of experience.[31] In saying that even the most elementary individuals have experience, Whitehead was, as he knew, reaffirming the position of Leibniz.[32]

Whitehead thereby rejected Cartesian dualism, with its two kinds of actual entities. He rejected vitalism, for example, on the grounds that it "involves an essential dualism somewhere."[33] The problem with an essential dualism is that the interaction, or at least *apparent* interaction, between essentially different types of actualities can be explained only by appeal to a supernatural agent, as Descartes saw and his followers Malebranche and Geulincx, with their "occasionalism," emphasized.[34] Among contemporary dualists, Richard Swinburne says that the mind-body problem can be solved only by appeal to divine omnipotence. Geoffrey Madell, who does not make this appeal, admits that the mind-body relation is "utterly mysterious."[35] Advocates of materialism—which is really, as Charles Hartshorne said, covert dualism[36]—face the same dilemma, with Colin McGinn admitting that "scientific naturalism runs out of steam" when it comes to the mind-body problem because "[i]t would take a supernatural magician to extract consciousness from [insentient] matter."[37] Whitehead avoided this

---

30 PR 113.

31 PR 29, 167.

32 SMW 155. Although in *Process and Reality*, which was based on Gifford Lectures given at Edinburgh, Whitehead portrayed Locke as the one who had most fully anticipated his position, he had stated the historical paternity more accurately in this passage in *Science and the Modern World*.

33 SMW 79.

34 On Descartes, see GORDON BAKER and KATHERINE J. MORRIS, *Descartes' Dualism* (London: Routledge, 1996), 153–4, 167–70. On Malebranche and Geulincx, see FREDERICK CHARLES COPLESTON, *A History of Philosophy*, vol. 4, *Descartes to Leibniz* (London: Burns and Oates, 1960), 117–9, 188–90.

35 RICHARD SWINBURNE, *The Evolution of the Soul* (Oxford: Clarendon, 1986), 198–9; GEOFFREY MADELL, *Mind and Materialism* (Edinburgh: The University Press, 1988), 2.

36 CHARLES HARTSHORNE, *Creative Synthesis and Philosophic Method* (LaSalle, Ill.: Open Court, 1970), 9, 27.

37 COLIN MCGINN, *The Problem of Consciousness: Essays Toward a Resolution* (Oxford: Basil Blackwell, 1991), 45.

problem by rejecting the assumption, common to materialists and dualists alike, that the matter of which the brain is composed is entirely insentient, devoid of feelings. As Hartshorne summarized the Whiteheadian solution to mind-body interaction: "cells can influence our human experiences because they have feelings that we can feel. To deal with the influences of human experiences upon cells, one turns this around. *We* have feelings that *cells* can feel."[38]

Although McGinn admits that this view would allow for a naturalistic solution to the mind-body problem,[39] he rejects panexperientialism as self-evidently absurd, alleging that it says that rocks have thoughts.[40] Although this and similar charges are common, they are misplaced, thanks to two other points that Whitehead shared with Leibniz. One of these is that, although all individuals have feelings, there are many different degrees of feeling. Conscious thoughts are a very high-level type, called "intellectual feelings."[41] Most individuals are incapable of such feelings. Very low-grade individuals, such as the molecules that make up rocks, have only very simple feelings, corresponding to what Leibniz called "petites perceptions."[42]

However, even if McGinn would accept that distinction, he would surely maintain, with Karl Popper and John Eccles, that it is absurd to think that things such as rocks and telephones have feelings of any sort.[43] But so would Whitehead, who followed Leibniz in distinguishing genuine individuals from nonindividualized groupings of individuals. Genuine individuals are of two types. Besides the most elementary entities, perhaps

---

38 Charles Hartshorne, *The Logic of Perfection and Other Essays in Neoclassical Metaphysics* (LaSalle, Ill.: Open Court, 1962), 229.

39 McGinn, *Problem*, 2 n, 28 n.

40 Colin McGinn, *The Character of Mind* (Oxford: Oxford University Press, 1982), 32.

41 PR 266.

42 Leibniz, noted Whitehead, used "perception" and "apperception" to distinguish between the lower and higher forms of taking account of other things. Similarly, Whitehead's "prehension," in distinction from "apperception," is meant to indicate an experiential "taking account of" that may or may not be conscious (AI 233–4). The idea that human experience must have some features in common with the most elementary units of nature, Whitehead points out, is an implication of "any doctrine which refuses to place human experience outside nature." It is then incumbent upon any philosophy holding such a doctrine to "find in descriptions of human experience factors which also enter into the descriptions of less specialized natural occurrences" (AI 184). Much of Whitehead's writing is devoted to this effort. I have described the justification for his approach in Griffin, *Unsnarling*, ch. 8, "Matter, Consciousness, and Misplaced Concreteness."

43 Karl R. Popper and John C. Eccles, *The Self and its Brain: An Argument for Interactionism* (Heidelberg: Springer-Verlag, 1977), 55, 517.

quarks or still simpler entities, there are what Hartshorne called "compound individuals."[44] Compound individuals are higher-level individuals compounded out of lower-level ones, as when atoms and molecules are compounded out of subatomic particles, living cells out of macromolecules, and animals out of cells. Only genuine individuals have feelings. Although the molecules making up rocks and telephones have feelings, a rock or a telephone as such does not. But an animal as such feels, by virtue of the fact that it is organized so as to give rise to "that supreme entity which is the one animal considered as one experiencing subject."[45] It is important to call Whitehead's doctrine "panexperientialism *with organizational duality*" because of this distinction between things that are and are not organized so as to give rise to a supreme entity, which Whitehead, echoing Leibniz's language of the "dominant monad," called the "dominant member."[46]

The importance of these two Leibnizian distinctions was stressed by Hartshorne, who said that, in formulating the contrast

> between active singulars and groups of these, only the former of which literally feel; and that between low and high levels or degrees of feeling, ... Leibniz took the greatest single step in the second millennium of philosophy ... toward a rational analysis of the concept of physical reality.[47]

Of course, as Hartshorne also pointed out, Leibniz "hid the importance of his insight by interweaving it with some of the most extraordinary fantasies in intellectual history." I will later discuss Whitehead's way of overcoming some of Leibniz's fantasies. For now, the point is that Whitehead's version of these Leibnizian distinctions enabled him, within a fully naturalistic system, to do justice to the presuppositions of science. I turn now to some ways in which panexperientialism with organizational duality can also do justice to our moral presuppositions.

---

44 Charles Hartshorne, "The Compound Individual," in *Whitehead's Philosophy: Selected Essays 1935–1970* (Lincoln: University of Nebraska Press, 1972), 41–61.

45 MT 23.

46 PR 102; AI 206.

47 Charles Hartshorne, "Physics and Psychics: The Place of Mind in Nature," in *Mind in Nature: Essays on the Interface of Science and Philosophy*, ed. John B. Cobb, Jr., and David Ray Griffin (Washington, D. C.: University Press of America, 1977), 89–96, at 95.

IV. Moral Freedom

"It is the primary aim of philosophy," said Henry Sidgwick, "to unify completely, bring into clear coherence, all departments of rational thought, and this aim cannot be realised by any philosophy that leaves out of its view the important body of judgments and reasonings which form the subject matter of ethics."[48] Whitehead, in quoting this statement with approval,[49] showed that he wanted his system to be adequate to our moral presuppositions. One of these presuppositions is human freedom. Against those who denied it, Whitehead said that our sense of freely making decisions is "too large to be put aside merely as misconstruction. It governs the whole tone of human life."[50] Furthermore, having included "freedom and necessity" among "the pairs of opposites [that] are in experience with a certain ultimate directness of intuition,"[51] Whitehead declared:

> Here we find an example of the value of a systematic philosophy. For we have either to explain the diverse senses in which freedom and necessity can coexist, or we have to explain away one or other of the most obvious presuppositions of our daily thoughts.[52]

This task of systematic philosophy is so urgent, Whitehead pointed out, because modern science-based thought involves a "radical inconsistency." While science-based thinkers share with everyone else "an unwavering belief in the world of men and of the higher animals as being composed of self-determining organisms," they articulate a mechanistic philosophy. In modern thought, accordingly, there is a "discrepancy between the materialistic mechanism of science and the moral intuitions which are presupposed in the concrete affairs of life."[53]

This radical inconsistency is illustrated by several contemporary philosophers. John Searle, for example, agrees that freedom is a hard-core commonsense idea. Pointing out that we *can* give up some commonsense ideas (which belong to those I classify as *soft*-core), Searle says that

---

48 Henry Sidgwick, *Henry Sidgwick: A Memoir* (London: Macmillan, 1906), Appendix I.

49 SMW 142.

50 PR 47.

51 PR 341.

52 MT 7.

53 SMW 76, 80.

> we can't similarly give up the conviction of freedom because that conviction is built into every normal, conscious intentional action .... [W]e can't act otherwise than on the assumption of freedom, no matter how much we learn about how the world works as a determined physical system.[54]

Like most other materialists, however, Searle believes that science "allows no place for freedom of the will," because science holds that "everything can be accounted for in terms of [material] particles and their relations."[55] "I would like to be able to keep both my commonsense conceptions and my scientific beliefs," Searle says candidly, but "when it comes to the question of freedom and determinism, I am ... unable to reconcile the two."[56] Some materialists, to be sure, redefine freedom so that it is compatible with complete determinism. As Searle says, however, freedom in this compatibilist sense is *not* freedom as we presuppose it.[57]

One reason why materialists cannot affirm freedom involves the "stuff" of which things are composed, which Aristotle had called the "material cause" and which he regarded as passive. According to Descartes' dualism, minds are composed of one stuff, "consciousness," which is capable of self-determination, while bodies are composed of another stuff, "matter," which is passive stuff, so that bodies are capable of being moved only from without. Materialists, in rejecting this dualism in favor of the view that all things are composed of matter, were left with the view that nothing is self-moving. Even when matter is described as a dynamic matter-energy, self-determination is not included in the dynamism.

Whitehead likewise held there to be one stuff of which all actual entities are composed. But he called this stuff "creativity," specifically pointing out that it is "divested of the notion of passive receptivity."[58] Whitehead's "creativity" is a larger concept than the physicist's "energy." Whereas the physicist's energy is limited to efficient causation, creativity, which is "emotional and purposeful," involves final or purposive causation as well.[59] Accordingly, rather than calling the stuff of which all things are composed simply "creativity," we should follow Hartshorne in calling it "creative

54 John R. Searle, *Minds, Brains, and Science: The 1984 Reith Lectures* (London: British Broadcasting Corporation, 1984), 97.
55 Ibid., 92, 86.
56 Ibid., 86.
57 Ibid., 87, 92, 95.
58 PR 31.
59 SMW 153; AI 186.

experience."[60] To say that all actual entities are embodiments of creative experience is to say that they are experiences capable of exercising final causation, or self-determination, as well as efficient causation.

A second reason why materialists cannot affirm freedom is their doctrine of identism, according to which the mind is numerically identical with the brain. Materialists do not, in other words, regard what we call "the mind" as a distinct entity: Searle says of the human head that "the brain is the only thing in there"; Daniel Dennett says that there are "[billions of] microagents (with no single Boss)" and "that's all that's going on."[61] Materialists cannot, therefore, regard the mind as a locus of power with the capacity for self-determination. Dualists, by contrast, do regard the mind as a distinct entity, and they typically affirm freedom. But they cannot explain, without presupposing supernatural assistance, how the mind, after exerting its self-determining freedom, is then able to influence its body. Naturalistic dualists, therefore, cannot really explain what morality presupposes—that we are at least partly responsible for our bodily acts.

This problem is overcome by Whitehead's position. On the one hand, his system, by virtue of its organizational duality, affirms a numerical distinction between mind and brain, so the mind can be regarded as a locus of self-determining power. On the other hand, this numerical distinction between mind and brain is not an ontological dualism, so the mind's influence on its brain cells is not unintelligible. And because the mind, as the supreme entity in the body, has far more freedom than any of our bodily cells and is in position to influence them all, we can understand how it can be the dominant factor in determining our bodily acts. The behavior of a compound individual is hence different in kind from that of a rock, in which any spontaneities in its individual members cancel each other out, so that the rock's behavior is "a mere aggregation of effects."[62] In an animal, the spontaneities of the bodily cells are coordinated by the *anima*, so that a description of the resulting behavior must take its purposes into account.[63]

---

60 Hartshorne, *Creative Synthesis*, 7: "in this philosophy there is nothing in the world but creative experience."

61 John R. Searle, *The Rediscovery of the Mind* (Cambridge: MIT Press, 1992), 248; Daniel E. Dennett, *Consciousness Explained* (Boston: Little, Brown & Co., 1991), 458, 459.

62 SMW 110; MT 28.

63 Thanks to this duality, Whitehead could reject the "wild overstressing of the notion of 'final causes' in the Christian middle ages"—medieval Aristotelians notoriously said that a rock falls because it seeks a state of rest—without succumbing to "the correlative overstressing of the notion of 'efficient causes' in the modern scientific period" (PR 84).

As explained thus far, Whitehead's affirmation of human freedom is based on the Leibnizian idea of organizational duality, according to which "diverse modes of functioning ... are produced by diverse modes of organization."[64] Leibniz himself, however, could not affirm genuine freedom, partly because of his retention of the traditional doctrine of divine omnipotence and partly because of his idea that all worldly monads are "windowless." Although Leibniz regarded each monad as a unity of perceptions, any real perception was denied. This denial meant that the so-called dominant monad could not really exert dominating influence over its body, because it could not be perceived by the bodily monads. Leibniz's explanation of our mind-body coordination—that my intention to raise my hand is followed by my hand's rising—took the form of his notorious fantasy of pre-established harmony, according to which the movements of all things were preordained by God at creation.[65] Although some of the molecules in my hand came from California carrots, others from Idaho potatoes, and still others from various other places, God had ordained that they would all rise simultaneously, just after my equally preordained wish for them to rise. Although Leibniz meant this doctrine to protect human freedom, it allowed only cosmological freedom, not theological freedom.[66] That is, while denying that our mind's decisions are determined by efficient causation from the world, Leibniz's doctrine allowed no freedom vis-à-vis God.[67]

Whitehead, besides rejecting the traditional doctrine of divine omnipotence, also revised Leibniz's panexperientialism by installing windows. This revision brings us to Whitehead's doctrine that the world is made of creative, experiential *events*. Pointing out that Leibniz "did not discrimi-

---

64 MT 157.

65 The fact that Leibniz was here simply bringing out the implications of traditional theism does not make his doctrine any less fantastic.

66 I have distinguished between cosmological and theological freedom in a discussion of Augustine and Pelagianism in ch. 7 of David Ray Griffin, *God and Religion in the Postmodern World* (Albany: State University of New York Press, 1989).

67 It was this Leibnizian solution to the problem created by windowless monads that evoked Whitehead's complaint about "a *deus ex machina* who was capable of rising superior to the difficulties of metaphysics" (SMW 156). As Whitehead elaborated elsewhere, Leibniz, having defined monads as incapable of communicating with each other, "boldly excepted God from the scope of the doctrine," thereby allowing "an indirect communication between monads by the mediation of God .... But no reason can be given why the supreme monad, God, is exempted from the common fate of isolation. Monads, according to this doctrine, are windowless for each other. Why have they windows towards God, and Why has God windows towards them?" (AI 133–4)

nate the event, as the unit of experience, from the enduring organism as its stabilisation into importance,"[68] Whitehead built his own system around this distinction by "toning down [Leibniz's] monads into the unified events in space and time."[69] That which is actual in the fullest sense of the term is a momentary event, called an "actual occasion" or an "occasion of experience."[70] Each individual that endures through time, such as an electron or a human mind, is a temporal *society* of these momentary experiences.

> The real actual things that endure are all societies. They are not actual occasions. It is the mistake that has thwarted European metaphysics from the time of the Greeks, namely, to confuse societies with the completely real things, which are the actual occasions.[71]

One reason this distinction is so important is that it allows us to fulfill "[o]ne task of a sound metaphysics," which is to "explain the interweaving of efficient and of final causation."[72] Leibniz's failure to explain this interweaving is only one example of modern philosophy's failure in this respect. Whereas Leibniz reduced all causation in finite things to final causation, materialists reduce it all to efficient causation. We must, said Whitehead, "hold the balance" between these two kinds of causation but without returning to Descartes' dualism between physical and mental actualities.[73]

Whitehead's explanation of the interweaving of efficient and final causation—which is one of his greatest achievements—involves the idea that the world's basic events are *physical-mental* events. Each enduring individual is a temporal society of actual occasions, each of which has phys-

68 SMW 155.

69 SMW 70.

70 Whitehead also used "events" more broadly, so that "[o]ne actual occasion is a limiting type of event" (PR 80). I am here using "events" exclusively for actual occasions, which are Whitehead's monads (PR 80).

71 AI 204. The fact that Whitehead violated this principle with regard to God, alluded to in note 27, meant that his brilliant solution to the relation of efficient and final causation, discussed below in the text, could not be applied to God and hence to the God-world relation. As a result, Whitehead implied that God was an exception to several metaphysical principles, thereby threatening the coherence of his whole system. Whitehead should have consistently applied his dictum that although the "notion of an actual entity which ... remains numerically one amidst the changes of accidental relations" is useful for some purposes, "in metaphysics [it] is sheer error" (PR 79).

72 PR 84; FR 28.

73 PR 19.

ical and mental poles.[74] An occasion begins as an open window to the past, receiving efficient causation from prior occasions. This reception of data from prior occasions constitutes the occasion's physical pole, which is simply *given* to the occasion.[75] At this point, the window is closed, so there is no more influence from the outside world, while the occasion decides precisely how to respond to the influences received.[76] This self-determination is the occasion's mental pole.[77] During these two phases, the occasion is a "subject," having experience. But with the completion of the subject's self-determination, the occasion becomes an "object" or "superject," exerting efficient causation on future occasions, especially the next occasion in the enduring individual in question. Accordingly, in an enduring individual such as an electron or a human mind, there is a perpetual oscillation between efficient and final causation, between givenness and self-creation.[78] (The term "perpetual oscillation" conveys Whitehead's meaning better than his own term "perpetual perishing," which has caused perpetual confusion.[79])

---

74 PR 230.

75 Given Whitehead's empirical procedure, whereby the features attributed to low-grade events must be based on analogy with features of our own experience, his account of efficient causation between low-grade events depends upon his description of human experience as containing an element of givenness. Much of Whitehead's epistemology is devoted to this point, which is emphasized by his "reformed subjectivist principle." I have discussed Whitehead's affirmation of givenness in my introduction to DAVID RAY GRIFFIN ET AL., *Founders of Constructive Postmodern Philosophy: Peirce, James, Bergson, Whitehead, and Hartshorne* (Albany: State University of New York Press, 1993), esp. 14–23. Whitehead's reformed subjectivist principle is discussed in OLAV BRYANT SMITH and DAVID RAY GRIFFIN, "The Mystery of the Subjectivist Principle," *Process Studies* 32, no. 1 (Spring / Summer 2003): 3–36.

76 The closing of the window is expressed in Whitehead's statement that in describing the becoming of an actual occasion after the past world has constituted its physical pole, "There must be no further reference to other actual entities" (PR 153).

77 "In each concrescence," Whitehead says, "there is a twofold aspect of the creative urge. In one aspect there is the origination of simple causal feelings; and in the other aspect there is the origination of conceptual feelings. These contrasted aspects will be called the physical and the mental poles of an actual entity" (PR 230). These two poles of an occasion of experience are parallel to the receptive and spontaneous dimensions of a moment of experience as described by Schleiermacher.

78 "The creative process is rhythmic: it swings from the publicity of many things to the individual privacy; and it swings back from the private individual to the publicity of the objectified individual. The former swing is dominated by the final cause, which is the ideal; and the latter swing is dominated by the efficient cause, which is actual" (PR 151).

79 PR 29. Whitehead's point was that after an actual occasion had completed its moment of subjective becoming, its *subjectivity*, with its *final* causation, perished, with the re-

With this doctrine of perpetual oscillation, Whitehead explained how freedom is possible in a fully naturalistic universe, in which every event is enmeshed in an unbroken causal web. Each event begins as an effect of efficient causation from past events, then becomes an efficient cause on future events. But the efficient causation it exerts is based on its own final causation, or self-determination. The event's freedom, in other words, is exercised *between* its reception of causation from the past and its exertion of causation on the future. Each "occasion arises as an effect facing its past and ends as a cause facing its future. In between there lies the teleology of the universe."[80]

To flesh out this idea, we need to look more closely at Whitehead's notion of feelings. Every feeling has both a datum—which is *what* is felt—and a subjective form—which is *how* that datum is felt. The term "feeling," as a synonym for "positive prehension" (there are also *negative* prehensions, which "eliminate from feeling"[81]), brings out "this double significance of subjective form and of the apprehension of an object."[82]

An occasion of experience as "an effect facing its past" is constituted by "physical feelings," which are feelings of the feelings of previous actual occasions. The datum of a physical feeling must be distinguished into the "initial datum," which is a prior actual occasion as a whole, and the "objective datum," which is one of that prior occasion's own feelings (the role of negative prehension is to effect the transition from the initial to the objective datum[83]).

A physical feeling is also called a "conformal feeling," because its subjective form conforms to that of the prior feeling. This point is so important

sult that the occasion now had *objectivity*, with its *efficient* causation. But many interpreters took the term "perish" too literally, assuming that it meant that the occasion was no longer *actual*. This would mean—by the ontological principle that only actualities can act—that the occasion would be incapable of exerting causation, which is the exact opposite of Whitehead's position. I have discussed this confusion in *Reenchantment*, 114–5.

80 AI 194. It seems to me that this idea is implicit in Schleiermacher's position.

81 PR 23.

82 AI 233. "In Cartesian language," said Whitehead (PR 41), "the essence of an actual entity consists solely in the fact that it is a prehending thing (i. e., a substance whose whole essence or nature is to prehend)," an "Ens prehendens." He could equally have said that it is "a feeling thing," which is why a "critique of pure feeling" should be able to deal with every type of human experience and hence every branch of knowledge.

83 PR 221, 236.

that Whitehead gives it a name, "the Doctrine of Conformation of Feeling."[84] He also uses the language of "sympathy," saying:

> The primitive form of physical experience is emotional—blind emotion—received as felt elsewhere in another occasion and conformally appropriated as a subjective passion. In the language appropriate to the higher stages of experience, the primitive element is *sympathy*, that is, feeling the feeling *in another* and feeling conformally *with* another.[85]

These physical, conformal feelings are also called *causal* feelings, because they are the means by which all efficient causation in the universe occurs. Efficient causation and physical feeling are, in other words, simply two sides of the same transaction: for B to feel A is for A to exert causal efficacy on B, which it does by transferring its feeling, with its subjective form, into B.[86]

If this were the totality of Whitehead's doctrine of feeling and causation, it would be a system of complete determinism, because the occasion's physical pole is simply given to it. But each occasion's physical pole is followed by a mental pole, which involves final causation, meaning self-determination. In this pole, the subjective forms of the inherited feelings can be modified: "[a feeling's] subjective form, though it must always have reproductive reference to the data, is not wholly determined by them."[87] Rather than total conformation, there is only "initial conformation."[88] Whitehead's doctrine with regard to an actual occasion as a whole, therefore, should be called "the doctrine of the *initial* conformation of subjective form," as indicated by his statement that "the continuity of subjective form is the initial sympathy of B for A."[89] Following this initial conformation, the occasion, in its mental or conceptual pole, determines the final subjective forms of its various feelings in the process of determining its overall aim, called the "subjective aim."[90]

---

84 PR 237–8; AI 183.

85 PR 162.

86 PR 116, 236.

87 PR 232. Cf. PR 85: "the *how* of feeling, though it is germane to the data, is not fully determined by the data."

88 AI 253.

89 AI 183.

90 PR 27, 235, 245–6.

This subjective aim includes the subject's anticipation of its effects on future occasions.[91] It is, of course, partly because human occasions of experience will have such effects that we evaluate them morally. But it is also partly because we presuppose that these effects flow, at least in part, from the subject's self-determination. Whitehead provided the basis for this presupposition with his principle that an actual occasion's effects on the future are a function of its "being" and that its being is determined by its "becoming," which involves the self-determination occurring in its mental pole. The importance of this principle for Whitehead's process philosophy is shown by the fact that he called it the "principle of process."[92] The centrality of this principle for Whitehead's justification of our moral presuppositions is brought out by his account of the relation between a subject and its feelings:

> The feelings are what they are in order that their subject may be what it is. Then transcendently, since the subject is what it is in virtue of its feelings, it is only by means of its feelings that the subject objectively conditions the creativity transcendent beyond itself. In our own relatively high grade of human existence, this doctrine of feelings and their subject is best illustrated by our notion of moral responsibility. The subject is responsible for being what it is in virtue of its feelings. It is also derivatively responsible for the consequences of its existence because they flow from its feelings.[93]

The most immediate consequences of a dominant occasion's decisions include their influences on its brain cells, which then transmit the desired influences to the relevant parts of the body. Whitehead's critique of pure feeling thereby explains how our bodily action can, as we all presuppose, reflect our free choices.

Although I have thus far discussed freedom without referring to God, a complete account requires this reference. The mental pole is called "mental" because its prehensions are of possible, rather than actual, entities. It is only insofar as an occasion of experience can choose among various possibilities that it has any freedom. But possibilities can neither exist nor exert

---

91 PR 27; AI 192–3.

92 PR 23.

93 PR 222. Cf. PR 255: "the actual entity, in a state of process during which it is not fully definite, determines its own ultimate definiteness. This is the whole point of moral responsibility."

causal influence on their own. This is Whitehead's "ontological principle," according to which possibilities can exist and exert agency only by virtue of existing within actualities.[94] Given the fact that there are possibilities that have never been actualized in the world of finite actualities, there must be another actuality, of cosmic scope, in which these possibilities are contained and through which they are made effective.[95] It was this insight that first led Whitehead to a *theistic* version of naturalism. This theism is *naturalistic* in that divine influence is part of, rather than an interruption of, the world's normal causal processes, because every finite occasion of experience feels God as well as the past finite world.

V. Normative Values and Ultimate Importance

Our moral beliefs, besides presupposing a degree of freedom, presuppose that there are normative values and that our decisions have ultimate importance.[96] In seeking to justify our basic moral beliefs, Whitehead sought to explain, in his critique of pure feeling, both the origin and the truth of these presuppositions. Whitehead's concern with the *origin* of these presuppositions was rooted in his acceptance of conceptual empiricism, according to which any notion used in a philosophical system must be "discoverable as an element in subjective experience."[97]

Whitehead's defense of this empiricism depended on his distinction between two modes of perception. Sensory perception, in delivering clear and distinct sense-data, is called "perception in the mode of presentational immediacy," because the data are immediately present to the mind, containing no inherent reference to a past or a future.[98] It was Hume's equation of perception with this mode that led him to deny that we have any direct experience of causation.[99] We know of causation because of another, more fundamental mode of perception, which Whitehead called "perception in the mode of causal efficacy." This mode is the basis for our notion

94 PR 46.
95 PR 40, 249–50.
96 Another presupposition about value presupposed by morality—the presupposition that at least some actual things have intrinsic value—had to be passed over here. See Whitehead's discussion at MT 110–11, 115–9, 158–9.
97 PR 166.
98 PR 179.
99 PR 123.

not only of causation but also of the external world, time, and the past, and hence for the basic categories that science uses to interpret sensory data.[100] But "perception in the mode of causal efficacy" is simply another name for "physical prehension," and that term is, as we have seen, largely equatable with "physical feeling." Whitehead thus showed how a critique of pure feeling can reveal the source of many of the categories necessary for scientific reasoning. Kant, having with Hume equated perception with *sense* perception, thought such categories must be a priori forms of the mind—an explanation that presupposes supernatural implantation. Whitehead provides a fully naturalistic explanation of these categories. Carrying out his claim that a critique of pure feeling should supersede *all* of Kant's critiques, he showed how physical feelings also provide the basis for our assumptions about normative values and ultimate importance.

1. Normative Values

One of our inevitable presuppositions, Whitehead held, involves the existence of values in the normative sense, often called normative ideals. A central feature of human experience is "the intuition of immediate occasions as failing or succeeding in reference to the ideal relevant to them."[101] For this reason, "good and evil" and "better and worse" are among the pairs of opposites in our direct experience.[102] Many other philosophers have agreed that we presuppose the existence of objective norms. For example, John Mackie, while denying that moral norms really exist, conceded that "most people in making moral judgments implicitly claim ... to be pointing to something objectively prescriptive."[103]

But modern moral philosophers, presupposing the sensationist doctrine of perception, have been unable to affirm the rootedness of ideals in our perceptual experience. The idea that we know of moral norms through "intuition," they believe, would require the existence of a special faculty—sometimes called a "moral sense."[104] Such a faculty would be explainable

100 PR 169–70; S 35; AI 182–3, 225; MT 133.

101 RM 60.

102 PR 341.

103 John Mackie, *Ethics: Inventing Right and Wrong* (New York: Penguin, 1977), 35, 299 n. For a similar view, see Gilbert Harman, *The Nature of Morality: An Introduction to Ethics* (New York: Oxford University Press, 1977), 11–13, 34.

104 See Mackie, *Ethics*, 38–9; Harman, *The Nature of Morality*, 9–10; Bernard Williams, *Ethics and the Limits of Philosophy* (Cambridge: Cambridge University Press, 1985), 149.

only as a supernatural implantation—as posited by deists such by Francis Hutcheson, Adam Ferguson, and Thomas Jefferson, who sought thereby to reconcile our knowledge of moral principles with Locke's sensationist doctrine of perception.[105] Among those who assume that viable explanations need to be naturalistic, moral assertions are, therefore, widely taken to be noncognitive, incapable of being true.[106]

Whitehead overcame this problem with his doctrine of perception in the mode of causal efficacy, which is a nonsensory mode of perception. Just as he rejected the idea of a "special religious sense,"[107] he could defend moral intuition without an ad hoc appeal to a special moral sense. Moral norms can be known in the same way that we know of time, the past, causation, and the external world.

Modern philosophy's difficulty with supporting our experience of normative ideals, however, is not due entirely to its sensationist doctrine of perception. Equally important are what can be called the Platonic problem and the Benacerraf problem. The Platonic problem is how ideal or abstract entities, traditionally called "Platonic forms," can exist. This problem presupposes the widespread intuition, sometimes called the Aristotelian principle,[108] that nonactual things can exist only in actual ones. In medieval

---

105 See the chapter entitled "Endowed by Their Creator" in GARRY WILLS, *Inventing America: Jefferson's Declaration of Independence* (New York: Vintage Books, 1978). Kant's moral a priori would arguably require the same kind of explanation. Rudolf Otto, being rooted in Kant and Schleiermacher, explained the existence of religion in terms of a divinely implanted category of the holy; see RUDOLF OTTO, *The Idea of the Holy*, trans. John H. Harvey (New York: Oxford University Press, 1958), 175. Mircea Eliade seemed to be presupposing Otto's view in saying that "the 'sacred' is an element in the structure of consciousness"; see MIRCEA ELIADE, *History of Religious Ideas*, trans. Willard R. Trask, vol. 1 (Chicago: University of Chicago Press, 1978), XIII. A criticism of the supernaturalism implicit in this position is expressed in J. Samuel Preus's query as to how, within an evolutionary world, Eliade's "remarkable 'element in the structure of consciousness' could have gotten there"; see J. SAMUEL PREUS, *Explaining Religion: Criticism and Theory from Bodin to Freud* (New Haven: Yale University Press, 1987), XIX.

106 See LEWIS EDWIN HAHN and PAUL ARTHUR SCHILPP, eds., *The Philosophy of W. V. Quine*, Library of Living Philosophers, vol. 18 (LaSalle, Ill.: Open Court, 1986), 663–5; J. J. C. SMART, "Religion and Science," in *Philosophy of Religion: A Global Approach*, ed. Stephen H. Phillips (Fort Worth, Tex.: Harcourt Brace, 1996), 217–24, at 223 (reprinted from PAUL EDWARDS, ed., *Encyclopedia of Philosophy* [New York: Macmillan Press, 1967], vol. 7).

107 RM 123.

108 Although Whitehead usually called it (one form of) the ontological principle, he also called it "the general Aristotelian principle" (PR 40).

thought, the ideal forms existed in God. Given the atheistic cosmology presupposed by most modern philosophers, however, there is no home for ideal entities. A few philosophers have thought that Platonic forms could exist on their own,[109] but most seem to agree with Reuben Hersh, who says: "Platonism without God is like the grin on Lewis Carroll's Cheshire cat .... The grin remained without the cat."[110] The other problem, pointed out by Paul Benacerraf, is that true beliefs can be considered *knowledge* only if that which makes the belief true is somehow *causally* responsible for the belief.[111] Accordingly, even if ideal entities could somehow exist on their own, we could not have knowledge of them, because they cannot be thought to exert any causal influence. Benacerraf's own argument focused on the question of how we can have knowledge of the ideal entities involved in mathematics.

Most philosophers, in dealing with mathematical knowledge, have found a way around the Platonic and the Benacerraf problems,[112] or else simply ignored them.[113] But they do use these problems as reasons for denying that we can have knowledge of moral norms.[114]

The theistic dimension of Whitehead's philosophy is most obviously relevant to his moral theory in relation to these two problems. It was, in fact, largely to solve these two problems that Whitehead came to affirm a version of theism. In a formulation of the ontological principle that addresses the Platonic problem, he said:

> Everything must be somewhere; and here "somewhere" means "some actual entity." Accordingly the general potentiality of

109 For an example of nontheistic Platonism, see CHARLES LARMORE, *The Morals of Modernity* (Cambridge: Cambridge University Press), 1996.

110 REUBEN HERSCH, *What is Mathematics, Really?* (New York: Oxford University Press, 1997), 12.

111 PAUL BENACERRAF, "Mathematical Truth" (orig. 1973), in *Philosophy of Mathematics*, ed. Paul Benacerraf and Hilary Putnam, 2nd ed. (Cambridge: Cambridge University Press, 1983), 403–20. Whitehead, who also held causation and perception to be two sides of the same relationship, would have agreed.

112 PENELOPE MADDY, *Realism in Mathematics* (Oxford: Clarendon Press, 1990), 44, 59, 78.

113 See HILARY PUTNAM, *Words and Life*, ed. James Conant (Cambridge: Harvard University Press, 1994), 153, discussing Willard Quine.

114 MACKIE, *Ethics*, 24, 38, 40; GILBERT HARMAN, "Is There a Single True Morality?," in *Relativism: Interpretation and Confrontation*, ed. Michael Krausz (Notre Dame: University of Notre Dame Press, 1989), 363–86, at 365, 366; IDEM, *The Nature of Morality*, 8.

> the universe must be somewhere .... The notion of "subsistence" is merely the notion of how eternal objects can be components of the primordial nature of God .... [E]ternal objects, as in God's primordial nature, constitute the Platonic world of ideas.[115]

The Benacerraf problem is addressed in another formulation of the ontological principle, which says that "apart from things that are actual, there is nothing—nothing either in fact or in efficacy."[116] "[T]he agency whereby ideas obtain efficiency in the creative advance," Whitehead suggested, is "a basic Psyche whose active grasp of ideas conditions impartially the whole process of the Universe."[117] Accordingly, after making a phenomenological claim—"There are experiences of ideals—of ideals entertained, of ideals aimed at, of ideals achieved, of ideals defaced"—Whitehead offers this interpretation: "This is the experience of the deity of the universe."[118]

Ideals as actively grasped by the divine psyche can influence us because we feel it in the same way that we feel other actual entities, so the same causal principles are operative.[119] Besides the fact that God prehends ideal possibilities with the subjective form of appetition that they be realized in the finite world,[120] there is a "prehension by every creature of the ... appetitions constituting the primordial nature of God."[121] In feeling the divine feeling of an ideal, a creature feels it, at least initially, with a confor-

115 PR 46.

116 PR 40.

117 AI 147.

118 MT 103. In *The Oxford Dictionary of Philosophy*, SIMON BLACKBURN says that "a naturalist will be ... opposed to accepting real moral duties and rights as absolute and self-standing facets of the natural order" ("Naturalism," in *The Oxford Dictionary of Philosophy* [Oxford: Oxford University Press, 1996], 255). Whitehead, by not regarding moral principles as "self-standing," can be said to fit that description. But whereas the naturalists Blackburn has in mind do not think of moral duties as "facets of the natural order" in *any* sense, Whitehead, with his theistically enlarged understanding of "the natural order," does. This is one of the crucial differences between theistic and nontheistic naturalism.

119 This assertion about the same causal principles presupposes that the remedy to Whitehead's own doctrine, discussed in notes 27 and 71, has been made. That is, God is here understood to be a personally ordered society of divine occasions of experience. With this revision of Whitehead's position, the worldly occasions can prehend God without God's being an exception to the metaphysical principle that an actual entity cannot be prehended until its subjectivity, with its final causation, has perished.

120 PR 32.

121 PR 207.

mal subjective form—that is, with appetition to realize it.[122] This divinely derived feeling is called the "initial subjective aim," or simply the "initial aim."[123] It is only the *initial* subjective aim, rather than the subjective aim as such, because the causality from God, like the causality from other actual entities, is not all-determining. The finite subject, once constituted by its prehension of God and the past world, is then "autonomous master" of its own becoming, deciding precisely what to make of its endowments.[124] The subject, therefore, "is conditioned, though not determined, by [the] initial subjective aim."[125] It is our freedom to depart from the divine ideal that explains why we have the experience not only of "ideals aimed at" and "ideals achieved" but also of "ideals defaced."[126] Whitehead thereby shows how a critique of pure feeling, with its category of the initial conformation of subjective form, can explain human conscience.

2. Ultimate Importance

Another presupposition of morality is that our decisions and actions are important in some ultimate sense. We presuppose that "there exists importance beyond the finite importance for the finite actuality," because "importance for the finite" requires "importance for the infinite," also called "importance for the universe."[127] Apart from some form of theism, however, it is difficult to see how this presupposition could be true. For example, Bernard Williams agrees that, besides the notion of *relative* importance, "we have another notion, of something's being, simply, important (important *überhaupt* ... or important *period*)." However, Williams continues, "It is not at all clear what it is for something to be, simply, important. It does not mean that it is important for the universe; in that sense, nothing is important."[128] We have here another example of a modern philosopher's willingness to rest content with a worldview that cannot do justice to presuppositions acknowledged to be inescapable.

122 PR 225.
123 PR 108, 244.
124 PR 245.
125 PR 108.
126 Whitehead's inclusion of both freedom and evil among our inevitable presuppositions provides the basis for his rejection of another of Leibniz's extraordinary fantasies, his doctrine that this is the "best of all possible worlds," which Whitehead called "an audacious fudge produced in order to save the face of a Creator constructed by contemporary, and antecedent, theologians" (PR 47).
127 MT 86.
128 Williams, *Ethics and the Limits of Philosophy*, 182.

Whitehead's solution to this problem again depends on his theism. But this solution involves not God's influence on the world but the world's influence back on God, which Whitehead calls God's "consequent nature." Our refusal to regard our lives as merely "transient enjoyment, transiently useful," our haunting sense that there is an everlasting order, and our "insistent craving that zest for existence be refreshed by the ever-present, unfading importance of our immediate actions," suggested Whitehead, reflect our sense that our experiences are taken up into the divine experience.[129] This sense lies behind our "insistent notion of Right and Wrong."[130]

Whitehead's thinking on this subject changed. When he wrote *Religion in the Making*, he thought that the idea of God as a personal being who responds to the world could be based only on metaphysical inference, not on direct religious experience.[131] But in *Process and Reality*, he said of God's experience of the world that "each [finite] actuality includes it as an immediate fact of relevant experience," so that God is experienced as "the great companion—the fellow-sufferer who understands."[132] In *Modes of Thought*, Whitehead said that our value experience includes three dimensions: one's present value experience, "the many other value experiences," and "the totality of value experience."[133] In other words, the experience of actuality as "something that matters" involves "intrinsic importance for itself, for the others, and for the whole."[134] Our "intuition of holiness," Whitehead held, is "our sense of the value of the details for the totality."[135]

129 PR 340, 351.

130 ESP 94.

131 RM 61–6.

132 PR 351. Whitehead thus provided a basis for supporting Abraham Joshua Heschel's description of the Hebrew prophets as having "a fellowship with the feelings of God, a sympathy with the divine pathos" (Abraham Joshua Heschel, *The Prophets* [New York: Harper & Row, 1962], 26).

133 MT 110.

134 MT 116–7.

135 MT 120. This point brings up the question of whether there is any sense in which Whitehead could have agreed with Schleiermacher's view that our feeling of God is a "feeling of absolute dependence." He could not, of course, have agreed with Schleiermacher's view that we are aware that we, while being dependent on God, do not influence God in return (CF § 4 [14–18]; § 33,2 [135]; § 38,1 [146]; § 51,1 [201]; § 54 [211–4]; § 55,1 [220]). In fact, Whitehead, in saying that importance for us implies importance for the infinite, explicitly rejects the idea that "one final perfection with static existence constitutes a notion which is relevant to our experience" (MT 86). However, if Schleiermacher's point were formulated simply to say that we are aware of divine causation upon which our existence is dependent and upon which we have no influence, Whitehead could agree. His position, I have argued in chapter 5 of *Reenchantment*, contains both a metaphysical and

Whitehead thereby grounded the presuppositions of morality in our direct feeling of God's feelings of both types: God's appetitive feeling of normative ideals and God's sympathetic feeling of the world. This twofold feeling of God was also central to his treatment of the relation between religion and morality.

VI. Self-Interest, the Common Good, and World-Loyalty

In his discussion of the problem of morality in contemporary civilization, Whitehead focused on the humanitarian ideal, with its respect for human rights, and the tension between self-interest and concern for the common good. Whitehead found the basis for dealing with these issues in the idea of reverence for God, which implies world-loyalty.

At the heart of civilization at its best, Whitehead believed, was the rise of respect for humans qua humans, which led to the idea of "the essential rights of human beings, arising from their sheer humanity."[136] In the Western world, the intellectual justification for inculcating emotions of respect and friendliness for other humans as such was provided by philosophy and religion, especially Stoicism, Platonism, and Christianity. Modernity, however, has undermined this traditional justification, and without "a reconstructed justification for the doctrine of regard to man, as man," the natural tribalism of human beings, with their naturally limited sympathies, will be unchecked.[137] Whitehead's reconstructed justification involves both belief in God and our actual feeling of God's feelings.

What gives the life of a human being its distinctive value is "the way in which unrealized ideals shape its purposes." Although the difference between humans and other animals is only one of degree, "the extent of

---

a cosmological argument for the existence of God. The metaphysical principles applying to all possible worlds can exist and be influential only by virtue of their subsistence in God, and the contingent "natural laws" of our particular cosmos depend upon a divine decision at the outset of our cosmic epoch. A Whiteheadian could agree, therefore, that the feeling described by Schleiermacher accurately reflects the *structure* of the whole of which we are parts. This structure, which consists of God-with-this-world, is that upon which I am completely dependent, and it is in no way dependent on me. Even though I affect God as a concrete, experiencing individual, nothing I do can affect the *nature* of the God-world totality.

136 AI 13.

137 AI 37–8.

the degree makes all the difference. The Rubicon has been crossed."[138] Put otherwise, "The worth of men consists in their liability to persuasion. They can persuade and can be persuaded by the disclosure of alternatives, the better and the worse."[139] This distinctive characteristic of human beings provides a basis for a "bond of sympathy" greater than those provided by other features of social life.

> This bond is growth of reverence for that power in virtue of which nature harbours ideal ends, and produces individual beings capable of conscious discrimination of such ends. This reverence is the foundation of the respect for man as man.[140]

Lying behind this statement is Whitehead's agreement with Hume that morality is based primarily on sympathy, in which one has concern for the feelings of others, and his further agreement that this sympathy is naturally limited to a quite restricted circle of fellow beings. Hume denied that this limitation could be overcome, saying that "there is no such passion in human minds as the love of mankind, merely as such."[141] But Whitehead, pointing to people who *have*, through the influence of religion, "cared for 'mankind, merely as such,'"[142] held that through reverence for God, our concern for others could be extended to humanity as such.[143]

Turning now to the natural tension between self-interest and the general good: Insofar as we accept the view that all human beings, or even all human beings within some particular society, have similar needs and rights, we know that we should promote the general good. We all have a strong tendency, however, to be more interested in our own satisfaction than in that of others. Whitehead, in fact, saw this tendency as metaphysically grounded, saying that each occasion of experience aims at intensity of feeling in itself and in the relevant future, with the "relevant future" defined in terms of the present experience's anticipation of its influence on future

138 MT 27.

139 AI 83.

140 AI 86.

141 DAVID HUME, *A Treatise of Human Nature (1739)*, in *The Philosophical Works of David Hume*, vols. 1 & 2 (Boston: Little, Brown, and Company, 1854), Book III, Part II, Section I; quoted by Whitehead, AI 30.

142 Ibid.

143 Although Whitehead, like virtually everyone else of his time, thought of morality almost entirely in terms of human-to-human relationships, his doctrine that all actual entities have intrinsic value—see note 82, above—provides the basis for an ecological ethic, which many have developed.

experiences.[144] Although an occasion's "relevant future" is usually quite limited, it can be extended indefinitely, so that the natural tendency toward selfishness can be overcome. On this basis—saying that "[t]he greater part of morality hinges on the determination of relevance in the future,"[145] he suggested the possibility of an ideal resolution of the tension between the self-interest and the general good:

> The antithesis between the general good and the individual interest can be abolished only when the individual is such that its interest is the general good.[146]

Such individuals will still act in terms of their interests, but they will now have "a wider sweep of interest." With regard to what might bring about this widening, Whitehead said that religion "is directed to the end of stretching individual interest beyond its self-defeating particularity."[147]

Whitehead related God to this stretching in two ways. First, God is working towards this end whether we consciously believe in God or not. A subject's feeling of another actuality brings that actuality, as an object, into the present subject, and every finite subject feels God. There is, accordingly, an "inclusion of God in every creature."[148] Although God is present only as an object of experience, not a subject,[149] this divine presence is not inert. "The only intelligible doctrine of causation," insisted Whitehead, "is founded on the doctrine of immanence. Each occasion presupposes the antecedent world as active in its own nature."[150] Applied to God, this account means that God is actively present in each finite experience. God's giving of an initial aim, in other words, is not simply God's proffering a possibility for our existence. It is prevenient, enabling grace. As thus actively present in us, God expands the scope of our interests:

---

144 PR 27.

145 Ibid.

146 PR 15.

147 Ibid.

148 RM 94. Cf. PR 156: "The world lives by its incarnation of God in itself."

149 This point follows from the fact, discussed in prior notes, that an actual entity can be prehended only after its subjectivity has perished, so that, as an object for other subjects, it is no longer a subject for itself (even though it is not a *mere* object but an object-that-had-been-a-subject). Whitehead would, therefore, have agreed with Schleiermacher that God was present in Jesus only objectively, not as an experiencing subject. I discussed Schleiermacher's Christology in "God's Objective Presence in Jesus," in David Ray Griffin, *A Process Christology* (Philadelphia: Westminster, 1973).

150 MT 165.

> God is that function in the world by reason of which our purposes are directed to ends which in our own consciousness are impartial as to our own interests .... He is that element in virtue of which our purposes extend beyond values for ourselves to values for others ... [and] such a value for others transforms itself into value for ourselves.[151]

Besides being brought about by our physical feelings *of* God, this expansion of our interests can also be brought about by our intellectual feelings *about* God. Whitehead did not mean, of course, simply any concept of God. He rejected a tribalistic concept of God in favor of one that embodies world-consciousness,[152] and he criticized traditional Christian theology for directing human worship towards a cosmic barbarian.[153] Whitehead approved of a concept of God in which the "concept of the goodness of God replaces the older emphasis on the will of God," a concept of a God whose goodness we imitate in order to become like God.[154] God thus conceived stands in contrast with our limited vision: "The consciousness which is individual in us, is universal in him: the love which is partial in us is all-embracing in him."[155]

Given such a concept of the object of worship, "Religion is world-loyalty," in which the human spirit "has merged its individual claim with that of the objective universe."[156] Insofar as this merger occurs, the tension between the individual's interest and the general good, which is *God's* interest, is overcome. Whitehead discussed this overcoming in terms of the final quality of civilized existence, "peace," in which "the soul's preoccupation with itself" is replaced by "a wider sweep of conscious interest."[157] Peace involves "the harmony of the soul's activities with ideal aims that lie beyond any personal satisfaction."[158] "One of its fruits is that passion whose existence Hume denied, the love of mankind as such."[159] Reverence for God, accordingly, can overcome the tension between self-interest and

---

151 RM 158.
152 RM 40–1.
153 RM 55; PR 342–3; AI 26, 51, 83.
154 RM 41.
155 RM 158.
156 RM 60.
157 AI 285.
158 AI 288.
159 AI 286.

the common good as well as the problem of the basis of respect for fellow human beings as such.

In Whitehead's kind of religion, which directs "conduct towards a unified purpose commanding ethical approval" and seeks to stretch "individual interest beyond its self-defeating particularity,"[160] religion and morality are obviously closely related. Both "arise as aspects of [the] human impetus towards the best in each occasion."[161] Each one involves our feelings of God as both influential in, and influenced by, the world.[162] But religion and morality are not identical. "Morality emphasizes the detailed occasion; while religion emphasizes the unity of ideal inherent in the universe."[163] Religion, in other words, is worship of the Holy Reality and loyalty to the divine purposes, whereas morality seeks to realize these purposes in relation to particular events. Each requires the other. Religion without world-inclusive morality is not religion oriented towards the truly divine; morality without such religion will lose motivation and universal scope.

160 RM 31; PR 15.

161 MT 28.

162 Whitehead concludes *Adventures of Ideas* by discussing our experience of the "Great Fact," which includes both the "initial Eros," which is "the living urge towards all possibilities," and the "final Beauty," which is the divine experience of the world. "It is the immanence of the Great Fact," says Whitehead, "which constitutes the zest of self-forgetful transcendence belonging to Civilization at its height" (AI 295–6).

163 MT 28.

Process and God in Whitehead and Schleiermacher

Eilert Herms, Eberhard Karls Universität Tübingen

Schleiermacher neither met Whitehead nor read a single line of his philosophy, and Whitehead, as far as we know, never read a single line of Schleiermacher's texts. So why compare their theories? We do so for a number of reasons. Both contradict the Kantian claim that metaphysics is not only impossible but also unnecessary. Both saw that a theory of the "on he on," that is, of being as such, must be a theory of the universal conditions of becoming. Furthermore, both saw that this conviction—the universal conditions of becoming—beats unavoidably at the very heart of religion, and that the subject of such a theory of universal becoming must provide the universal conditions for the world's becoming, which includes both the universal conditions of the becoming of human persons and refers to the work of God. For these reasons, the substance of both authors' arguments can be compared, especially if we consider the fact that both worked completely independently from one other. Such independence shows that their respective undertakings and achievements are a trustworthy witness to their common subject matter.

So much for the convergences between our two classic authors. Just as informative, however, might be the observation that there are certain marked differences between them. These differences pertain to their respective starting-points and methods, which consequently result in unmistakable differences in their work, and especially the culmination of their respective work: the relation between the world's becoming and God's work and, therefore, the understanding of God's own being and essence. These differences have an undeniable impact on religion. The comparison between our two heroes, therefore, will lead us to see and to understand how and why decisions regarding starting-point and method are decisive for the determination of metaphysical and religious content. Furthermore, the comparison will help us to determine that and why discussions of different worldviews are empty and without foundation as long as no account can be given of the situation whose mode of givenness provides us with the possibility of reaching some trustworthy conception of the universal conditions of the world's becoming and God's work.

In this chapter, therefore, I first delineate the basic elements of Whitehead's metaphysical procedure and his resulting view concerning the relation between the process of creativity and God.[1] Second, I hint at some problematic features in Whitehead's method, which infect the resulting view of world and God in the process of creativity. Third, I sketch the starting-point, procedure, and outcome of Schleiermacher's view of the world's becoming *as* God's work, which seems to me to avoid the problematic steps of Whitehead's view. I conclude with some remarks on why I think that the differences between Whitehead's and Schleiermacher's worldviews really make a difference for life.

I. Basic Traits of Procedure and Result in Whitehead's Thought

The epistemological turn of Western thought in Hume and Kant focused our attention on the difference between the realm of insight over and against the realm of arbitrary assumption and on the boundary between these two realms. This difference was drawn by giving the necessary and sufficient conditions for the first realm, that is, insight. Whatever is more than arbitrary assumption, is dependent on—as well as limited and controlled by—the fact that a situation is given *for* us. The situation shows itself *to* us, *as* one that we are required to understand and to master. This showing itself to us or this givenness to us *as* to be understood and mastered by us, necessitates us at once to understand and to somehow master the situation in the most satisfactory way, as well as providing us with the ability to do so. Almost immediately after Hume and Kant, it became clear that this situation, constituted by a type of givenness to us, was nothing else and could not be anything other than the situation of the immediate self-consciousness of

1 For a fine elucidation of Whitehead's argument in *Process and Reality*, see Kirsten Huxel, "A. N. Whiteheads Theorie des Bewußtseins," in *Befreiende Wahrheit: Festschrift für Eilert Herms zum 60. Geburtstag*, ed. Wilfried Härle, Matthias Heesch, and Reiner Preul, Marburger theologische Studien, no. 60 (Marburg: N. G. Elwert, 2000), 257–84.

persons living in the world. These persons found themselves to be determined in an absolutely passive way that simultaneously determined them as relatively free. Such was the situation already confronting Augustine in the fifth century and Descartes in the seventeenth century. Finally at the end of the nineteenth and the beginning of the twentieth century, it also became clear that all scientific knowledge falls within the boundaries of the situation of that which is primordially and insurmountably given for us. The decisive question is, therefore, whether or not this situation implies an allowance and demand for metaphysics. Or in more precise terms, whether or not there can be insight into the permanent conditions of this primordial and insurmountable situation that is universally valid for its every possible determination.

The remarkable feature of Whitehead's thought is his wholehearted acknowledgment of this situation as the starting-point for all human insight, whether empirical or metaphysical. Whitehead combines with this basic insight the basic terminological decision to call this situation the situation of experience. From this starting-point, the following series of steps leads him to his organistic metaphysics of creativity.

The first step (1.) is the step from empirical to metaphysical insight. This move is governed by a logical conclusion from the basic insight mentioned above: that nothing exists beyond the sphere of experience, which, as a negative all-proposition, includes the positive all-proposition that whatever exists, exists within this sphere. This means, however, that whatever exists, exists under conditions that are valid for experience. Consequently, the conditions of experience are the conditions of whatever there is, which means, of being as such.

Second (2.), a step of abstraction follows from the first. If we cannot identify being as such, called experience, with human being and human experience, the conception of the universal conditions of being / experience must be abstracted from all specific human traits of being / experience. And if, on the other hand, we cannot deny that human being / experience is a special kind of being / experience, the universal conditions of being / experience must be thought of as allowing for that special kind of being / experience and as producing it. The central difficulty in this step of abstraction is obviously the following. On the one hand, there are indeed special traits of human being / experience; it is necessary to abstract from these traits in order to reach the universal conditions of being / experience. On the other hand, it cannot be denied that in a certain sense "we" are an integral element in the primordial and insurmountable situation; by abstracting from the "we," the very essence of this situation would be destroyed. The decisive question is the following. In what sense is abstraction demanded of us and in what sense is it prohibited? We will now investigate the question as to whether or not Whitehead deemed this an important question and whether or not he answered it coherently.

Now to step three (3.). This step is the elucidation of the creative character of all experience by describing it as apprehensive feeling. Apprehension lies at the very heart of creativity. The continuous process of creativity results in a universe, and the presence of creativity does not lose anything of its past in passing over to its ever richer future. The task at step three, however, is at least threefold. The first task to be accomplished is to elucidate the relationship between feeling and apprehension (3. 1.). The second task is to describe feeling in the apprehension constituting its creativity (3. 2.). Finally, the third task is to face the consequences for the notion of subject as its own superject or as creative event in itself (3. 3.).

As far as task 3. 1. is concerned, Whitehead accepts the primordial situation. This primordial situation is one in which something is immediately given to a subject or felt by a subject as to be apprehended somehow. In this situation, there is no separation of feeling (immediate self-presentation) and apprehension (causality). Rather, there is always a distinction that assigns the task of determining the relation between feeling and apprehension: Does immediate self-presentation (feeling) allow for and demand causality (apprehension) or vice versa? Here, Whitehead votes for the priority of causality. It is apprehension that constitutes feeling and its apprehension.

Then—task 3. 2.—the creative continuum of apprehensive feeling and its elements can be described as follows. A subject feeling is affected by a constellation of felt data. There are different kinds of data: physical feelings on the one hand, eternal objects and the so-called hybrid feelings and propositional feelings on the other hand. The subject apprehends feelingly an objective datum from this felt data. The evaluation of physical feelings guides feeling through the constellation of primordial data. By this procedure, the subject feeling is motivated to select one physical feeling from among others.

Finally—task 3. 3.—the consequence of this selection for the concept of subject must be drawn. This subject is always the subject feeling. According to Whitehead's vote for causality (apprehension) as primordial over and against immediate self-presentation (feeling), this subject feeling cannot be anything other than the result. The subject feeling is the objective datum, either of a companion feeling or of its own; it is its own superject, or in other words, a creative event in itself.

Such is the felt apprehension. Its creativity is continuously working and results in a creative presence. This creative presence gathers the universe of past creative events together with their results, including human being and consciousness, without any loss. Furthermore, it passes over into ever richer future states of the universe. Process and reality are viewed in such a way that reality in itself is nothing but the continuous process of apprehensive feelings that are creativity. This understanding of creativity implies that God and world are nothing but two aspects of the same metaphysical process. God is both primordial nature, meaning the primordial situation of this creative working, and consequent nature, meaning the continuously growing result of this creative working; the world is the continuous flux leading from God's primordial to God's consequent nature. It is the world that connects God's primordial to God's consequent nature. Whitehead describes the way in which God's two natures are connected by the world in a famous passage. "The primordial nature (of God) is conceptual, the consequent nature is the weaving of God's physical feelings upon his primordial concepts."[2]

II. Problematic Steps and Results in Whitehead's Philosophy

It can now be shown, I think, that the elements supporting the methodological structure of Whitehead's philosophy and the resulting picture of the process of creativity can be the objects of critical observations and remarks.

First, there is the problem of reaching the level of metaphysical insight over and against the empirical one. It is of course a formal condition of metaphysical insight that it be logically universal. The formal condition must, furthermore, be articulated in all-propositions that distinguish between what is impossible for all possible beings (for example, as is the case with Whitehead, being outside the sphere of experience) and what is valid for every possible being. This formal condition of metaphysical insight is, nevertheless, only a necessary, not a sufficient condition. For any formal condition to be a sufficient one, it must be shown—since Kant—that the universal conditions of being are given to us as possible objects of our

2 PR 345.

knowledge as well as the kind of objects that are given to us to be known. Whitehead acknowledges the situation of immediate self-consciousness. (This is the situation that is given for us or shows itself *to* us *as* one that we must understand and master. This showing itself to us or this givenness to us *as* to be understood and mastered by us, necessitates us at once to understand and to somehow master this situation.) Whitehead, I think, acknowledges that this situation is primordial and insurmountable for all possible beings and knowledge. The question can now be posed as to whether or not only empirical facts are presented by this primordial and insurmountable immediate givenness or whether or not the universal conditions of being are presented by this immediate giveness over and against empirical facts. One does not reach metaphysical insight merely by drawing the logical conclusion from the denial that metaphysical insight is possible. What is indispensable is an analysis of the primordial and insurmountable situation of immediate givenness. This analysis would show that this givenness presents at once both empirical facts *and* the universal conditions of all possible beings as those conditions to be "apprehended"—somehow understood and mastered—by the receiver of that presentation. During Whitehead's lifetime, Husserl and Heidegger successfully attempted such an analysis of immediate primordial givenness. I do not find any evidence that Whitehead undertook such an analysis.

That leads to some problems in step 2: the abstraction from the human traits of experience. In order for this abstraction to be performed, we must hold on to a few assumptions. First, we must hold on to the acknowledgment that the above-mentioned situation of immediate givenness of something is given to its receiver *as* to be somehow apprehended by this receiver. Second, we must hold on to the acknowledgment that human beings must be reckoned among such receivers. If this is the case, then the human traits of experience can be distinguished from one another. Some traits are inseparable from the very constitution of the primordial situation, others are not. Therefore the following question can be formulated. *Which* traits are inseparable and *which* are not? I do not see that Whitehead poses this question, let alone answer it.

This question concerning the separability or the inseparability of human traits from the primordial situation is, in turn, closely connected to step 3. 1. This step is Whitehead's vote for the priority of causality (apprehension) over and against immediate self-presentation (feeling) in the inseparable connection between feeling and apprehension. Whitehead vindicates this decision by referring to bodily experience. For this kind of experience, there is no feeling unless it is caused by some outer affection. But let us look more closely at the example of bodily experience. We must ask whether or not there is or can be any bodily experience outside of the primordial situation of immediate givenness. I do not see how this claim can be asserted. We are then drawn back to the analysis of the primordial situation itself. And here we see that feeling as immediate self-presentation by its own constitution—namely immediate givenness of something *to* a receiver *as* to be somehow apprehended (understood and mastered) by this very receiver—does not exist by itself and can never occur without apprehension. It is feeling (immediate self-presentation) that contains all the conditions—including the sufficient conditions—for apprehension.

The claim concerning feeling as containing all the conditions for apprehension can be frankly acknowledged without saying that feeling as immediate self-presentation should be the Absolute. Feeling does not necessarily imply the absolute, all-constituting creativity and thereby the whole process of reality. On the contrary, the following question is possible and justified. Does the situation of immediate givenness, which for us is primordial and insurmountable, show itself to be constituted passively in some way as the effect of some higher "working," or does it show itself to be constituted by itself? The description of this whole constellation would, however, become thoroughly blurred and confused if we did not make a careful distinction. We must distinguish between that working, on the one hand, whose *effect* is the primordial situation of feeling, and that other working, on the other hand, whose necessary and sufficient conditions are implied by that primordial situation of feeling. These are two different kinds of working because they proceed from different conditions. This distinction might be quite plausible to a stubborn Whiteheadian because Whitehead himself insisted on the claim that feeling presents all the necessary and sufficient conditions of apprehension.

In step 3. 2., Whitehead shows us how the subject feeling is always conditioned by the relations between its felt data—at least the physical feelings and eternal objects—and how these relations determine the apprehension of feeling. On this point, Whitehead asserts that the subject feeling is always affected by both different kinds of felt data and the relationship between the different kinds of data. Whitehead's assertion, however, can be deemed what Hegel calls a "trockene Versicherung." It is a "dry insurance" as long as Whitehead cannot show how the very constitution of feeling includes its determination by different kinds of givens, called "data," and as long as he cannot clearly analyze the character of a datum by an analysis of both feeling and the different kinds of passivity constituting it. I do not find anything of this sort in Whitehead's thought.

Finally, step 3.3.: The conception of the subject feeling as subject-superject and as creative event only draws the consequences of steps 3.1. and 3.2. If our critical remarks regarding them are justified, however, then this final conception would merely be the sum of the problems in steps 3.1. and 3.2. Apart from these problems, we must ask whether or not this conception suffices to meet the requirements of causality, which is the very heart of creativity. Let us take proceed with our investigation of causality as follows. Let us take causality as given. According to Whitehead, causality cannot be known from the beginning. If causality could be known from the beginning, it would fulfil the condition of its own knowability. This condition is the distinction between cause and effect. Although neither separated nor separable, cause and effect must be distinguished in order that many effects in different circumstances can be attributed to one cause and its workings. It is necessary to think of causes in the Whiteheadian context as having the character of some process; in fact it would be inconsistent on Whiteheadian premises to think of causes as things or substances. The knowability of this causative process inexorably demands its distinctness from its effects, its relative determination and its relative identity. Only then would it be possible to distinguish between several kinds of causes that are distinct from each other through their typical mode of apprehensive working. Such a distinction is forced upon us by everyday life. I do not see that Whitehead's concept of subject feeling does or can fulfil this requirement.

This criticism already addresses the metaphysical result of Whitehead's philosophical procedure. And this criticism, I think, extends to the whole Whiteheadian picture of reality as creative continuum. Whitehead's picture of reality does not take the type of determination of every subject feeling into account. This determination results from the constitution of its relative identity and distinctness from its own effect on the one hand and from its own "cause"—whatever that might be—on the other hand. Consequently in Whitehead's system, there is no determination of processes *within* the universal process and no determination of the whole process. Furthermore, there is no reflection on the determination of certain processes that might show which possible processes in this world, given all processes, are actualized and working in this world. Finally, there is no reflection on the determination of the universal process of this world from all possible worlds that are creatively apprehendable from the universe of all possible worlds.

I am now considering two claims. The first claim concerns the determined finite process-continua within this world. The second claim concerns this whole world as a special—determined and finite—process-continuum within that creative process-continuum whose creativity apprehends and determines one out of all possible world-processes. When these two claims are considered together, one cannot avoid making clear distinctions between different spheres and modes of creativity. The following distinctions are inevitably made between a) the creativity *within* certain process-continua in this world, b) the creativity *of* these process-continua in this world, c) the creativity of this world, and d) the creativity of the higher process-continuum, within which this world as a determined finite process-continuum is apprehended out of the sphere of all possible worlds. The result of this analysis is the conception of a relation between God and world that differs radically from Whitehead's conception.

III. Methodological Structure and Content of Schleiermacher's Worldview

Now let us turn to Schleiermacher. Unlike Whitehead, Schleiermacher did not write a comprehensive essay on metaphysical cosmology. In his main works, he makes several contributions to such a theory, which, when added together, are sufficiently coherent to form a comprehensive whole. The relevant texts are: *Reden über die Religion an die Gebildeten unter ihren Verächtern*, *Monologen*, the introductory parts in his lectures on speculative ethics as well as the late essays on the same topic that he delivered at the Prussian academy of sciences, furthermore, part one (the so-called transcendental part) of his lectures on *Dialectic*, his lectures on *Psychology* and last but not least, his dogmatics, the *Glaubenslehre* as well as the *Christliche Sittenlehre*.

1. What makes the comparison with Whitehead fruitful is, firstly, the fact that Schleiermacher seems to be one of the first thinkers who discovered—even earlier than Fichte, it seems—that the situation of immediate self-consciousness is the primordial and insurmountable situation for all possible human activity. Schleiermacher calls this situation "feeling." For Schleiermacher, feeling is the situation of the immediate givenness of something to the receivers of those givens (data) *as* to be somehow apprehended (understood and mastered, be it theoretically by means of symbolizing or practically by means of organizing) by these very receivers. This means

that the "something" that is given is the inner-worldly situation of the receivers themselves. The givenness is characterized by the immediate disclosure of this receiver-situation to every receiver *as* her own. This situation of immediate disclosure constitutes her as an inner-worldly agent. By this disclosure, she is enabled and necessitated to apprehend the situation somehow, in a free and responsible manner. In apprehending this situation, she shares it with others like herself, thereby interacting freely and responsibly with others. Whitehead obviously acknowledges this same situation as one that grounds all of human life's abilities and obligations. For this reason, it is an insurmountable situation for us. The acts occurring in human life are all reflective apprehensions. Human life proceeds as continuous self-determination by these reflective apprehensions, specifically by means of self-organization and self-symbolization. The latter kind of self-determination, self-symbolization, results in the system of human knowledge.

2. In Schleiermacher's case, this knowing—this is the second point to be made here—sticks tenaciously to what is given to us to be known in the primordial situation of immediate self-consciousness. This "given" is the universe of single events[3]—present and past—*as* individuals, irreplaceable in their individuality, each of a certain kind, and therefore always within the horizon of the class of events like them. To know them, therefore, is to apprehend them in their specific relation to the universal conditions of the class. In each case, each individual is an individual variation of those universal conditions characterizing those individuals who are like them. This apprehension, which is called knowing, combines two apprehensions which must be distinguished from each other. The distinction between the two apprehensions does not at all mean that the two can be or can occur in separation from each other. One apprehension intends and apprehends the individual in the specific determination highlighting its irreplaceability. The other apprehension intends and apprehends the universal conditions determining the class-horizon. In this case, the essence of individual variation can be unmistakeably apprehended within this horizon as both quantitatively and qualitatively distinguished from every possible other individual. To summarize: All knowledge functions critically, referring the empirical to the speculative and vice versa, so that the progress of speculative knowledge is the condition of progress for empirical knowledge and

3 Schleiermacher speaks of "acts" instead of "events," but does not limit this expression to human actions.

vice versa.[4] If speculative knowledge of this sort can be called philosophical or metaphysical, then metaphysical knowledge cannot be abandoned by the empirical domain. Metaphysical knowledge is the indispensable condition of empirical knowledge. And this level of metaphysical knowledge is not, as is the case with Whitehead, reached by logical operation. Metaphysical knowledge is obtained, as is the case with Schleiermacher, by reflective analysis of the primordial situation of immediate self-consciousness. This primordial situation constitutes in and by itself the possibility for the task of knowing. This task cannot be refused. And it is made possible by critically moving between the singular and the universal.

3. Thus we have reached the third point to be stressed. The universal, which Schleiermacher unfolds in his speculative theory, is not reached by abstraction from the human condition. It is reached by reflection on it as it is given to us to know—*nota bene*: not as known!—in immediate self-consciousness. Consequently, the universal conditions of becoming or acting, as Schleiermacher describes them, do not articulate an overall universality without any specifically determined traits. These universal conditions are the universality of the overall conditions of only a specific—that means, determined and finite—sphere of possible events or of possible becoming. It only shows us the universal conditions of the becoming of *this* world, given to us and inhabited by us. At first glance, this seems to be a classic example of the pseudo-metaphysics that Whitehead criticizes as resulting from the barren identification of the universe with ourselves. On closer examination, however, this criticism does not hold. By unerringly holding on to the primordial and insurmountable given, Schleiermacher insists only on the finite determination and individuality—scholastically speaking: the *haecceitas*—of the universe given to us and inhabited by us. Such a determinate individuality does not in any way imply that this world consists of nothing but us and those like us. On the contrary—and this is my point 4—: the universal conditions of humankind in this world comprehend the dependence of all human life and history on physical process and its organic productivity. This process is older than human history and is, probably at the same time, younger than it.

4 Compare these insights with Whitehead's finding of the "two levels of ideas which are required for successful civilization, namely, particularized ideas of low generality, and philosophic ideas of high generality. The former set are required to reap the fruit of the type of civilization immediately attained; the latter set are required to guide the adventures toward novelty, and to secure the immediate realization of the worth of such ideal aim" (AI, VIII).

4. I stress this as point 4. An honest account of the universal conditions of all becoming in this world given to us and inhabited by us and those like us shows a determined complex of *different modes of becoming*. Each mode is determined in itself and is finite; most of the modes are far older than human life. Schleiermacher's insistence on the given by no means narrows his perspective. It only focuses his view to the determined idiosyncrasy of the becoming in this world as well as to the internal differentiation of this becoming and its relative stable ordering. And the final important point is the following. A metaphysic of this world, which is unerringly faithful to the given, does not even end with the insight into the universal conditions of this world. Such a metaphysic knows these universal conditions only by referring to the horizon of that which is possible for those like it. Only within this horizon is this world given as *this* world. This world is not given as this world without relation to other possible worlds like it. Each of these possible worlds must show the trait of *haecceitas*; each must be a determined, finite, single world; each must be a "this world" in itself.

5. Let us now shift our view back to our world. It is a simple fact that Schleiermacher concentrates his speculative theory on only one sphere of becoming in it. In his speculative ethics, Schleiermacher concentrates on the universal conditions of human history. The basis of this concentration is a sharp distinction between the physical process of nature and the ethical process of history. The one is given to us only as the continuum of that physical process producing organic systems for human life and exercising its will upon it; the other is given to us as the personal and social systems producing workings of will that actualize human life in the physical process. The distinction between physical and ethical process is not made in isolation. Schleiermacher's statements on the *terminus a quo* show that he conceives of the ethical process as embedded in the physical process. History begins with the "Beseelung der Natur"; history begins with the situation that the physical process is given to us in immediate self-consciousness. Schleiermacher's statements on the *terminus ad quem* show the insurmountable dependence of the ethical process (of history) on the physical process (of nature). History ends with the dissolving of this unity. The unity comes to an end when the physical processes as necessary conditions of our inner-worldly life come to an end.

On the other hand, the embeddedness of history in and its dependence on nature only mean that nature is the necessary, not the sufficient condition of history. Consequently, there is no way to conceive of the ethical process as caused by the physical one. Rather, a new mode of causation enters the world with the start of the ethical process. It is the causation initiated by the responsible actualization of human will. And this is a mode of causation which—if we stick to the given—we find in human action and nowhere else. Does this, however, not mean that the unity of the world and the process-continuum in it is lost?

6. The answer is: No. I will explain this negative answer as point 6. The unity of our world is not guaranteed by the fact that there is one single mode of process in it. There is no one single mode of causality, leading from a present state to an altered one. On the contrary, it is a given fact that within our world, there are different modes of causality working together at the same in the transition from the present to the altered state. The unity of our world and the different modes of processing its overarching continuum are guaranteed by the openness of all these different modes of process to being mutually conditioned by the workings of all the rest. This harmony implies that there is no single mode of process which is absolute. No mode of process within this world given to us and inhabited by us is the unique ground for the existence of other modes; no one mode produces and guarantees the harmonious openness of all different modes of proceeding, and no one mode is a self-creating and self-supporting *causa sui*. Schleiermacher is faithful to what is given in the primordial situation of self-consciousness: to the determination of different modes of causality in our world, to the determination of the relations between these different causal modes, and to the determination of the whole process-continuum in our world. Schleiermacher is faithful to the fact that none of these determined modes of proceeding is given as determined by itself. They all are given as individual modes of proceeding from the sphere of other possible modes of those like them. As such, these individual modes of proceeding are given as determined by a process that occurs at a higher level, a process that apprehends the individual modes from their respective class of individual modes. For Schleiermacher, this higher level of proceeding (the apprehension of this determined finite world from all other possible worlds like it and the determined finite order of determined finite modes of proceeding in it) is the level of God's creativity. God's creativity determines the modes of process and their ordering, which are constitutive for the determinate finite process-continuum of this world.

7. Schleiermacher's claim considering the determinative capacity of God's creativity does imply, however, that God's creativity refers to the entire process-continuum of this world. I want to explain Schleiermacher's insight as my seventh point because it has been obscured by widespread

misunderstandings of § 4 of the *Glaubenslehre*. One finds the famous phrase in this paragraph of the CF, that God is the "Woher" (the "Whence") of our absolute dependence.[5] God is felt when we feel that we are necessitated to interact in the world in both relative freedom and relative dependence. This passage, if understood correctly, does not refer to some private motion of the heart, conceived of as a ghost in the machine. The passage refers to the situation of immediate self-consciousness, which, for Schleiermacher, is that primordial and insurmountable situation both constituting every human life as a life of responsible interaction and providing that life with all kinds of abilities, tasks, and obligations essential for such a life. The passage concerns the givenness of a human individual as a variation of the universal conditions of humankind. It, furthermore, speaks of the givenness of the universal conditions of humankind, including its dependence on physical process, and it, therefore, speaks of the givenness of this determined finite world, which is given to us and inhabited by us. To speak of God as the "Whence" (*Woher*) of the primordial and insurmountable situation of all possibilities, tasks, and obligations is, therefore, to speak of God as the "Whence" of our world as a whole.

8. But—and this is point 8—another misunderstanding can be seen to disappear along with the misunderstanding concerning the "Whence" of our absolute dependence. This second misunderstanding concerns our thinking about the world. What does it mean when we speak of our world as *this* world, which is given to us and inhabited by us? The ambiguity regarding our world as *this* world rests on the following two meanings. Either the term could mean the fact that this world is given to us and inhabited by us in a way that is accidental to this world. Such a world would be what it is—"this world"—whether or not it be given to us and inhabited by us. Or the term could mean that it is essential for the *haecceitas* of this world to be given to us and inhabited by us and our kind. The ambiguity between accidental and essential givenness is resolved by the claim concerning God's creativity as the "Whence" of the situation of the immediate givenness of all constituents of human life. As such, God's creativity is primordial and insurmountable for the situation of immediate givenness and refers to the whole of the determinate and finite world, which is given to

5 "As regards the identification of absolute dependence with 'relation to God' in our proposition: this is to be understood in the sense that the *Whence* of our receptive and active existence, as implied in this self-consciousness, is to be designated by the word 'God' and that this is for us the really original signification of that word" (CF § 4,4 [16]).

us. The world given to us *in toto* is apprehended by God's creativity as *this* world from possible alternatives of its kind. Divine creativity apprehends this world as given to us and inhabited by us. If the *haecceitas* of this world is determined by God's creativity, then its essential element consists of being given to us and inhabited by us and our kind. Schleiermacher makes this point clear when he—reaffirming classic Christian insight—holds that God's creativity follows and actualizes God's providence.

9. So much for the material insights concerning the relationship between the activity or process of our world and God, as we find those insights articulated in Schleiermacher's texts. I have treated them so far as speculative insights.

The question now arises as to whether or not this interpretation is adequate. Are not all God-propositions, for Schleiermacher, religious insights and propositions of religious faith? Are they propositions of faith, which, as such, do not achieve any theoretical status and do not claim any theoretical truth? With this question in view, we touch an area of remarkable confusion in the research on and in the interpretation of Schleiermacher. In order to clarify this confusion, let me state the following facts.

9. 1. All propositions in the series given above are taken from texts in which Schleiermacher explicitly articulates reflective insights on what is given to us as that which is to be apprehended, understood and mastered, symbolized and organized by us. The basic subject of reflection is—in the *Reden* (*Speeches*), in the *Monologen* (*Soliloquies*), and then again in the *Lectures on Psychology*—the human condition. Schleiermacher specifies the human condition in its fullness and wholeness as immediately given to us in order that it be understood and mastered by us. The givenness of the human condition occurs together with the givenness of the primordial and insurmountable situation of immediate self-consciousness, which comprehends the determined and inescapable whole of possibilities, tasks, and obligations of human life.

9. 2. A specific human activity delineates the content of reflective intuition (*Anschauung*) of the self. (The content of the self's reflective intuition is immediately disclosed and presented as a single variation of humankind, to be understood and mastered by the self.) For Schleiermacher, this activity is symbolizing activity. Symbolizing activity grasps the indissoluble relation between the individual and the universal, the empirical and the speculative. A sort of "speaking of life in its wholeness" grounds and embraces all scientific speech about the given. As a result of its critical movement, symbolizing activity concentrates on either the empirical domain (within

the horizon of the universal to be grasped speculatively) or on the speculative domain (the universal given as determined and finite together with the individual).

9.3. Consequently, the reflective intuition (*Anschauung*) of the self that is immediately given can also serve as the starting-point for both a speculative description of the universal traits of human existence and an empirical description of a single human individual and its essence. The latter path is, for example, taken in the *Monologen*, the former in the *Reden*. As a key text in Schleiermacher's corpus, the *Speeches* are a speculative description of the primordial and insurmountable situation of human life. In the *Speeches*, Schleiermacher describes this situation as constituted and as given in the immediacy of self-disclosure. His discussion of the situation offers a speculative theory regarding the constitution of human knowing (indissolubly connected with willing and feeling), and the actualization of the human will (action) (indissolubly connected with knowing and feeling). This speculative theory also accounts for human feeling as the mere suffering of an immediate self-disclosure that necessitates the human self to understand and to responsibly master its own existence through the interchange between willing and knowing.

9.4. In the *Reden* and in the psychological part of the Introduction to the *Glaubenslehre*, Schleiermacher shows that this human feeling—the suffering of immediate self-disclosure—is first in need of and is destined to undergo a series of formative phases. Each formative phase takes effect when the human person suffers the respective phase; the effect of enduring the respective phase is the explicit feeling of the complex content of immediate self-consciousness. The content of immediate self-consciousness is the structure of the inner-worldly situation as the differences and relations between the individual and its surroundings. These differences also include the differences between individual persons who, in light of immediate self-consciousness, interact responsibly with each other and with non-personal individuals. These differences also finally include the differences between the individual's felt relation to itself and to the world as well as the differences between the individuals felt relation to itself and to the divine creativity, which is the origin of the presence of this world of ours. For Schleiermacher, all these formative stages of feeling make up religion: from "Fetishism" to "Polytheism" to "Monotheism."[6]

6 CF § 8 (34–9).

9. 5. The thesis of Schleiermacher's speculative theory of humankind can be summarized as follows. Feeling, the immediacy of self-consciousness, is the constitutive, primordial, and insurmountable situation of human existence, and this feeling suffers the divine creativity necessitating humankind to its existence in responsible, relational freedom. Finally, all formative phases of this feeling of being necessitated by divine creativity to a responsible life in this world make possible all of our willing and knowing. This same willing and knowing is always grounded in and tinged by religion.

Schleiermacher's contribution to the speculative theory of human action consists of both a speculative theory of human willing as described by his speculative ethics[7] or a speculative theory of knowing as described in the transcendental part of the *Dialectic*.[8] In these passages of the *Dialectic*, Schleiermacher shows how the modes of human process are each grounded in and made possible by immediate self-consciousness. Immediate self-consciousness is the feeling of the suffering of divine creativity as the self-revelation of both God's existence and essence. This suffering is "religion," the felt suffering of the relation between this our world as apprehended, determined, presented, and given by God's providential creativity.

IV. The Significance of the Difference

All idiosyncrasies and differences in Schleiermacher's thought result from two basic decisions. The first decision is to unerringly stick to the given, the second decision is to take seriously the given determination and finitude of this world of ours.

These two decisions have precise consequences for Schleiermacher's view of the relation between world and God. For Schleiermacher, both spheres have a processual constitution; one is not given without the other, both form an indissoluble unit, but both are irreducibly distinct and related to one another. The determination of both the world's process and of all processes in the world must be distinguished from the higher process of the

7 Paragraphs 3 and 4 of the *Glaubenslehre* are situated in the context of propositions derived from and belonging to speculative ethics.

8 Schleiermacher describes religion in this section only according to its special function as the *fundamentum* of all possible knowing; in addition, Schleiermacher treats God only as the ground of the possibility of adequate knowing. This description, however, is only a partial view of religion and God. *In concreto*, religion and God are not only the ground of knowing, but are also the ground of our presence in this world.

creativity determining both the world's process and the processes in the world. To be taken into account in this distinction is the determination of different kinds of processes in the world that are more complicated cases of older kinds of process and that are manifest as determined by completely new principles of process. The difference between two types of these various inner-worldly processes is manifest, at least, as the difference between physical process and ethical process; ethical process is manifest in the actualization of responsible willing in the context of physical process.

To sum up: Schleiermacher's commitment to the primordial und insurmountable determination of the world's process and all processes in the world leads him to distinguish between divine creativity and all worldly and inner-worldly processes. This distinction has obvious consequences for understanding the position and role assigned to human life within the world's process. The human person is necessitated to proceed ethically, by actualizing reasonable (*besonnene*) modes of striving and willing. The human being is, therefore, necessitated to live in a manner responsible for sustaining the progress of humankind in the physical process. By this mandate for ethical striving, Schleiermacher holds to the traditional Christian and especially Reformation insight that humans are unable to choose freely what they will; they only will what they can reasonably will on the basis of their respective religions. And this means, on Schleiermacher's account, that humans will on the basis of their respective formative states of feeling.

I conclude what has, until now, been merely a descriptive contribution with a short and modest judgment concerning both of our two heroes.

Regarding Whitehead, I do not see that or how his cosmological and theological views can be rendered in a way compatible with the traditional Christian view concerning the origin, constitution, and destiny both of the world and of God's creativity.

It is not this incompatibility that leaves me with reservations regarding Whitehead's metaphysical scheme. My hesitation is due rather to Whitehead's failure to hold on to what is given. Clear-cut definitions and sharp thinking are good, but I would like to see a metaphysics that has more control of definition through the phenomena than is the case in Whitehead. Without a clear determination of the empirical domain in the metaphysical domain, Whitehead ends up with a worldview that does not explicate our responsibility in this world's process. The more I consider this point, the more I gain the impression that Whitehead's worldview may perhaps be a religious one. But his religious worldview might be, it seems to me, a manifestation of the kind of religion that Charles Renouvier characterized in a letter to William James more than a hundred years ago. "Speculative people, and religious people as well, are always engaged with much earnest in persuading themselves that there is something—some single thing, matter, force, God etc.—which is all that they are, and does all they do, so that they are themselves responsible for nothing."[9]

Regarding Schleiermacher, I do not find that his pre-scientific conception of the human condition and that his scientific, namely critical, empirical, and speculative delineations of the given are methodologically deficient in principle. Nevertheless, many of his insights must be explicated further than he did (for example, his theory of God) and some observations must be corrected.

When viewing his achievements as a whole, however, I do not see how human relational existence within the world *coram deo* can be reconstructed in a way inheriting the biblical and Christian view of what is given by disregarding and bypassing Schleiermacher's insights.

---

9 Cited in RALPH BARTON PERRY, *The Thought and Character of William James: as revealed in unpublished correspondence and notes, together with his published writings*, vol. 1 (London: Milford, 1935), 679.

Schleiermacher and Whitehead on Religious Pluralism

John B. Cobb, Jr., Claremont School of Theology

Contributors to this book have argued convincingly that Whitehead and Schleiermacher pursue systematic thought in ways that are not undercut by standard postmodern critiques of "system." Both present systems that are open and nondogmatic. Both systems encourage attention to the particular as particular. Both recognize the perspectival character of their work. Both believe, nevertheless, that systematic thinking is important for human well-being and understanding in general.

I would like to acknowledge Christine Helmer's dedication to the renewal of systematic theology. It is good to be part of a counter-offensive against those forms of post-modernism that so encourage suspicion of efforts to think comprehensively and coherently that they lead to rejection of all universal claims other than their own negative ones. It is good to work in alliance with others who care about systematic theology and recognize that from each particular point of view, something can be seen about the whole.

My own graduate studies were in "constructive" theology. The choice of this label at the University of Chicago Divinity School was not based on any bias against being systematic. The accent, however, was on the need to do more than systematize what we inherit from the tradition. To do theology of any sort, of course, one must understand the tradition and, in a deep sense, continue it. But our context is so different from that in which most of the tradition was formulated, that our task is to construct a fresh way of understanding our faith, recognizing that our way of doing so is not the only way.

Clearly, the constructive project does not drive any wedge between Schleiermacher and "constructive theology," which at Chicago in those days was often Whiteheadian. On the contrary, no major figure in Christian history has been more creatively constructive in the reformulation of the inherited tradition than Schleiermacher. He provides our best model for tackling the current need for construction.

One stimulus to Schleiermacher for constructing a new systematic theology was the growing awareness, in his day, of religious pluralism. For him, reflection about the meaning of Christian faith should be carried out in that context. His judgment has been strongly vindicated by ensuing developments. The growing awareness of the plurality of religious traditions has been one of the elements that has been used against systematic theology. That we have his model of building a systematic theology in the context of that awareness helps greatly in countering this type of argument against systematic theology.

Similarly, those of us who are deeply influenced by Whitehead pick up this awareness at a later stage of its development. We cannot construct theology apart from recognizing the particularity and the limitations of the Christian tradition. But for us, this is a challenge to construct systematic theology differently, not to avoid systematic thought. Indeed, the dialogical model to which Whiteheadian theologians are inescapably committed encourages participants from the diverse religious traditions to think more systematically about the points of view from which they come.

Because I think the context of religious pluralism is so important for the development of particular systematic theologies, I have chosen to look at how the awareness of religious diversity influenced Schleiermacher and Whitehead. My interest is not so much in the particularities of their understanding of religion in general, or their accounts of the several religions they chose to discuss, as in the systematic implications of their basic approaches for the current theological discussion of religious pluralism. If we are to launch an effective counter-attack against those who have dismissed systematic theology, we must build on our heroes but move ahead in our own systematic work.

That there are other religious traditions in the world besides one's own has been known throughout human history. In this sense, there is nothing new or particularly interesting about affirming religious pluralism. However, during most of history, most people did not suppose that understanding their own tradition required the examination of others. That is, they did not locate their own tradition in the context of a wider study of the history of religions. For Protestants, Schleiermacher played an important role in introducing this approach to theology. Its appropriateness has been, and now is, contested. But this debate is not our concern here. This book focuses on the relation of Schleiermacher to another thinker who took for granted the wisdom and necessity of this approach. Since I share the conviction that this is a desirable and even necessary way of understanding ourselves as Christians, I will not deal with the objections and alternatives that still flourish in the theological community.

To limit my task in comparing Schleiermacher and Whitehead on their treatment of religious plurality, I am restricting myself to Schleiermacher's *The Christian Faith* and Whitehead's *Religion in the Making*. I take these to be the most thematic treatments of this topic by the two thinkers. We could derive somewhat different interpretations from Schleiermacher's *Speeches* and Whitehead's *Process and Reality*, and the results of comparing those books would add richness to the discussion. But the similarities and differences of the two books I have selected provide adequate material for one short essay.

Schleiermacher was a true pioneer. He developed his own understanding of what religion is in general and of how this universal has been developed in diverse traditions. Locating Christianity in this context created a new genre of writing. Whitehead, writing a century later, followed in what was already a long and rich tradition. Nevertheless, his approach within this genre was also quite distinctive.

One difference between them lies in their perspectives. Schleiermacher was writing theology. To write the new kind of theology he believed needed, he borrowed from other disciplines, and the discussion of religion in general and of the other particular religions, which is central for this paper, is part of the borrowing. But the purpose and goal was theological. Whitehead wrote as a philosopher who was not committed to one or another religious community. The question for him was what could be learned from the several traditions and which, if any, were favored by the evidence to which all contributed.

Schleiermacher wrote at a time when Christendom took for granted the superiority of Christianity. One does not sense that he doubted this assumption.[1] He did think that this superiority should be displayed on a scientific basis, and he proceeded to do this to his own satisfaction. Whitehead wrote at a time when Christian superiority, while still widely affirmed, was by no means to be taken for granted. In England a good many intellectuals had been attracted to Buddhism. For Whitehead, this was the primary

1 CF § 3,3 (7–8).

religious alternative to Christianity, whereas Schleiermacher says nothing about it unless it is included under Indian polytheism.[2]

For both Schleiermacher and Whitehead, the focus in the understanding of global religion is experience. For both, experience is constituted by feelings. For both, experience and its interpretation must be distinguished, even if both recognize the dialectical relation between them. The definitions of religion in the two cases are both experiential.

I. Schleiermacher's Understanding of Religion

With respect to the distinctive nature of religious experience, Schleiermacher and Whitehead move in different directions. Schleiermacher is quite univocal in his definition and draws conclusions from it systematically. The essence of religion is the feeling of absolute dependence. Absolute dependence means, for Schleiermacher, a wholly one-sided relationship. As creatures, we are dependent on something that is in no way dependent on us. Schleiermacher thinks that this relationship is universally present, and, indeed, that this feeling is the ground for positing the reality of the world. Thus one can move quite directly from the distinctive religious feeling to judgments about that which is felt. The doctrine of God is directly entailed in the feeling. Indeed, the feeling of absolute dependence can equally be called the feeling of being in relation with God.[3]

Not only so, but one can evaluate the various religious traditions according to the way they embody or express the essence of religion. It is clear to Schleiermacher that the failure to see that the feeling of absolute dependence has a single, self-identical object is a mark of an inferior religion. Although all religions must be valued positively, since the highest forms develop from the lower ones, only the monotheistic forms of religion are candidates for full recognition.

Schleiermacher finds three religions that are equally monotheistic. To distinguish these he requires another principle. For this purpose, he introduces the distinction between the receptive or aesthetic dimension of human experience and the active or ethical. The former predominates at lower stages; so its presence in a monotheism shows that the lower stage has not been completely transcended. This, Schleiermacher holds, is the

---

2 CF § 8 (34–8).
3 CF § 4,4 (16).

situation with Islam.[4] In Christianity, the sense of absolute dependence is bound up with the moral (teleological) side of experience, showing the fullness of its transcendence of the primitive expressions of religion. Judaism, like Christianity, is a teleological monotheism, but it betrays elements of a lower level of religion in its limitation of focus on the "race of Abraham."[5]

Whitehead's understanding of religious experience, on the other hand, leaves open the issue of monotheism and, indeed, of theism generally. We will postpone consideration of his view of religion to the next section. Here we will discuss his thought only as it critically engages that of Schleiermacher. A Whiteheadian, approaching Schleiermacher, can try to understand the feeling of absolute dependence in two ways: (1) as a feeling of God, and (2) as a realization of creativity. I will consider both possibilities.

(1) For Whitehead, as for Schleiermacher, many of our feelings involve dependence. Indeed, dependence is involved in all feelings. What we become in each moment depends on what we feel. But when what we feel is actual, it not only affects what feels it, but also is affected by what it feels. Schleiermacher recognizes that this is characteristic of our feeling of other creatures. Whitehead believes it is characteristic of our feeling of all other actual entities whatsoever, including what he calls "God." Hence, the relation to God in Whitehead is not one of absolute dependence as Schleiermacher defines this.

There is, however, a deeper meaning of absolute dependence. It means dependence for our existence as such. This does apply, for Whitehead, uniquely to God among all actual entities. In the case of all other feelings, a creature is dependent on the datum for just what it becomes, but if that datum were absent, the result would be that it would become something different. The lack of that datum would not prevent it from becoming. In the case of God, the situation is different. Apart from God, it would not come into being at all.

One wonders whether this is not the really important meaning of "absolute dependence" for Schleiermacher. Is not the feeling of absolute dependence the sense of being dependent on another for our existence? Does it really contain an awareness that our being makes no difference to the one on whom we are dependent? I find it very difficult to imagine that anyone could discover such an awareness phenomenologically in the feeling of any actual entity. I find it much easier to imagine that Schleiermacher was influ-

---

4 CF § 8,3 (37–8).
5 CF § 8,4 (37).

enced in his interpretation of the feeling by the long history of traditional Christian thinking about divine impassibility or by his own metaphysics.

Even if we freed "absolute dependence" of this connotation of divine independence, a difference would remain between Whitehead and Schleiermacher. Whitehead, philosophically, has no reason to deny that the feeling, or, more technically "prehension"[6] of God, is characterized in part by a feeling of absolute dependence in the more limited sense. But religiously Whitehead did not believe that dependence is the most important feature of our relation to God. For him, the idea of God and the experience of God are indissolubly connected to the good, whereas viewing God as equally causative of all that happens obscures that connection.[7]

(2) There is another, quite different, way of interpreting the feeling of absolute dependence from a Whiteheadian perspective. This way distinguishes this feeling from the feeling or prehension of what Whitehead calls God. Nevertheless, the fit may be better. For Whitehead, the feeling of God has the same subject-object structure as all other feelings, that is, it is the prehension of a datum, thereby incorporating some aspect of the given into the becoming subject. But Schleiermacher does not seem to understand the feeling of absolute dependence in this way.

Whereas for Whitehead, there is a prehension of God, even if, for the most part, there is no consciousness that the datum involved is divine, in Schleiermacher's analysis, the feeling of absolute dependence does not seem to have a structure of this kind. The feeling seems to be of the subject itself as absolutely dependent. Schleiermacher describes this as a feeling of being in relation to God, rather than as a feeling of God. This feeling is found in the immediate self-consciousness of the subject, not as part of its receptive or active poles, but as that which holds these in unity. This fits better, in thinking of Whitehead, with awareness of being an instance of the creativity that constitutes the occasion in its feeling of its past and its decision about how to form itself so as to be suitably effective in the future. Creativity is the unification of the many into a new actuality, apart from which there would be nothing at all. Creativity is thus the process of becoming of every occasion.[8]

The problem is that, for Whitehead, an occasion cannot prehend the process of its own becoming. In the strictest sense, there cannot be any

6 PR 19, 23.
7 SMW 179.
8 PR 21.

completely reflexive self-consciousness, although Whitehead occasionally uses the term in a less rigorous way. Nevertheless, Whitehead's own analysis of a moment of experience or an "actual occasion"[9] is based in part on something closely approximating what Schleiermacher calls "immediate self-consciousness."

For Whitehead, a moment or occasion of human experience is conscious of some features of its world, including its own past. This consciousness is part of the subjective form of intellectual feelings of that world. These subjective forms constitute much of the conscious part of the subjective immediacy[10] of the occasion. Subjective immediacy is, thus, partly conscious even though it cannot be the datum of its own conscious feelings. Perhaps this is equivalent to Schleiermacher's immediate self-consciousness. This subjective immediacy clothes the creativity of which the occasion is an instance.

Whitehead speaks of every occasion, of the world itself, as in the inexorable grip of creativity. This may be a deduction from his metaphysics, but it may also express a sense of the absolute dependence of the occasion on creativity. What Schleiermacher discovers in immediate self-consciousness may well be the same as what Whitehead describes as the relationship of every occasion to creativity.

This approach does not necessarily bring Whitehead and Schleiermacher closer together. Whitehead encourages us not to hypostasize creativity. Also, he sees creativity as wholly neutral from a moral perspective. It is as fully expressed or embodied in vicious as in kindly acts. The explosion of an atomic bomb is as full an embodiment of creativity as a mother's tender care for her child. Whitehead discerns that there is that in the totality of things that gives a character to creativity, indeed, the character of love, and it is this that process theologians, following him, call God.

Perhaps we could bring Whitehead and Schleiermacher closer together if we posit that Schleiermacher's feeling of absolute dependence is the realization of creativity characterized by the divine character. Certainly he attributes to God characteristics other than those derived from absolute dependence as such. If so, the connection may be quite close.

Regardless of how we understand the feeling of absolute dependence, it does not work well in identifying the common element in all religious traditions. A feeling of radical dependence on God, shorn of its non-reciprocity,

---

9 PR 16, 18.
10 PR 155.

does play a central role in the Abrahamic traditions, but elsewhere it is hardly discernible. Among Hindus, Advaita Vendantists discover in immediate self-consciousness the identity of Atman and Brahman, the self and ultimate reality. Identity is quite different from dependence. Also Brahman is quite different from the Abrahamic God.

The Buddhist case is particularly significant. Some Buddhists seek to understand who or what they are through meditational practices that clarify immediate self-consciousness. They realize that they are instances of dependent origination. Thus the sense of dependence is important. But this dependence is understood to be on the causes and conditions, the circumstances, or the events, that constitute their coming into being. It is not on God.

That the feeling of absolute dependence fails to describe anything central to all religious communities does not mean that the effort to find a common element of feeling through the world's religions has been abandoned. Far from it. The basic program was taken up most influentially by Rudolf Otto.[11] He identified the distinctively religious element in experience as the feeling of the holy. This overlaps phenomenologically with the feeling of absolute dependence as described by Schleiermacher, and Otto develops his interpretations of Christianity and of other religions from this analysis in a way that parallels Schleiermacher. We can well treat Otto as an updating of Schleiermacher based on much greater knowledge of other religious traditions.

The broader tradition inspired by Schleiermacher includes thinkers with definitions of religion that are not so close to his. I take this broader tradition to be any that identifies an essence of religion that functions as a norm for evaluating religions. One influential example is Paul Tillich, who defined religion as "ultimate concern."[12] Like the definitions of Schleiermacher and Otto, this points quite directly to some kind of deity. Much more different, but still falling in this general genre, is the work of John Hick,[13] who understands religion in terms of transformation from self-centeredness to centeredness in the Real. In the development of Hick's thought, the Real is derivative from the theistic tradition, but Hick is at pains to show that it may be understood non-theistically. Unlike the oth-

11 Rudolf Otto, *The Idea of the Holy*, trans. John H. Harvey (London: Oxford University Press, 1936 [1917]).

12 Paul Tillich, *Dynamics of Faith* (New York: Harper & Row, 1957).

13 John Hick, *An Interpretation of Religion: Human Responses to the Transcendent* (London: The Macmillan Press, 1989).

ers, Hick finds the higher religions more or less equal in their attainment of the common religious goal.

In comparing Whitehead and Schleiermacher on religious *pluralism*, it is important to recognize that Schleiermacher's basic program can be, and has been, updated in this way. The issue in evaluating Schleiermacher, therefore, is not so much whether his formulation of the essence of religion fits our current knowledge of non-Abrahamic religions, as whether the identification of an essence of religion and evaluation of the various traditions in its light is the best way to proceed. Whitehead did not pursue that path.

II. Whitehead's Understanding of Religion

In evaluating Schleiermacher's contribution, I have stressed that we should not concentrate on the question of whether his definition of religion and his accounts of other religious traditions fit the present stage of knowledge of the religious traditions. They do not. It is anachronistic to expect such a fit. The problem is less acute with Whitehead because he wrote more than a century later. But his book is now more than three-quarters of a century old, and scholarship has advanced greatly since he wrote. Obviously, his judgments also need updating.

Whitehead's understanding of how religion came into being and his interpretation of Buddhism, while sometimes insightful, require adjustment in light of more recent scholarship. He lacked appreciation of the values of primal religions. His treatment of Judaism and the Christian Old Testament can easily support supersessionism. Although he emphasized that religion is a source of historical evil as well as good, he did not notice some of the forms taken by this evil. For example, he did not realize the close connection between the rise of what he called "rational religion" and alienation from the natural world. He was oblivious also to the deep bond between rational religion and patriarchy. However much we benefit from Whitehead's contribution, a Whiteheadian can no more follow him in detail than an admirer of Schleiermacher can follow him in that way. Today the debate should be between those who follow Schleiermacher's program of finding a normative essence in religion and those who follow Whitehead's program of defining religion more neutrally and evaluating its expressions with norms not derived from its definition.

Whitehead does not provide us with a single definition of religion as Schleiermacher does. In the section of *Religion in the Making* entitled "Religion Defined," he provides a number of statements that take the form of definitions. However, referring back to them later in the book, he characterizes them as "general descriptions."[14] In any case, they are consistent with one another and mutually supportive; so the lack of a single definition need not pose a serious problem for the reader.

The more serious problem is that Whitehead uses "religion" in two ways, one broader and one narrower.[15] His account of the history of the development of religion from pre-human animal antecedents to rational religion calls for a broad definition that he does not provide. The definition-like statements he does provide are directly applicable only to rational religion.[16] The latter is roughly comparable to what Karl Jaspers led us to understand as the religious traditions arising in the axial age. We will focus, as Whitehead does, on rational religion, but before we do so, it may be well to sketch his speculations about the earlier history.

Whitehead sees ritual as the context for the rise of religion. Ritual is found already among animals. It comes into being when there are energies that are not needed for survival and that can be expressed in ways that are immediately satisfying. The satisfying results are emotions, and gradually human beings develop rituals for the purpose of generating satisfying emotions. They then formulate myths to accompany and interpret the rituals, and eventually reason operates on these myths to rationalize them. Their rationalizing involves making them both internally coherent and also consistent with other knowledge. In this context, the individual becomes important, whereas, previously religion was a social phenomenon.

Whitehead's first definition of religion is "force of belief cleansing the inward parts."[17] The second definition is "the art and the theory of the internal life of man, so far as it depends on the man himself and on what is permanent in the nature of things."[18] The third is "what the individual does with his own solitariness."[19] And finally, Whitehead writes: "Thus religion is solitariness."[20]

The emphasis on interiority and solitariness clearly contrasts the rational religion in which, for Whitehead, the history of human religious develop-

14 RM 58.

15 RM 13–18.

16 That Whitehead's definitions are of *rational* religion in the *narrow* sense, rather than the world-wide history of religion, i. e., the *broad* sense, is made clear in RM 58.

17 RM 15.

18 RM 16.

19 Ibid.

20 RM 17.

ment culminates, with what precedes it. Just as ritual arises in the transcendence of animals over necessities for survival, rational religion arises in the capacity of individuals to transcend their immersion in society. To realize that one must make one's own decisions is a step toward the discovery of interiority. To realize that these decisions deal not only with external actions but also with the interior life, and to accept responsibility for that interior life, complete the process. Solitariness is that dimension of human life that transcends determination by the natural world, including the body, and by society.

The emphasis on solitariness can easily mislead a reader who is not familiar with Whitehead's thought more broadly. Solitariness for him is not isolation. What one is in any moment, however solitary one may be, is a synthesis of relations with one's own past, one's body, and other creatures. Late in the book, Whitehead clarifies that "the topic of religion is individuality in community."[21]

The feeling of absolute dependence and the experience of the holy certainly belong to this religious sphere. But the religion in which Whitehead is interested is not necessarily centered on any particular feeling. It is the active cultivation of the inner life informed by particular beliefs. From this point of view, whether the feelings and orientations identified by Schleiermacher, Otto, and Tillich are universal is a secondary question. Hick's transformation from self-centeredness to centeredness in the Real describes such cultivation. But Whitehead is open to other forms of cultivation of the inner life, not all of which are so positive. Hick's goal of transformation may, indeed, be held to be normative, but this must be argued on independent grounds. It cannot be taken as descriptively the primary goal of all rational religions.

How one cultivates the inner life depends on what one really believes rather than on particular experiences; so beliefs are of great importance. Beliefs may lead to cultivation in a way that is harmful to self and others. Whitehead emphasizes that religion has had many terrible consequences in human history, but that should lead, not to its rejection, but to the improvement of the beliefs that inform the cultivation. For him theology, broadly understood, is critical.

In the list above, I omitted his definition of religious beliefs. "A religion on its doctrinal side, can thus be defined as a system of general truths which have the effect of transforming character when they are sincerely held and

21 RM 88.

vividly apprehended."[22] Again, nothing is said as to the direction of change effected by the beliefs. Whitehead wrote before the discussion of the religious character of Communism, Nazism, and Fascism was widespread. To a large extent, no doubt, they functioned as pre-rational religions, that is, by manipulating behavior socially. But there is also strong evidence that they evoked inward devotion and changed the interior lives of many of their believers. Further, the many patterns of personal growth, recommended and practiced in recent decades, are, or at least can be, religious, even though they are more often thought of as psychological.

My depiction of Whitehead's understanding of religion as radically open to diverse patterns of ordering the inner life should be qualified. He affirms a commonality in religious experience "founded in the concurrence of three allied concepts in one moment of self-consciousness."[23]

1. That of the value of an individual for itself.

2. That of the value of the diverse individuals of the world for each other.

3. That of the value of the objective world which is a community derivative from the interrelations of its component individuals, and also necessary for the existence of each of these individuals.[24]

This leads Whitehead to formulate one further general description of religion. "Religion is world loyalty."[25] The solitary individual can find no answer to the question, "What, in the way of value, is the attainment of life?" that does not refer to the whole. In transcending tribal religion through solitariness, the individual sees the whole as the value context of the spiritual quest. This is closely connected to the "intuition of immediate occasions as failing or succeeding in reference to the ideal relevant to them."[26]

Do we have here a normative essence of rational religion that goes beyond general description? There is little doubt that the description is affected by Whitehead's own metaphysical vision, and that the choice of the word "loyalty" is influenced by the Abrahamic traditions. No description

22 RM 15.
23 RM 59.
24 Ibid.
25 RM 60.
26 Ibid.

is purely neutral. Nevertheless, the close connection between individuality and universality is widely attested. Whitehead does not use this account to rank the rational religious traditions. His intention is to describe what is common to them.

Furthermore, it would misrepresent Whitehead to see the fit between his account of religious experience and his metaphysics as simply reading his metaphysics into religious experience. He looks to religious experience as a major resource for metaphysics, and these insights of religious experience are data for that metaphysics. Value plays a central role in that metaphysics, and Whitehead cannot derive that from the sciences. If he is wrong in his account of religious experience and his ideas about the centrality of value are not supported by religious experience, then his metaphysics requires drastic revision.

The features of religious experience noted by Schleiermacher, Otto, and Tillich betray their origin in a theistic context. The one on whom we are absolutely dependent is God. So is the Holy One. Our ultimate concern should be something like God. This is not true of Whitehead's definition. Hindus, Buddhists, Taoists, and Confucians may respond more immediately than Christians to the understanding of religion as the cultivation of the inner life. For many practitioners of Asian religious traditions, meditation is central. This cultivation in no way depends on belief in God. On the other hand, beliefs influence the form of cultivation; so whether one believes in God, and if so, how God is understood, are matters of great importance. Much of Whitehead's book is devoted to these questions.

Clearly, the understanding of religion as the cultivation of the inner life leading to universal concern leaves open the question of whether there is common experience of God or experience that witnesses directly to the reality of God. Whitehead thought that we could not identify a common theistic experience, at least in any direct and obvious way. Belief in God arises through interpretation, rather than through direct analysis of experience.[27] Theistic belief can be suggested by "the intuition of immediate occasions as failing or succeeding by reference to the ideal relevant to them." For Whitehead, this implies that there is "a rightness attained or missed,"[28] or a "rightness in things partially conformed to and partially disregarded."[29]

27 In later writings this judgment is modified.
28 RM 60–1.
29 RM 66.

However, the interpretation of this shared experience can be immanental. Buddhists interpret it in terms of dharma and Buddha-nature. For many Chinese, it is the Tao or Li. Describing it as immanent does not mean that it cannot be distinguished from the actual course of events. In this minimum sense, at least, it is transcendent.

The Abrahamic traditions interpret this difference between what is and what ought to be through the idea of the purposes of God. Thus the emphasis is on transcendence. But unless the divine purpose is manifest to believers in the world, it has no possibility of effectiveness. In this minimum sense, at least, it is immanent.

Whitehead's own interpretation of what he takes to be the common element in religious experience lies between the two extremes. He thinks it is the experience of God, but that the God thus experienced is fully immanent. It is the immanence of God that is experienced in the awareness of better and worse. As in the case of the dharma, the Tao, or Li, that which is experienced as immanent cannot be identified with the ongoing events in the world, and thus also transcends it.

How are alternative interpretations to be evaluated? Whitehead offers three criteria. Do they help to provide a self-consistent doctrine? Are they consistent with what we know of the world through the natural and social sciences and through history? What effects do they have on the inner life that is shaped by belief in them?

Clearly the selection of these criteria is already a judgment that may not be shared by all the traditions one intends to evaluate. Further, the focus on the rightness in things, even if there should be agreement as to the universality of this experience, may play only a secondary role in some traditions. Even if there should be agreement among the traditions on these points, the judgment of relative strength would remain a very complex and difficult one. All the great traditions are capable of relatively self-consistent expression or, more accurately, multiple forms of such expression. All of them are capable of formulations that are consistent with what is reliably learned from the sciences. All of them are likely to claim that the way they shape the inner life is best, and there are no neutral bases for evaluating these claims. This suggests that for a Whiteheadian, the definition of religion does not function as a norm for ranking religious traditions in general, or for judging them to be more or less equal. It may, however, function to encourage particular formulations of particular traditions, including Christianity.

Whitehead's own judgment was that, of the traditional religions, Christianity and Buddhism have the most to offer. He saw no need to choose between them. He thought that both are in decline,[30] and that both could gain from interacting with one another. A quotation will indicate his general view, one that has been important for many of his followers.

> The decay of Christianity and Buddhism, as determinative influences in modern thought, is partly due to the fact that each religion has unduly sheltered itself from the other. The self-sufficient pedantry of learning and the confidence of ignorant zealots have combined to shut up each religion in its own forms of thought. Instead of looking to each other for deeper meanings, they have remained self-satisfied and unfertilized.[31]

Whitehead goes on to make another point about Christianity and Buddhism indicative of his general view of the role of religion and its place in the world. "Both have suffered from the rise of the third tradition, which is science, because neither of them had retained the requisite flexibility of adaptation."[32] Whitehead held that because science suggests a cosmology, it also has religious implications.[33] Nevertheless, science in itself is not a third religious tradition. He understood science to deal with the external world, the world known primarily through sense experience. This is different from the internal world, which is the province of religion. But this difference, for Whitehead, does not allow for separation. What we know of the external world affects the beliefs that shape the internal world. When scientific findings are ignored by a religious tradition, it loses effectiveness and credibility. The larger truth Whitehead seeks builds on what is learned religiously about the inner life and what is learned scientifically about the outer world, and seeks to integrate these into a cosmology. Every cosmology has implications for religion.

III. Implications for Theology

In comparing Schleiermacher and Whitehead we are comparing a theologian and a philosopher. Of course, the difference is not as great as that

30 RM 44.
31 RM 146.
32 Ibid.
33 RM 141.

may make it seem. Schleiermacher often wore the hat of a philosopher and did respected work in that field. When as a theologian he borrows from philosophy, especially ethics, he is borrowing from his own creative work as a philosopher, developed for the specific purpose of being used in theology. Whitehead's philosophy explicitly draws on religious experience, and his understanding of religion is certainly influenced by his Christian upbringing. He is keenly interested in Christianity and makes many fruitful comments about it and about how it should develop. One may either complain that he crosses boundaries illegitimately, or agree with him that boundaries are, in general, obstacles to creative thought. Although the differences on this point between Schleiermacher and Whitehead are not so great after all, it is worthwhile to ask how they distinguish the approaches of philosophy and theology, and make the transition between them.

Schleiermacher uses his scientific account of religious experience to show that Christianity is the purest and fullest embodiment of this experience. Given this conclusion, it is natural and proper that one proceed to spell out the Christian teaching that best displays this purity and adequacy. This means that one turns from philosophy to theology.

The theology that results takes the distinctive religious experience ideally embodied in Christianity as the principle through which all Christian theology is expounded. The doctrines of God, of Christ, and of the church are all shaped in relation to the distinctively religious feeling of absolute dependence. The remarkable success of Schleiermacher in developing a comprehensive theology on the basis of this single feature of experience attests to his genius and has led to recognition of *The Christian Faith* as the greatest achievement in systematic theology between John Calvin and Karl Barth.

The strength of this achievement is not invalidated by the judgment that the ongoing study of world religions shows that absolute dependence is an Abrahamic idea and experience. Nevertheless, this change does undercut part of Schleiermacher's claim and opens the door to quite different theological developments. There are two options. Christianity may develop its theology based on distinctively Christian history and experience without connecting it to any idea of a universal religious experience. Alternately, theology may be developed in relation to a different central theme that can lay greater claim to universality. As noted, we can see in Rudolf Otto a partial fulfillment of this second option, and we can interpret Tillich's theology as a fuller development of this alternative. On the other hand, if the first option is chosen, then one may decide that the distinctively Christian theme around which theology should be developed is quite different from that of any other religious tradition. The Neo-orthodox theologians chose this route. In a very different way, Whitehead also points in this direction.

Whitehead as philosopher seeks to view the global religious scene more neutrally. His neutrality, however, is quite different from that cultivated in many religious studies programs. For Whitehead, being a philosopher does not count against judgments of truth and value. It would not in principle prevent him from judging his own tradition, Christianity, as superior. It does prevent him from assuming this or seeking the definition of religion through the generalization of Christian experience or teaching.

Whether in the end he succeeds in avoiding a Christian bias is doubtful. Even his selection of Buddhism as the chief conversation partner for Christians reflects a widespread Christian bias, one which I share. His preference for a theistic interpretation of value experience, albeit highly immanentalist, is almost certainly influenced by his Christian background. In short, no more than anyone else, could Whitehead avoid the particularity of his own experience.

This is not, however, a very serious criticism. Whitehead knew that all thought is historically and culturally conditioned and did not suppose that his philosophy or any other escaped this conditioning. His awareness of the value of other cultures and traditions influenced his judgments considerably more than was common in his day—or even in ours. He emphasized the interpretative character of his own theories. He called his philosophy "speculative" in order to make clear that it is provisional and subject to endless testing. This certainly applies also to his judgments about religion and religions.

Whitehead called for theologians to adopt this open attitude, arguing that their failure to do this was the source of many of their problems. He emphasized that this openness should be particularly directed toward Buddhism and toward science. In principle it should be open toward any place where the theologian judges there may be wisdom and insight. Hence the task of theologians can never be simply to transmit the inherited teaching of their traditions.

Nevertheless, Whitehead viewed the task of Christian theology as quite different from the philosophical one, even as he pursued it. He thought that theology should ground itself on the founding events of the faith and the meanings that arise from them rather than on the generalities sought by philosophers. Christian theologians should attend especially to the events recounted in the New Testament. Whitehead goes on to comment on the meanings he finds in those events, thus crossing the line into theology. For him these events count heavily against the association of God with unilateral power so celebrated in traditional theology and still suggested by Schleiermacher's focus on "absolute dependence."

IV. The Current Scene

Among those who take religious pluralism seriously, several approaches are prominent. Some view difference as so important that they encourage each community to develop its own symbolic system with little regard to others. Others look for commonality in mystical or esoteric experience, viewing the ordinary expressions of all of the traditions as secondary.

Among those who continue to regard understanding multiple religious communities as an important context for understanding one's own, many follow Schleiermacher's model. They seek some essence shared by all religions that can also function as a norm for evaluating their relative success. Since most of these today are influenced by the current state of knowledge about religions, they differ from Schleiermacher in their accounts of the essence of religion. They are less likely today to affirm that any one religious tradition is markedly superior to the others. John Hick is typical in his judgment that the great traditions are roughly equal in the extent to which they achieve the goal of religion.

A smaller, but growing, number follow Whitehead in rejecting this normative use of essence. In this perspective it is not important to have a clear definition of religion, even one as clear as Whitehead's. Whitehead's definition can lead to setting Christianity in the context of systems only some of which are usually considered to be religions. The World Council of Churches spoke of "faiths and ideologies," thereby recognizing the importance of Communism, Nazism, and extreme nationalism, as well as secular humanism and scientism, as alternative ways of shaping life, including interior experience. Still following Whitehead, one might also set Christianity in the context of psychotherapeutic systems. This works fruitfully for a number of the Asian traditions as well.

One may follow the general pattern of Whitehead's approach, while arguing that the organization of social relations is as important to the rational religions as the interior aspects highlighted by Whitehead. One can compare the several traditions in this regard as well, and also compare them with other bases on which social relations are ordered. Of course, social order affects the interior life, but this is a dimension of the religious traditions that Whitehead does not discuss. In general, the focus on "religion" directs attention inwardly and cultically, and leads to slighting social and political features of the major "religious" traditions.

These broadly Whiteheadian approaches open the door for a wider range of interactions and mutual learning among the great traditions of humankind than do the approaches of the essentialists. For the essentialists, religions can learn how other religions work toward the goal that is common to all. This may lead to improvements, but it is not likely to lead to radical transformation.

For those whose approach is more like that of Whitehead, there is no reason to posit a common essence or goal. Learning of diverse goals and how they are realized can lead to much more fundamental changes. Christians can learn how Buddhists cultivate their inner lives to attain enlightenment, and some may decide to follow Buddhist methods in pursuit of that goal, without giving up the quite different goals of traditional Christianity. Buddhists can be inspired by Christians to cultivate their hopes and desires for social justice and their commitment thereto. Both can learn methods of disciplined inquiry from the sciences and can order their inner lives toward achieving greater objectivity in their judgments.

The judgment of those who take this approach is likely to be that each tradition is relatively successful in attaining the goals it identifies as its own. Since these goals differ, there is no one hierarchy in which they can be ranked. Since each tradition is likely to regard its goals as the most important, there is no neutral perspective from which their relative worth can be evaluated. It is best to abandon any effort along these lines. But this does not mean that we should abandon critical value judgments. All our traditions have dark sides. All are ambiguous in their effects. Intense mutual criticism in the context of fundamental mutual appreciation can benefit all the great historic traditions in terms of their own norms.

Indices

Index of Names

Index of Subjects

Contributors

ANNA CASE-WINTERS is Professor of Theology at McCormick Theological Seminary in Chicago, Illinois. She is the author of *Divine Power: Traditional Understandings and Contemporary Challenges* (Westminster John Knox 1990) and other works related to the interface between traditional and contemporary theologies, such as process, feminist, and eco-justice theologies.

PHILIP CLAYTON is Ingraham Professor of Theology at Claremont School of Theology and Professor of Philosophy and Religion at Claremont Graduate University in Claremont, California. He is also Co-Director of the Center for Process Studies. Among numerous publications in the areas of philosophy, religion and science, and theology, his most recent books are *God and Contemporary Science* (Edinburgh University Press 1997), *The Problem of God* (Eerdmans 2000), and the forthcoming *Mind and Emergence* (Oxford University Press 2004).

JOHN B. COBB, JR. is Professor Emeritus of Theology at Claremont School of Theology and Professor Emeritus of Religion at Claremont Graduate University in Claremont, California. He is the founding Director of the Center for Process Studies. His many books employ Whitehead's conceptuality for contemporary philosophy and theology. Among them are *A Christian Natural Theology* (Westminster 1965), *Christ in a Pluralistic Age* (Westminster 1975), *Beyond Dialogue: Toward a Mutual Transformation of Christianity and Buddhism* (Fortress 1982), and co-authored with David Ray Griffin, *Process Theology: An Introductory Exposition* (Westminster 1976).

KRISTA DUTTENHAVER is a Ph. D. student in systematic theology at University of Notre Dame's Department of Theology in Notre Dame, Indiana. She received her A. B. in religion at Princeton University and her M. A. T. S. in philosophy of religion and theology at Claremont School of Theology.

ROLAND FABER is Professor of Systematic Theology at the University of Vienna, Austria. His publications range from topics on process thought and mystical theology, science, philosophy and theology, to Buddhist-Christian dialogue. They recently include *Prozesstheologie: Zu ihrer Würdigung und kritischen Erneuerung* (Matthias-Grünewald 2000) and *Gott als Poet der Welt: Anliegen und Perspektiven der Prozesstheologie* (Wissenschaftliche Buchgesellschaft 2003).

Katie Goetz is a Master of Divinity student at Claremont School of Theology in Claremont, California and a candidate for ordination in the United Methodist Church. She assisted in editing *Truth: Interdisciplinary Dialogues in a Pluralist Age* (Peeters 2003), edited by Christine Helmer and Kristin De Troyer.

David Ray Griffin is Professor of Philosophy of Religion and Theology at Claremont School of Theology and Professor of Religion at Claremont Graduate University in Claremont, California. He is also Co-Director of the Center for Process Studies. His most recent books are *Religion and Scientific Naturalism* (State University of New York Press 2000), *Reenchantment without Supernaturalism: A Process Philosophy of Religion* (Cornell University Press 2001), and *Two Great Truths: A New Synthesis of Scientific Naturalism and Christian Faith* (Westminster John Knox 2004).

Christine Helmer is Associate Professor of Theology at Claremont School of Theology and Associate Professor of Religion at Claremont Graduate University in Claremont, California. Her publications include *The Trinity and Martin Luther: A Study on the Relationship between Genre, Language and the Trinity in Luther's Works (1523–1546)* (Philipp von Zabern 1999) and a volume co-edited with Christiane Kranich and Birgit Rehme-Iffert, *Schleiermachers Dialektik: Die Liebe zum Wissen in Philosophie und Theologie* (Mohr Siebeck 2003).

Eilert Herms is Professor of Systematic Theology at Eberhard Karls Universität Tübingen, Germany. He is also Director of the Institut für Ethik in Tübingen. He was Chair of the Wissenschaftliche Gesellschaft für Theologie e. V., 1996–2002. Among numerous publications in the areas of Luther research, as well as fundamental theology, dogmatics, and ethics (forthcoming: *Die Wirtschaft des Menschen* [Mohr Siebeck]), especially church and society (e. g., *Von der Glaubenseinheit zur Kirchengemeinschaft*, vol. 2 [N. G. Elwert 2003]), his most recent book on Schleiermacher is *Menschsein im Werden: Studien zu Schleiermacher* (Mohr Siebeck 2003).

J. R. Hustwit is a Ph. D. student in philosophy of religion and theology at Claremont Graduate University in Claremont, California, where he is working on a dissertation concerning the hermeneutics of inter-religious dialogue. His research interests include process thought, religious pluralism, hermeneutics, and the interface between theology and culture.

JULIA A. LAMM is Associate Professor of Theology at Georgetown University in Washington, D. C. She specializes in Schleiermacher's theology and in the history of Christian thought. Her first book is *The Living God: Schleiermacher's Theological Appropriation of Spinoza* (Pennsylvania State University Press 1996) and she is currently working on another book, *Schleiermacher's Plato: The Construction of a New Platonism.*

JACQUELINE MARIÑA is Associate Professor of Philosophy at Purdue University in West Lafayette, Indiana. She specializes in Kant, Schleiermacher, and on the relation of ethics to religion. She is the author of numerous articles appearing in journals such as *Journal of the History of Philosophy*, *Journal of Religion*, and *Religious Studies*. She is currently editing the *Cambridge Companion to Schleiermacher*, and is working on a book, *Love of God and Neighbor and What It Means to Be a Person.*

JOHN QUIRING is Program Director at the Center for Process Studies. He is Adjunct Professor of Philosophy at Victor Valley College in Victorville, California and also teaches theology at Westwood Presbyterian Church in Los Angeles, California.

MARJORIE HEWITT SUCHOCKI is Director of the Process & Faith Program at the Center for Process Studies, Former Dean and Professor Emerita of Theology at Claremont School of Theology, and Former Professor of Religion at Claremont Graduate University in Claremont, California. She has published widely in the areas of process thought and feminism, doctrines of sin, and religious pluralism. She is author, among other books, of *God, Christ, Church: A Practical Guide to Process Theology* (Crossroad 1989) and *Divinity & Diversity: A Christian Affirmation of Religious Pluralism* (Abingdon 2003).

JACK C. VERHEYDEN is Professor Emeritus of Theology at Claremont School of Theology and Former Professor of Religion at Claremont Graduate University in Claremont, California. He edited Schleiermacher's *Life of Jesus* (Fortress 1975) and has written on a variety of figures and topics in philosophy of religion and theology from the nineteenth century.

Bei Fragen zur Produktsicherheit wenden Sie sich bitte an:
If you have any questions regarding product safety,
please contact:

Walter de Gruyter GmbH
Genthiner Straße 13
10785 Berlin
productsafety@degruytergbrill.com